HANDBOOK OF ELECTRONICS INDUSTRY COST ESTIMATING DATA

HANDBOOK OF ELECTRONICS INDUSTRY COST ESTIMATING DATA

THEODORE TAYLOR
Corporate-Tech Planning
Waltham, Massachusetts

With contributions by
Robert K. Jenner and James A. Burbank, II

Based on
ELECTRONICS INDUSTRY COST ESTIMATING DATA
by Fred C. Hartmeyer

A Wiley-Interscience Publication
JOHN WILEY & SONS
New York • Chichester • Brisbane • Toronto • Singapore

Copyright © 1985 by John Wiley & Sons, Inc.

All rights reserved. Published simultaneously in Canada.

Reproduction or translation of any part of this work
beyond that permitted by Section 107 or 108 of the
1976 United States Copyright Act without the permission
of the copyright owner is unlawful. Requests for
permission or further information should be addressed to
the Permissions Department, John Wiley & Sons, Inc.

Library of Congress Cataloging in Publication Data:

Taylor, Theodore, 1929–
   Handbook of electronics industry cost estimating data.

   Based on: Electronics industry cost estimating data/Fred C. Hartmeyer.
   "A Wiley-Interscience publication."
   Includes index.
   1. Electronic industries—Estimates—Handbooks, manuals, etc. I. Hartmeyer,
Fred C. Electronics industry cost estimating data. II. Title.

TK7835.T28   1985   338.4′3621381   85-6294
ISBN 0-471-82264-7

Printed in the United States of America

10 9 8 7 6 5 4

PREFACE

This book is a unique accumulation of time standards, manufacturing methods, and the overall "rules of thumb" which are used for cost-estimating electronic equipment and systems. Originally published in 1964 by Fred C. Hartmeyer, it has been extensively revised, updated, and expanded to reflect today's electronic technology and manufacturing practices. All operations from machining and sheet metal fabrication through wiring, circuit board and integrated component assembly, electrical testing, and packing are covered. In addition, learning curves and hourly rates and burden along with the complementary fields of production scheduling, manufacturing and facility planning ratios, and concept estimating are described.

In short, this book is for the person working in, associated with, or utilizing the results of the planning and cost estimating of electronic products. For the business manager it provides the basic techniques and methods that are used in cost estimating and program planning For the practitioner of cost estimating it serves as a ready reference volume.

Although the book bears some resemblance to the original Hartmeyer version (retention of chapter titles where applicable), all data have been revised. The text has undergone major revisions and expansion, and in many instances has been completely rewritten to reflect 20 years of technological changes since the original version. A few exceptions are in the areas of coil winding and electroplating, where the manufacturing processes have not varied that significantly. Five new chapters have been added covering the subjects of plastics, fiber optics, automatic testing, integrated circuits, and computer aided design and manufacturing. Finally, the contents has been reorganized so that the chapters more logically follow the sequence in which cost estimates are developed. This was further enhanced by grouping the subjects under the categories of labor (fabrication, assembly, inspection and test), materials, manufacturing support, and program planning.

The data in this book are based on the experiences of many companies and individuals. On-site interviews were used extensively in updating the data and developing new subject areas. Consequently, the book does not necessarily represent the policies and practices of any one source.

The Defense Logistics Agency (with the permission of John Wiley and Sons) played a key role in bringing this book to fruition by sponsoring the updating work which the author directed. This new edition, consequently, represents an unabridged version of the results of that effort. It is one of the standard references used by the Defense Contract Administration Services in performing technical cost analysis of contract proposals for electronic systems. In addition to small-volume military type systems, the book also applies to cost estimating any electronic product, be it military or commercial.

The author wishes to express his appreciation to the numerous companies and individuals who provided sources of data, helpful suggestions, and reviews of the material in the book. Special recognition is given Robert K. Jenner and James A. Burbank, II who, as colleagues and participants on the updating program, made major contributions to many of the sections of the book, and to Katherine E. Blythe for her diligent library research and subject indexing.

THEODORE TAYLOR

Waltham, Massachusetts
July 1985

CONTENTS

PART I INTRODUCTION AND SUMMARY LEVEL DATA

1. Introduction to Estimating Standards   3

    1.1. Merits of Standard Data   4
    1.2. Basis of Standard Data in This Book   5

2. Organizational Overview and Estimating Formats   6

    2.1. Organization   6
    2.2. Estimating Forms   8

3. Summary Cost Data   14

Sec. 7	Machining Operations	15
Sec. 8	Sheet Metal Operations	19
Sec. 9	Electroplating and Chemical Surface Treatment of Metals	23
Sec. 10	Plastics	24
Sec. 11	Printed Circuit Boards	24
Sec. 12	Terminal Board Fabrication	25
Sec. 13	Painting Operations	26
Sec. 14	Silk Screen Printing and Engraving	27
Sec. 15	Riveting, Mechanical Assembly, Fastening	27
Sec. 16	Coil Winding Operations	30
Secs. 17–18	Wiring	30
Secs. 21–22	Inspection and Test	32
Sec. 24	Packing and Packaging	32

Sec. 25	Learning Curve	34
Sec. 26	Labor Allowances and Multipliers	34
Sec. 27	Material Cost and Allowances	35
Sec. 29	Material Discount Curves	35
Sec. 30	Special Tooling & Test Equipment	36
Sec. 31	Manufacturing Engineering	37
Sec. 33	Hourly Labor and Burden Rates, Earnings	38

PART II PROGRAM PLANNING

4. Schedule Determination — 41

- 4.1. Manufacturing Phasing Charts — 41
- 4.2. Delivery Schedule Estimating — 45

5. Personnel and Facility Planning Ratios — 49

- 5.1. Direct Labor Proportions — 49
- 5.2. Burden Expense Ratios — 50
- 5.3. Space Planning Ratios — 51
- 5.4. Sales and Facility Ratios of Selected Electronic Companies — 51

6. Concept Estimating — 56

- 6.1. Concept Estimating Parameters for Hardware — 57
- 6.2. Concept Cost Estimating Parameters — 63

PART III COST ESTIMATING ANALYSIS

FABRICATION

7. Machine Shop Operations — 67

- 7.1. Cut Raw Material — 68
- 7.2. Turret Lathe – 1-in. Diameter Stock — 70
- 7.3. Engine Lathe – 6-in Diameter Stock — 73
- 7.4. Milling — 77
- 7.5. Drilling — 79
- 7.6. Broach — 81
- 7.7. Grind, Centerless — 82
- 7.8. Grind, External Cylindrical — 84
- 7.9. Grind, Internal Cylindrical — 87

CONTENTS

7.10.	Grind, Surface	88
7.11.	Gear Hobbing	89
7.12.	Speeds and Feeds	94
7.13.	Tolerances and Surface Finishes	94

8. Sheet Metal Operations — 101

8.1.	Cut Blank	102
8.2.	Notching	103
8.3.	Holes	103
8.4.	Press Operations	106
8.5.	Trim, Profile, Rout	113
8.6.	Burr Removal — Bench, Belt, Vibrator	115
8.7.	Arc Welding	116
8.8.	Spot Weld	118
8.9.	Rivet	120
8.10.	Surface Preparation	121
8.11.	Paint	121
8.12.	Silk Screen and Engraving	123

9. Electroplating and Chemical Surface Treatment of Metals — 124

9.1.	Plating Method	125
9.2.	Analysis of Minutes Per Lot, Batch and Part Values	127
9.3.	Process Descriptions	138

10. Plastics — 149

10.1.	Types of Plastics	150
10.2.	Fabrication Methods	155
10.3.	Costs	160

11. Printed Circuit Board Fabrication — 164

11.1.	Double-Sided Etched Circuit Board Fabrication	165
11.2.	Multi-Layer Etched Circuit Board Fabrication	171
11.3.	Electrical Testing	175

12. Terminal Board Fabrication — 177

12.1.	Calculation Method	177

13. Painting Operations — 179

13.1.	Surface Preparation	180

	13.2.	Mask and Demask Part for Painting	180
	13.3.	Set-Up Time per Paint Type	181
	13.4.	Primer	181
	13.5.	Surfacer	182
	13.6.	Lacquer	182
	13.7.	Enamel and Modified Vinyl	183
	13.8.	Plastic Protective Film, Strippable	183
	13.9.	Fungicide(Spray Application)	184
	13.10.	Miscellaneous Detail Values	184
	13.11.	Calculation Method	188
14.	**Silk Screen Printing and Engraving**		**189**
	14.1.	Silk Screen Printing Process Description	189
	14.2.	Photographic Operations to Prepare Silk Screen Stencils	191
	14.3.	Silk Screen Printing, Including Decals	193
	14.4.	Engraving	194
	14.5.	Calculation Method	196

ASSEMBLY

15.	**Riveting and Mechanical Assembly**		**198**
	15.1.	Parts Handling	198
	15.2.	Rivet, Stake, Eyelet	201
	15.3.	Screw and Nut Operations	202
	15.4.	Miscellaneous Assembly Operations	204
	15.5.	Tool Handling Values	206
	15.6.	Tape and Tags	206
	15.7.	Cement and Glyptol	207
	15.8.	Walking	207
	15.9.	Dynamic Balance Operations	207
16.	**Coil Winding Operations**		**210**
	16.1.	Coil Winding Terms	210
	16.2.	Example	211
	16.3.	Set-Up Analysis	212
	16.4.	Run Time Analysis	213
17.	**Wire Preparation**		**218**
	17.1.	Preparation Insulated Wire, Machine	219
	17.2.	Preparation Insulated Wire, Hand	219

CONTENTS

17.3.	Stake Taper Pin to Wire	221
17.4.	Cut to Length Bus Wire and Sleeving	222
17.5.	Cut and Bend Resistors, Capacitors, and Other Components	222
17.6.	Cut to Length Coaxial and Shielded Cable	223
17.7.	Preparation Shielded Cable with Ground Lead	224
17.8.	Preparation Coaxial Cable	225
17.9.	Stamp and Assemble Identification Tapes	227
17.10.	Assemble Plug and Connector to Harness	227
17.11.	Buzz Wire to Identify Continuity	229
17.12.	Pull Tubing over Cable	230
17.13.	Twist Wires	230
17.14.	Spot Tie Harness	230
17.15.	Harness Fabrication	231
17.16.	Fabricate Harness Nail Board	234

18. Wiring and Component Insertion 235

18.1.	Insulated Wire: Prepare and Install (2) Ends	236
18.2.	Bus-Wire, Crimp, Solder (2) Ends	240
18.3.	Resistor, Capacitor, Transistor: Prepare and Install	240
18.4.	Sleeving: Prepare and Assembly	241
18.5.	Shielded Cable: Prepare and Install (2) Ends	241
18.6.	Coaxial Cable: Prepare and Install (2) Ends	242
18.7.	Crimp Wire to Terminal: Detail Values	243
18.8.	Crimping Obstruction Allowances	245
18.9.	Wire Dressing Values	246
18.10.	Select Wire from Group	246
18.11.	Tool Handling Values	246
18.12.	Develop Wire List from Schematic Diagram	247
18.13.	Set Up Components at Work Station	247

19. Soldering 248

19.1.	Solder Wire to Terminals	248
19.2.	Wire Gauge and Diameter Data	249
19.3.	Dip and Wave Soldering of Circuit Boards	250
19.4.	Spot and Seam Soldering	253

20. Fiber Optics and Optoelectronics 254

20.1.	Background	255
20.2.	Components	256
20.3.	Applications	269
20.4.	Typical Costs	271

INSPECTION AND TEST

21. Inspection   272

    21.1. Inspection Estimating Ratios   272
    21.2. Visual Inspection and Record Keeping   273
    21.3. Machine Shop Gauging Times   274
    21.4. Gauging Frequencies   275
    21.5. AQL and Inspection by Random Sampling   276

22. Test   277

    22.1. Test Estimating Ratios   277
    22.2. Elemental Test Time Standards   279
    22.3. Troubleshoot and Retest Allowances   280

23. Automatic Test Equipment   281

    23.1. Products   282
    23.2. Market and Suppliers   284
    23.3. Processes   289
    23.4. Cost   293

SHIPPING

24. Packaging And Packing   304

    24.1. Packaging and Packing Methods   305
    24.2. Time and Material Requirements   306
    24.3. Calculation Method   306
    24.4. Packing Material Price List   311
    24.5. Reusable Shipping Containers   316

LABOR ADJUSTMENTS

25. Learning Curves   317

    25.1. Basic Principles   318
    25.2. Industrial "Learning"   318
    25.3. Applications of Learning Curves   320
    25.4. Mechanics of the Learning Curve   321
    25.5. Learning Curve Selection   324
    25.6. Use of Learning Curves   329
    25.7. Completion vs. Expenditure, Considering Learning   332

26. Labor Allowances and Multipliers   333

26.1.	Standard Hour Allowances	334
26.2.	Standard Hour Multipliers	339

MATERIALS

27. Material Costs and Allowances — 340

27.1.	Raw Material Prices	340
27.2.	Line Stock and Raw Material Ratios	342
27.3.	Total Bid Material Allowances	343

28. Integrated Circuit Packaging — 345

28.1.	Background	345
28.2.	Integrated Circuit Packages	349
28.3.	Integrated Circuit Packaging Process	353
28.4.	Typical Costs	359

29. Material Discount Curves — 362

29.1.	Uses of Material Discount Curves	362
29.2.	Application of Material Discount Curves	363

SUPPORT

30. Special Tooling and Test Equipment — 366

30.1.	Tooling, Test Equipment and Production Planning Program Cost Ratios	367
30.2.	Comparative Tooling and Manufacturing Methods	368
30.3.	Tooling Cost Factors	368
30.4.	Special Test Equipment Cost Factors	372

31. Manufacturing Engineering — 373

31.1.	Basis for Estimating Manufacturing Engineering	374
31.2.	Small Production Quantities—Minimum Manufacturing Engineering	374
31.3.	Moderate Production Quantities—100% Planning 100% Manufacturing Engineering	375
31.4.	Cost Comparison Example	379

32. Computer Aided Design and Manufacturing — 382

32.1.	Computer Aided Design	385

32.2.	Computer Aided Manufacturing	389
32.3.	Benefits and Costs	390

HOURLY RATES AND BURDEN

33. Hourly Labor and Burden Rates, Earnings   394

33.1.	Hourly Labor Rates	394
33.2.	Burden Rates	395
33.3.	Earnings	396

INDEX   **399**

PART ONE

INTRODUCTION AND SUMMARY LEVEL DATA

SECTION ONE

INTRODUCTION TO ESTIMATING STANDARDS

1.1. Merits of Standard Data	4
1.2. Basis of Standard Data in This Book	5

The estimating data of this book consist of Labor Time Standards; shop methods upon which the standards are based; and the allowances, learning curves, burden rates, and so forth, which convert labor hours and material dollars into total manufacturing costs. In addition, there are a great number of experience ratios by which any cost item, from line stock material to total program start up costs, can be estimated. The cost information under each subject addresses some or all of the following categories: a description of product applications; manufacturing methods, including a description of the process and major operational steps; labor content in minutes; cost and calculation methods; cost estimating work sheets; and general rules of thumb used in the estimating procedure.

A one-page summary table has been developed for each area of cost so that a complete estimate of an electronic system can be accomplished in abbreviated form. The summary tables are contained in Section 3 and a sample set of cost estimate forms in Section 2. These two sections contain a condensed working model of the entire book. Each subsequent section gives the detailed analysis and descriptive information for the summary tables.

Electronic manufacturing comprises the skills of a wide range of markedly different industry groups. A unit may perform an electronic function, but the physical hardware is produced by the standard manufacturing processes of machining, sheet metal fabrication, electroplating, painting and finishing, and various assembly operations. Much of this estimating data is therefore applicable to other industries that utilize the same manufacturing processes.

The estimating standards are aimed at practical application rather than detailed precision. The usual pages of detail time tables have been minimized because, although they are extremely accurate, few people have the time to use them. The summarized time standards coupled with the estimator's judgment is a much more practical approach.

1.1. MERITS OF STANDARD DATA

Estimating can be done by three general methods. The first is comparison. By this method a skilled or unskilled person can make gross comparisons of unknown items to known items. The second method is by having intimate experience with the operation and, thus, knowing approximately how many hours it will take to do the job. The third is by using standard time data. By this method all of the possible elements of work are measured, assigned a standard time, and classified in a catalog. When a specific operation is to be estimated, the required number of standard time values are added together to determine the total time. This method has certain advantages over other estimating methods:

1. It is more accurate. It is based on the work *content* rather than on the double estimate of how much work is to be done, and how long it will take to do it.

2. It is easier to justify. A series of individual elements and operations adding up to 100 hours is easier to justify than one overall judgment of 100 hours.

3. Standard data promote consistency between estimates and estimators. No two people think alike. A review of estimates of similar units made by different people without the use of common standards will prove this point. Conversely, standard data *will* show where there is a legitimate difference between similar equipment.

4. Standard data coupled with learning curves can be used to estimate labor for any production quantity. Where the experience method is used, the shop normally producing in the 5 to 50 range finds it hard to estimate the same job at 1000 units. Standard data plus learning curves will cover the entire quantity spectrum.

5. Personal experience with the specific operation is not mandatory when using standard data, although it is certainly desirable. But in comparison, the experience-dependent estimator *cannot* estimate operations that are outside his or her previous experience.

6. Standard data estimating emphasizes the method rather than various individuals' estimates of how much time an operation should take. Standard

data has already resolved the problem of how much time an element is worth. The remaining problem is to determine the correct elements that are required to accomplish the job.

Standard data can be used to estimate labor requirements for cost estimates, schedules, labor forecasts, make-or-buy decisions, and as a norm by which to compare actual shop performance.

1.2. BASIS OF STANDARD DATA IN THIS BOOK

The standard data in this book are based on predetermined time standards, time studies, and rough synthesis by methods analysis. The predetermined time standards measure body motions in 0.001 and 0.0001 minute, while the rough synthesis technique measures overall operations in 0.05 to 0.50 minute. The data are based on the experience of many companies and industries. The standards, allowances, and learning curves represent a system of estimating which has been in industrial use for many years.

The standards are based on a level of efficiency which is attainable by a job shop working in the 1 to 1000 quantity range. Producers in the 10,000 to 100,000 quantity range would require standards of greater precision than are presented here. Because short production runs (less than 1000 units) do not benefit from the economy of scale that is derived from automation of large production quantities, the time standards still reflect a high degree of manual operation. A few exceptions are in those classes of manufacturing operations where a third party or captive shop can achieve higher volume by combining many small orders of a similar product. Printed circuit boards fall into this category. It is common practice to fabricate several small circuits as a single panel and separate them after the operation is completed rather than run them through individually.

In the updating process, a number of the subject areas have progressed from cost estimating by process step to one of "commodity" estimating. This is particularly the case for such areas as integrated circuits (chips) and printed circuit boards (PCB). These products are now considered purchased materials in the same manner applied to discrete components (transistors, resistors, and capacitors). Cost estimating is done on a per lot or per square inch basis (in the case of circuit boards), and not built up by labor time standard at the detailed process step level. The latter estimating procedure has now moved to the next higher level of assembly—the packaging of integrated circuit chips and/or assembly onto PCBs.

SECTION TWO

ORGANIZATIONAL OVERVIEW AND ESTIMATING FORMATS

2.1. Organization	6
2.2. Estimating Forms	8

2.1. ORGANIZATION

Figure 2.1 provides an overview of the cost estimating process illustrating the relationship of the various cost estimating sections in the book. In addition to preliminary "Program Planning," there are three basic paths that together cover the total cost estimate of the particular product. The first is labor oriented involving the estimates of labor time standards for fabrication, assembly, integration, test and inspection and shipping. The estimate is built up in successive steps through application of labor allowances, learning curves and hourly rates and burden. A second path treats the raw materials and purchased components and their associated allowances and discounts. Manufacturing, tooling, and production support costs are covered as a third path using the appropriate ratios. Other costs and profit are then added to the sum of the three paths to arrive at total costs.

The user should refer to Section 33—Hourly Labor and Burden Rates, and Earnings—for more detailed treatment of the cost element terminology. Figure 33.1 is particularly helpful in this regard.

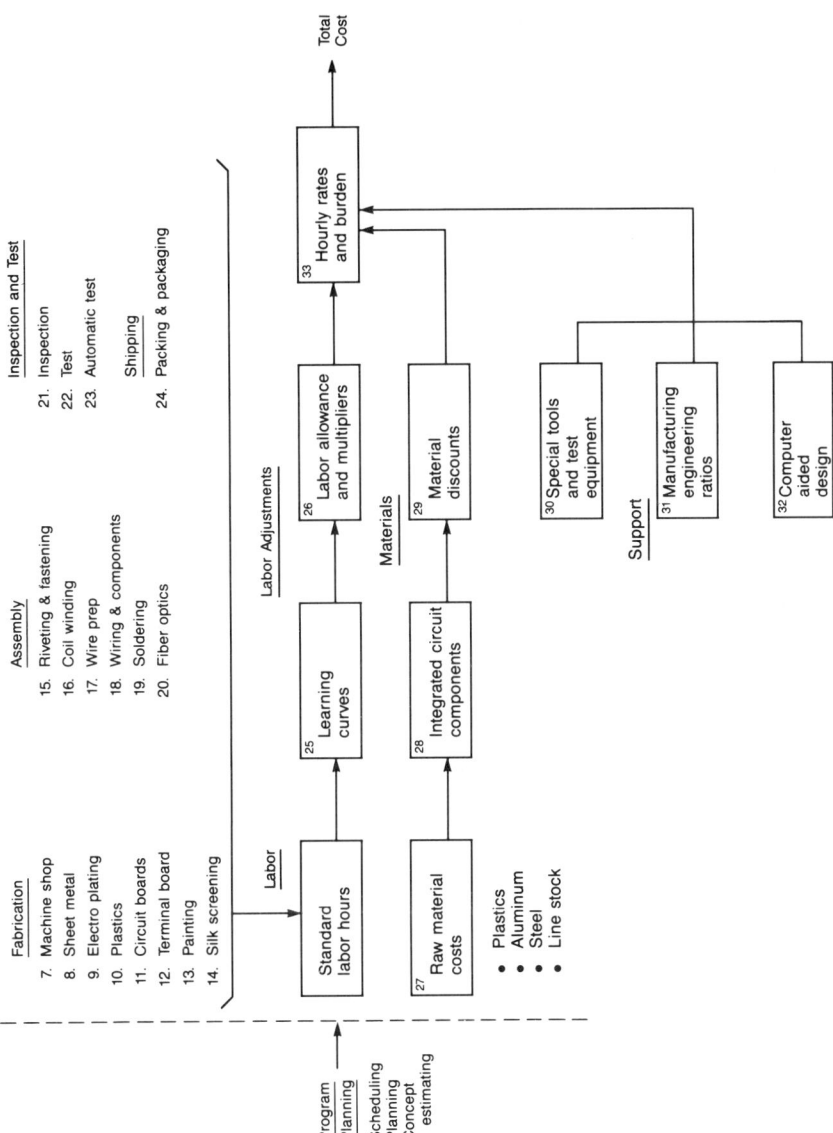

FIGURE 2.1. Cost estimating process.

8 ORGANIZATIONAL OVERVIEW AND ESTIMATING FORMATS

2.2. ESTIMATING FORMS

A set of cost estimating forms is provided as a framework to build detailed estimate data into total sales price. The forms are for convenience, but more important, they provide a consistent method of accumulating and factoring the raw estimate data. These forms are augmented by separate more detailed cost estimating worksheets which may be found in many of the sections.

COST ESTIMATE INFORMATION SHEET

Project Title _____ Proposal No. _____

About This Proposal:

Bid due: _____
Potential customer: _____
Proposal requested by: _____

Type of Business: __Military
                                     __Commercial

Type request: __Verbal
                                     __Written request for quote

Type pricing: __Fixed Price
                                     __CPFF
                                     __Budgetary (not legally binding)
                                     __Company internal only

Type program: __Production
                                     __R&D plus production follow on
                                     __R&D plus prototype

Quantity of units: A. _____
                                     B. _____
                                     C. _____

Schedule or period: A. _____
                                     B. _____
                                     C. _____

Prior History:

Previously built by: _____
Quantity: _____ Rate: _____ Date: _____
Type contract: _____ Fixed price: _____ CPFF _____
Cost per unit: _____
Other information: _____
_____

Cost Estimate Ground Rules

Design status   __Predisgn.   __In process   __Complete
Basis for estimate: __Verbal desc.   __Prelim. dwgs.   __Prod. dwgs

Other: _____
_____
_____

ESTIMATING FORMS

COST ESTIMATE SUMMARY MANUFACTURING

To _____ Class Estimate: _____ Project Name:

Sample Cost Estimate _____

_____ Fixed Price ☒ _____

_____ Incentive   ☒ _____

_____ CPFF   ☐ _____

_____ Budgetary   ☐ Date:

Item	Quantities		
	10	100	1000
1.	Max. 2/Mo.	Max. 10/Mo.	Max. 50/Mo.
2.			
3.			
4. Recurring cost			
5. Mfg. cost per unit	XXX	XXX	XXX
6. Mfg. cost per lot	XXXX	XXX XXX	X XXX XXX
7.			
8. Non-recurring cost			
9. Production planning	XX	XXXX	XXXX
10. Special tooling		XXXX	XXXX
11. Special test equipment		XXXX	XXXX
12. Subtotal	XX	XXXXX	XXXXXX
13.			
14. Total program	XXXX	XXXXXX	X XXX XXX
15.			
16.			
17.			
18.			
19.			
20.			
21.			
22.			
23.			
24.			
25.			

Remarks: Originator:
               Approvals: (1) _____
                           (2) _____
                           (3) _____
                           (4) _____

10 ORGANIZATIONAL OVERVIEW AND ESTIMATING FORMATS

MANUFACTURING COST ESTIMATE BREAKDOWN

P. No. _____
Date _____

To _____ Class Estimate: Project Name:
_____ Fixed Price   ☒   Sample Cost Estimate _____
_____ Incentive   ☒   Mfg. Cost per Unit _____
_____ CPFF   ☐   Lot Quantity _____ 100 ____
_____ Budgetary   ☐   Max. Monthly Rate _____ 10 ____

Manufacturing Cost	Rate	Hours per Unit	$/ Unit	$/ Lot
1. Direct labor and overhead				
2. Fabrication	X	X	X	
3. Assembly	X	X	X	
4. Inspection	X	X	X	
5. Test	X	X	X	
6. Manufacturing engineering	X	X	X	
7.				
8.				
9. Total direct labor		X	X	
10. Overhead	(X%)		X	
11. Total direct labor and overhead			X	
12. Other direct costs				
13. Material cost	(X%)	X	X	
14.				
15.				
16.				
17. Travel				
18. Fares				
19. Per diem				
20. Total other direct costs			X	
21. Total direct costs			X	
22. G and A expense	(X%)		X	
23. Total cost			X	
24. Earnings	(X%)		X	
25. Total price			X	X
26.				
27.				
28.				

Remarks   Originator: _____
                                               Approvals: _____

LABOR HOURS SUMMARY

Date _____ Equipment _____
Originator _____ Unit _____
Drawing No. _____

Alternate Quote Quantities	Allowance Range	Fabrication	Assembly and Packing	Inspection	Test	Manufacturing Engineering	Total
A. Quantity _____							
1. Standard hours		X	X		X		
2. Learning factor		X	X		X		
3. Realized hours		X	X		X		
4. Variance from measured labor	10–20%						
5. Internal rework and repair	10–20%						
6. Total labor hours		X	X		X		
7. Engineering change allowance	0–15%						
8. Design growth allowance	0–30%						
9. Other							
10. Total allowances							
11. Grand total hours		X	X	X	X	X	
B. Quantity _____							
1. Standard hours							
2. Learning factor							
3. Realized hours							
4. Variance from measured labor	10–20%						
5. Internal rework and repair	10–20%						
6. Total labor hours							
7. Engineering change allowance	0–15%						
8. Design growth allowance	0–30%						
9. Other							
10. Total allowances							
11. Grand total hours							
C. Quantity _____							
1. Standard hours							
2. Learning factors							
3. Realized hours							
4. Variance from measured labor	10–20%						
5. Internal rework and repair	10–20%						
6. Total labor hours							
7. Engineering change allowance	0–15%						
8. Design growth allowance	0–30%						
9. Other							
10. Total allowances							
11. Grand total hours							

Note: Any one, all, or none of the above allowances may be used for a given estimate.

MATERIAL COST SUMMARY

Date _____   Equipment _____
Originator _____   Unit _____
                                 Dwg. No. _____

Cost Items	Allowance Range	Cost/Unit per Quantity			
		Estimate		Quotation Quantity	
		X	X		
		Quantity Discount			
Standard Purchased Parts		X%	X%	%	%
1. Basic purchased parts cost		X	X		
2. Packing material		X	X		
3. Total basic cost		X	X		
4. Line stock – hardware, solder, steel, brass, Al, etc.	1–4%				
5. Attrition	1–5%				
6. Purchasing variance	0–5%				
7. Engineering change allowance	0–10%				
8. Design growth allowance	0–30%				
9. Inflationary price increase	3%/yr.				
10. Other					
11. Total allowances			X		
12. Total standard parts cost			X		
Subcontract Parts or Services	Ref. Page				
1 _____			X		
2 _____					
3 _____					
4 _____					
5 _____					
6 _____					
7 _____					
8 _____					
9 _____					
10 _____					
11 _____					
12 _____					
13 _____					
14 _____					
15 _____					
Total Material Cost Unit			X		

Note: Any one, all, or none of the above allowances may be used for a given estimate.

ESTIMATE DATA SHEET

Originator _____
Date _____
Equipment _____
Unit _____
Dwg. No. _____

Assembly Pos. 1 2 3 4 5 6	Part Number	Chg. Ltr.	Qty. Rqd.	Description	Labor: Standard Hr./Required Qty.							Tooling		Material	
					Fabrication		Assembly		Test		Code	$	$ Per Unit	Total $
					SU	Oper.	SU	Oper.	SU	Oper.				
X	X	X	X	X	X	X		X				X	X	X
	X	X	X	X	X	X		X				X	X	X
	X	X	X	X	X	X		X				X	X	X
X	X	X	X	X	X	X		X				X	X	X
	X	X	X	X	X	X		X				X	X	X
X														
X														
				Total	X	X		X				X		X

13

SECTION THREE

SUMMARY DATA

This section presents a summary of the detailed analyses contained in Part III. It can be used to perform quick, but less thorough estimates of a particular cost situation where "ballpark" information is desired. Because of the abbreviated nature of this data, it is recommended that the user acquire some familiarity of the full scope of the book before using the Summary Data.

The tables listed use the same section numbers as their respective analyses sections.

Sec. 7	Machining Operations	15
Sec. 8	Sheet Metal Operations	19
Sec. 9	Electroplating and Chemical Surface Treatment of Metals	23
Sec. 10	Plastics	24
Sec. 11	Printed Circuit Boards	25
Sec. 12	Terminal Board Fabrication	25
Sec. 13	Painting Operations	26
Sec. 14	Silk Screen Printing and Engraving	27
Sec. 15	Riveting, Mechanical Assembly, and Fastening	27
Sec. 16	Coil Winding Operations	30
Secs. 17, 18	Wiring	30
Secs. 21, 22	Inspection and Test	32
Sec. 24	Packing and Packaging	32
Sec. 25	Learning Curve	34
Sec. 26	Labor Allowances and Multipliers	34
Sec. 27	Material Cost and Allowances	35
Sec. 29	Material Discount Curves	35
Sec. 30	Special Tooling and Test Equipment	36
Sec. 31	Manufacturing Engineering	37
Sec. 33	Hourly Labor and, Burden Rates, Earnings	38

SUMMARY DATA FOR SEC. 7. Machining Operations

Reference Section	Operation	Time Allowance
7.1	Power hack saw or band saw, set up	8.08 minutes/OCC
	Add: to weld blade for internal cuts	8.03 minutes/OCC
	Parts Handling—(Incorporates Input & Output Stock Handling)	0.30 minutes/Part Produced
	Allow additional handling time for contour cuts	
	Saw time for flat stock-straight cuts	

	Minutes per In. of Cut		
	Aluminum	Mild Steel	S/S
⅛ in. Thick-minutes per in. of cut length	0.02	0.04	0.12
½ in. Thick	0.06	0.16	0.22
Saw time for bar stock-minutes per inch of diameter	0.60	1.60	3.03

Reference Section	Operation	Time Allowance
7.2	Turret Lathe (No. 3 Size) Set-up	85.25 minutes per job
7.2.1	Parts Handling (Incorporated Input and Output Stock Handling)	1.74 minutes per part Produced
7.2.2	Turn O.D. or Bore I.D.	

	Minutes per In. of Cut		
	Aluminum	Mild Steel	S/S
Making a roughing cut, up to 0.150 in. deep, 1 in. long on 1 in. diameter stock	0.21	0.26	0.30
Make 0.025 in. deep finish cut, 1 in. long	0.23	0.32	0.37

Reference Section	Operation	Time Allowance
7.2.3 and 7.2.4	Tap or thread with tap or die ¼ in. × 32 NF	
	Additional set-up for tap or thread	1.32 minutes/hole (average)

	Minutes per In.		
	Aluminum	Mild Steel	S/S
Tap or thread per in. or less of thread length	0.289	0.435	0.514

Reference Section	Operation	Time Allowance
7.2.5	Taper	
	Additional set-up for taper	1.53 minute/taper angle

(continued)

SUMMARY DATA FOR SEC. 7. *(Continued)*

Reference Section	Operation	Time Allowance		
	Turn taper	Minutes per Linear In.		
		Aluminum	Mild Steel	S/S
		0.244	0.370	0.488
7.3	Engine lathe (10 in. swing) set-up	39.35 minutes/job set-up		
7.3.1	Parts handling (Incorported input and output stock handling)	1.41 minutes/part produced		
7.3.3	Turn O.D. or bore I.D.	Minutes per In. of Length		
		Aluminum	Mild Steel	S/S
	Make a 0.175 in. deep roughing cut 1 in. long, 6 in. diameter stock	0.09	0.31	0.34
	Make a 0.025 in. deep finish cut 1 in. long on 6 in. diameter stock	0.23	0.32	0.37
7.3.5	Turn a Taper up to 7°			
	Additional set-up for taper	3.37 minutes/taper angle		
		Minutes per In. of Taper		
		Aluminum	Mild Steel	S/S
	Turn taper, per in. of taper length	0.24	0.37	0.49
7.4	Milling			
	Set-up	45.81 minutes/set-up		
7.4.1	Parts handling	1.52 minutes/parts produced		
7.4.2	Advance, back off, adjust table	0.27 minutes per cut		
7.4.3	Profile or end mill ½ in. deep × ¾ in. wide	Minutes per In. of Cut		
		Aluminum	Mild Steel	S/S
		1.03	2.80	7.35
7.4.4	Surface or face mill ½ in. deep × 1 in. wide	Minutes per In. of Cut		
		Aluminum	Mild Steel	S/S
		0.09	0.23	0.29

SUMMARY DATA FOR SEC. 7. *(Continued)*

Reference Section	Operation		Time Allowance			
7.4.5	Side mill, straddle mill or slotting mill ½ in. deep × 1 in. wide		Minutes per In. of Cut			
			Aluminum	Mild Steel		S/S
			0.06	0.20		0.35
7.4.6	Corners (½ in. Radius) groove or chamfer (½ × ½ in.) slot (1 in. deep × ⅜ in. wide)		Minutes per In. of Cut			
			Aluminum	Mild Steel		S/S
			0.06	0.20		0.35
7.5	Drill press		23.82 minutes per job			
7.5.1	Operation	Handling Time Minutes/Hole	Minutes per In. of Depth			
			Aluminum	Mild Steel		S/S
	Position stock and drill ⅛ in. hole	0.55	0.09	0.25		0.55
	Position stock and drill 2 in. hole	1.10	0.33	0.68		1.34
7.5.2	Countersink ⅛ in. diameter hole	0.55	0.05	0.15		0.30
7.5.2	Operation	Handling Time Minutes/Hole	Minutes per In. of Depth			
			Aluminum	Mild Steel		S/S
	Countersink 2 in. diameter hole	1.10	0.26	0.37		0.69
	Machine tap ⅛ × NS40	0.55	0.06	0.10		0.18
	Machine tap 2 × 4½ TPI	1.10	0.06	0.10		0.18
7.6	Broach—Set-up		16.35 minutes per job			
7.6.1	Handle Parts and Operate Broach	Weight	Minutes per Piece			
			0–20	21–40	41–80	Over 80
		Time	4.29	4.61	5.01	8.33

(continued)

SUMMARY DATA FOR SEC. 7. *(Continued)*

Reference Section	Operation	Time Allowance

7.6.2	Machine Time	Keyway		Minutes per In. of Keyway Length		
		Depth	Width	Aluminum	Mild Steel	S/S
		$1/16$	$1/8$	0.16	0.31	0.61
		$1/4$	$1/2$	0.42	0.92	1.82
		$1/2$	1	0.52	1.02	2.00

7.7	Grind, centerless—set-up	42.33 minutes per job

7.7.1	Grind Time		Handling time minutes/part	Minutes per Linear In.		
				Aluminum	Mild Steel	S/S
	Through feed-1 in. diameter, .013 in. stock removed		—	0.08	0.08	0.08
7.7.2	In feed—1 in. diameter		1.55	0.10	0.10	0.10

7.8	Grind—external cylindrical set-up = 25.09 minutes per job

7.8.1	Handle Parts and operate grinder		Minutes per Piece			
		Weight	0–20	21–40	41–80	Over 80
		Time	2.87	3.17	3.57	6.30

7.8.3	Grind	Diameter (in.)	Stock Removed (in.)	Minutes per Linear In. of Grind		
				Aluminum	Mild Steel	S/S
		$1/2$	0.010	0.023	0.075	0.100
		1	0.010	1.20	1.50	0.162
		4	0.010	1.16	0.934	0.352

7.9	Grind, internal cylindrical set-up =	34.01 minutes per job

7.9.1	Handle parts and operate grinder		Minutes per Piece			
		Weight	0–20	21–40	41–80	Over 80
		Time	1.20	1.50	1.90	4.73

SUMMARY DATA FOR SEC. 7. *(Continued)*

Reference Section	Operation			Time Allowance		
7.9.3	Grind time—inside of hole			Minutes per Hole		
	Hole Depth (in.)	Diameter (in.)	Stock Removed (in.)	Aluminum	Mild Steel	S/S
	1	½	0.006	0.252	0.780	0.780
	2	1	0.010	0.420	1.300	1.300
	3	4	0.018	1.300	3.950	3.950

7.10 Grind surface set-up = 19.08 minutes per job
Handle parts and operate grinder

	Minutes per Piece				
Weight	0−10	11−20	21−40	41−80	Over 80
Time	0.15	0.48	1.27	1.67	4.50

7.10.1 Grind 0.008 minutes per in.^{2} of surface ground

SUMMARY DATA FOR SEC. 8. **Sheet Metal Operations**

Reference Section	Operation	Set-up Minutes	Minutes/Operation		
			Small 3 × 3 in.	Medium 18 × 18 in.	Large 30 × 30 in.
8.1	Set-up power shear—per set-up	6.20			
8.1.1	Cut blank with power shear per piece produced		0.36	1.16	1.81
8.2	Set-up power notcher per set up	6.94			
8.2.1	Notch blanks, including parts handling per notch		0.16	0.27	0.49
8.3.1	Set-up turret punch per set-up	32.77			
8.3.2	Handle blanks for punching per completed blank		0.29	0.47	0.84

(continued)

SUMMARY DATA FOR SEC. 8. *(Continued)*

Reference Section	Operation	Set-up Minutes	Minutes/Operation		
			Small 3 × 3 in.	Medium 18 × 18 in.	Large 30 × 30 in.
8.3.3	Punch hole per hole punched		0.06	0.06	0.06
	Set-up drill press per set-up	23.82			
	Handle blanks for drilling if part weighs over 40 pounds add 0.110 minutes for each hole-to-hole move with no template or 0.087 minutes for each hole-to-hole move if a template is used. Per completed blank.		0.29	0.47	0.84

8.3.3
Hole Drilling Time Values

Hole Diameter	Hole Depth (in.)	Location Guide	Move Time Hole-to-Hole		Machine Time		
			Through 40 pounds	Over 40 pounds	Mild Steel	S.S	Aluminum
1/8	1/8	Scribed Lines	0.078	0.110	0.05	0.15	0.02
1/8	1/8	Template	0.061	0.087	0.05	0.15	0.02
1/4	1/4	Template	0.061	0.087	0.13	0.109	0.11
3/4	3/8	Template	0.061	0.087	0.23	0.42	0.05
1	1/2	Template	0.061	0.087	0.35	0.61	0.07
2 to 4 in. Flycutter	1/8	Scribed Lines	0.078	0.110	2.16	3.76	0.25

8.4	Press operations						
8.4.1	Set-up to blank or pierce Minutes per die				21.15	24.85	30.85
	Blank and pierce Minutes per part produced				0.36	0.42	1.53
8.4.2	Dimple or joggle Set-up press				21.15	24.85	30.85
	Dimple or joggle Per piece produced				0.36	0.42	1.53

SUMMARY DATA FOR SEC. 8. *(Continued)*

Reference Section	Operation	Set-up Minutes	Minutes/Operation		
			Small 3 × 3 in.	Medium 18 × 18 in.	Large 30 × 30 in.
8.4.3	Brake form				
	Set-up press	8.89			
	Change die (Top)	4.87			
	(Bottom)	7.65			
	Brake form parts				
	Per piece produced		0.23	0.38	0.72
8.4.4	Roll form				
	Set-up	4.82			
	Handle parts				
	Per piece produced		0.18	0.43	0.77
	Per in. of blank length		0.02	0.03	0.06
8.4.5	Deep draw				
	Set-up		21.05	24.85	30.85
	Run time				
	Minutes per part produced		0.25	0.40	0.74
8.4.6	Anneal				
	Set-up furnace				
	Minutes per temperature change	7.88			
	Run time				
	Batch furnace				
	Per blank		0.15	0.60	1.88
	Run time				
	Conveyor furnace				
	Per blank		0.13	0.50	1.00
8.4.7	Hydroform				
	Set-up press				
	Per set-up	28.87			
	Run time				
	Per blank produced		0.25	0.40	0.74
8.5	Trim, profile, route, burr edge				
	Set-up trim, profile or route				
	Per set-up	28.50			
8.5.1	Handle and jib part				
	Per part handles		0.26	0.54	1.14
8.5.2	Trim, profile route and burr edge				
	Aluminum, brass, phenolic or epoxy—				

(continued)

SUMMARY DATA FOR SEC. 8. (Continued)

Reference Section	Operation	Set-up Minutes	Minutes/Operation		
			Small 3×3 in.	Medium 18×18 in.	Large 30×30 in.
	¼ in. thick Per in. of cut		0.07	0.07	0.07
8.6	Burr removal with abrasive tool such as sander, vibrator, file Set-up	6.23			
8.6.1	Handle part Per part		0.13	0.28	0.62
8.6.2	Remove burr from edge Per in.		0.02	0.02	0.02
8.6.3	Remove burr from flat surface Per ft.^{2}		0.20	0.50	0.50
8.6.4	Remove burr from hole Per hole		0.03	0.03	0.03
8.7	Arc weld				
8.7.2	Set-up welding machine	12.91			
	Set-up welding fixture (If required)	31.82			
8.7.3	Handle parts to welding fixture		0.29	0.47	0.84
	Weld (MIG Process) Per in. of weld				
	0.062 in. Stock		0.25	0.25	0.25
	0.125 in. Stock		0.40	0.40	0.40
	0.250 in. Stock		0.75	0.75	0.75
8.7.4	Stress relieve Set-up	6.00			
	Cycle time		0.29	0.47	0.84
8.7.5	Grind weld smooth Set-up sander or grinder	6.00			
	Grind or sand edge Per linear in.		0.03	0.04	0.05
8.8	Spot weld Set-up	12.50			
	Handle parts Sum time for each input part by size		0.16	0.17	0.49
	Weld Per spot		0.07	0.07	0.10
8.9	Rivet (low volume operation only) See				

SUMMARY DATA FOR SEC. 8. *(Continued)*

Reference Section	Operation	Set-up Minutes	Minutes/Operation		
			Small 3 × 3 in.	Medium 18 × 18 in.	Large 30 × 30 in.
8.9.1	Section 9 also Set-up	19.77			
	Handle parts				
	Per completed part		0.13	0.28	0.62
	Hand fed rivet, Machine driven				
	Per rivet		0.18	0.18	0.18
8.9.2	Machine fed rivet				
	Per rivet		0.05	0.05	0.05
	See Section 9 for surface protection operations.				

SUMMARY DATA FOR SEC. 9. Electroplating and Chemical Surface Treatment of Metals

Process	Minutes to Process One Batch	
	First	Successive
Anodize Aluminum, Clear	28	19
Anodize Aluminum, Dye	31	22
Anodize Aluminum, Hard, Clear	25	16
Anodize Aluminum, Hard, Dye	28	19
Black Oxide Coat Iron and Steel[a]	26	17
Cadmium Plate Steel[b]	23	14
Chemical Conv. Coat Aluminum	22	13
Chrome Plate Copper[a]	23	14
Chrome Plate Steel[b]	25	16
Copper Plate Copper[a]	22	13
Gold Plate Aluminum	31	22
Gold Plate Copper[a]	23	14
Nickel Plate[a]	21	12
Nickel, Electroless[a]	27	18
Passivate Stainless Steel[a]	19	10
Phosphate Treat Steel[a]	23	14
Rhodium[a]	25	16
Silver Plate Aluminum	31	22
Silver Plate Copper[a]	23	14
Tin Plate Aluminum, Hot Oil Fuse	35	26
Tin Plate Copper[b]	22	13
Zinc	24	15

[a] Heat treated parts require abrasive cleaning
[b] Heat treated parts require abrasive cleaning and baking

SUMMARY DATA FOR SEC. 10. Plastics

Reference Section	Operation	Typical	Range
10.3 (line 1)	In-process scrap ratios	5%	3–8%
	Molded part rejects	2%	1–3%
10.3 (line 2 and Fig. 10.5)	Material costs (in absence of quotes)		
	Injection molding	$2.00/lb.	$0.78–$4/lb.
	Fluoroplastics (Teflon)	—	$10–$20/lb.
	Polyamide-imides (Torlon)	—	$10–$20/lb.
10.3 (line 3)	Labor (in minutes)		
	Thermoplastic cycle time	0.4	
	Thermosets cycle time	1.2	
	Preform shop press cycle	0.1	
10.3 (line 4)	Tooling costs		
	Simplest die set	$10,000	—
	Small part dies	—	$10,000–$20,000
	Large complex dies	—	$100,000–$150,000

SUMMARY DATA FOR SEC. 11. Printed Circuit Boards

Double-sided
Pre-Production
  Release for manufacture
    New orders   120.0 minutes
    Repeat orders   24.0 minutes
  Prepare N/C tapes (new orders only)   4.0 minute/in.
  per inch of perimeter
Production   0.30 minute/in.^{2a}

Multi-layer
Pre-Production
  Release for manufacture
    New Order   90.0 minute
                                             + 30.0 minute/circuit layer
    Repeat Order   24.0 minute
  Prepare N/C tapes (new orders only)   2.0 minute/in.
  per inch of perimeter, per circuit layer
Pre-Lamination
  Inner circuit (2 sides)   0.072 minute/in.^{2a}/sheet
  Prepreg   0.005 minute/in.^{2a}/sheet
  Outer circuit   0.005 minute/in.^{2a}/sheet

SUMMARY DATA FOR SEC. 11. *(Continued)*

Laminate
    Assemble books   0.13 minute/sheet
    Laminate   0.001 minute/in.^{2}

Post-Lamination
    Inspect for registration   1.06 minute/board
    Process as double-sided board   0.29 minute in.^{2a}

Test

Test for shorts and opens
    Set-up and program automatic test   3.75 minute/point
        equipment, per contact point
    Test each board   0.50 minute/board

[a] Area of one side. For example, use 80 in.^{2} for an 8 by 10 in. board.

SUMMARY DATA FOR SEC. 12. Terminal Board Fabrication

Reference Section	Operation	Set-Up Minutes	Minutes/ Operation	Number Operation	Total Minutes
7.1	Cut blank to size, 2 edges per in.	8.08	0.02		
8.6	Deburr 4 edges, per in.	6.23	0.02		
11.6	Drill holes				
	Set-up, per hole size	23.82	—		
	Drill, per hole		0.10		
8.6	Deburr holes, per hole, each side	6.23	0.06		
14.1	Silk Screen	63.30	0.73		
15.2	Stake Terminal	8.05	0.04		
15.1	Handle (5 operations/ board)		0.06	5	0.30
	Total set-up	115.71			
	Total oper. per board × ___ boards/lot	= ___			
	Total time/lot	= ___			

SUMMARY DATA FOR SEC. 13. Painting Operations

Reference Section	Operation	Set-Up or Drying-Min.	Minutes/Unit/Spray Coat — Part Size			
			To 3"	To 8"	To 20"	To 30"
13.1	Surface preparation (typical)					
	Steel or Aluminum-phosphate treat					
	Aluminum-chemical film		See Section 3, Table 9			
	Magnesium-dichromate					
13.2	Mask and de-mask part for painting (See also 13.11, below)					
	Minor		0.50	1.70	3.40	6.00
	Average		1.00	3.40	6.40	10.00
	Major		2.50	8.50	19.00	25.00
13.3	Set-up time per paint type	SU 22.10				
13.4	Primer					
	Wash primer	Dry 12	0.50	0.77	1.04	1.27
	Zinc chromate	Dry 12	0.80	1.22	1.74	2.17
13.5	Surfacer (includes power buff)	Dry 12	2.40	3.52	4.14	6.02
13.6	Lacquer					
	Flat	Dry 18	0.80	1.22	1.74	2.17
	Gloss	Dry 18	1.10	1.82	2.54	3.22
13.7	Enamel or Modified Vinyl					
	Flat	Dry 30	0.80	1.22	1.74	2.17
	Gloss	Dry 60	1.10	1.82	2.54	3.22
13.9	Plastic protective film, strippable	Dry 60	0.70	1.12	1.59	1.92
13.10	Fungicide (spray application)	Dry 30	0.90	1.42	2.04	2.52
13.11	Miscellaneous detail values					
	1. Handling time	/part	0.20	0.26	0.32	0.42
	2. File or burr edge	/in.	0.02	0.02	0.02	0.02
	3. Sand or grind by power	/sq. ft.	1.09	1.09	1.09	1.09
	4. Sand by hand	/sq. ft.	2.18	2.18	2.18	2.18
	5. Blow off surface	/sq. ft.	0.05	0.05	0.05	0.05
	6. Wash surface with solvent	/sq. ft.	0.05	0.05	0.05	0.05
	7. Spray paint	/sq. ft.	0.06	0.06	0.06	0.06
	8. Brush paint, flat	/sq. ft.	0.24	0.24	0.24	0.24
	edges, lips	/sq. ft.	1.00	1.00	1.00	1.00
	9. Apply & remove masking tape	per in.	0.02	0.02	0.02	0.02
	Trim tape	per in.	0.11	0.11	0.11	0.11
	10. Assemble & remove masking plugs and stencils	per item	0.10	0.10	0.10	0.10

Note: SU = set-up time.

SUMMARY DATA FOR SEC. 14.   Silk Screen Printing and Engraving

Reference Section	Operation	Time (Min.)
14.2	Photographic operations to prepare silk screen stencil...	
14.2.1,3,4	from line copy	48.8
14.2.2,3,4	from continuous tone copy	57.1
14.3	Silk screen printing, including decals	
14.3.1	Set-up for production, per color	14.5
14.3.2	Print, per color, per print	0.73
14.3.3	Add 2 cycles (14.2.1 & 14.2.2) for decals (to apply clear lacquer layers)	
14.4	Engraving	
14.4.1	Set-up	10.2
	per letter	0.3
14.4.2	Engrave and lacquer...	
	per piece	1.1
	per letter, per piece	
	Letter size	
	to $1/8''$	0.17
	to $1/4''$	0.21
	to $3/8''$	0.24
	to $1/2''$	0.28
	to $5/8''$	0.32

SUMMARY DATA FOR SEC. 15.   Riveting, Mechanical Assembly, and Fastening

Reference Section	Operations (Values include tool handling)	Set-Up Minutes	Min./ Oper.	No. Oper.	Total Minutes
	Mechanical Assembly				
15.1	Pick up and position parts and aside				
	Washers, brackets, switches, etc. up to 2 pounds		0.06		
	Subassemblies, transformers, etc. 3 to 8 pounds		0.11		
	large chassis, panels, etc. over 8 to 40 pounds		0.19		
15.3	Threaded fasteners—assemble and tighten with hand tools				
	1 piece (nut on stud, bolt into tapped hole, etc.)		0.29		
	2 pieces (nut and bolt, etc.)		0.35		

(continued)

SUMMARY DATA

SUMMARY DATA FOR SEC. 15. *(Continued)*

Reference Section	Operations (Values include tool handling)	Set-Up Minutes	Min./ Oper.	No. Oper.	Total Minutes
15.2	Rivet or stake				
	Use part handling time from section 15.1 to cover positioning parts for riveting or staking, set up to rivet or stake	8.05	0.04		
	Stake or rivet includes position part, activate machine, and move to next hole. Rivets automatically fed. Holes previously prepared. Mechanically driven riveter.				
	Add time to stake a previously assembled nut or terminal		0.04		
15.2	Arbor press – manually operated				
	Handle part to press and aside				
	Washers, brackets, switches, etc. up to 2 pounds		0.06		
	Subassemblies, transformers, etc. 3 to 8 pounds		0.11		
	Large chassis, panels, etc. over 8 to 40 pounds		0.19		
	Set-up press, adjust back-up plate, position stops, install ram nose	12.74			
	Press part and move to next hole		0.54		
15.2	Air operated press				
	Handle part to press and aside				
	Washers, brackets, switches, etc. up to 2 pounds		0.06		
	Subassemblies, transformers, etc. 3 to 8 pounds		0.11		
	Large chassis, panels, etc. over 8 to 40 pounds		0.19		
	Set-up press, adjust back-up plate, position stops, set pressure, install ram nose	12.74			
	Press part and move to next hole		0.27		
15.4	Rubber stamp employee ID	0.10	0.04		
15.4	Glyptol – apply to 1 point glyptol or cement	0.27	0.05		
	Apply to screw, nut or 1 square inch of surface	0.27	0.09		
15.4	Grommet – install				
	Includes using guide wire, and arbor press		1.23		
	Snap ring – Install				
	Includes using hand or power pliers		0.43		
	Cotter pin – Install				
	Includes installing cotter pin, and bend back using hand tools		0.74		

SUMMARY DATA FOR SEC. 15. *(Continued)*

Reference Section	Operations (Values include tool handling)	Set-Up Minutes	Min./ Oper.	No. Oper.	Total Minutes
	Drive pin or roll pin—Install				
	Includes driving pin in previously prepared hole, using hand tools		0.20		
	Knob—Install or shaft				
	Friction fit		0.13		
	Tighten set screw		0.17		
	Decal—Install—pressure sensitive adhesive				
	No backing		0.20		
	Peel backing and apply		0.28		
	Decal—Install, non-pressure sensitive				
	Remove decal from water, slip off backing, position, smooth with cloth		0.24		
15.6	Wrap cable with electrical tape—				
	1 layer per linear inch	0.14	0.09		
	Wrap wire junction with electrical tape	0.14	0.54		
15.8	Walk—minutes for 100 feet		0.52		
	Drill hole with portable electric				
	per hole 3/16" dia hole in .09" mild steel	0.64	0.57		
	per hole 3/16" dia hole in 1/8" aluminum	0.64	0.24		
15.9	Dynamic Balance of Rotating Part				
	For Military Specification				
	Weight to 40 pounds		135.00		
	Weight 41 to 600 pounds		235.00		
	Weight over 600 pounds		480.00		
	For Commercial Specification				
	Weight to 40 pounds		65.00		
	Weight 41 to 600 pounds		110.00		
	Weight over 600 pounds		240.00		

SUMMARY DATA FOR SEC. 16. Coil Winding Operations

Reference Section	Operations (Values Include Tool Handling)	Set-Up Minutes	Min./ Oper.	No. Oper.	Total Minutes
16	*Coil Winding*				
16.2	Manual time per winding-handle form to and from chuck, tape or solder start and finish leads		0.55		
16.4.14	Winding time				
	Space per turn	0.4	0.002		
	Close, universal, groove per turn	0.4	0.003		
	Hand wind per turn	0.1	0.013		
	Optional operations				
16.4.17	Make tap and anchor with tape		0.20		
16.4.35,37	Cut and cement Kraft paper to coil		0.15		
16.4.39	Saw multiple paper windings/ inches diameter		0.15		
16.4.40,21	Pull tap and tape		0.25		
16.4.34,25,28	Prepare and attach insulated lead		0.20		
16.4.37	Brush coil with compound/				
	½ inch		0.04		
	handle brush		0.03		

SUMMARY DATA FOR SECS. 17 & 18. Wiring

Reference Section	Operation (Run Time Values in Minutes)	Machine Prep.			Hand Prep.	
		Set-Up Min.	Length		Length	
			15"	60"	15"	60"
18.1	Insulated wire-prepare and install (2) ends					
18.1.1	Crimp and solder					
	Point to point	5.04	0.72	1.11	2.00	2.22
	Lay in U channel		0.77	1.29	2.05	2.40
	Lace harness	5.04	1.07	1.83	2.35	2.94
18.1.2	Taper pin (solderless)					
	Point to point		0.58	0.98	1.10	1.32
	Lay in U channel		0.77	1.29	2.05	2.40
	Lace harness	5.04	1.07	1.83	2.35	2.94
18.1.3	Pneumatic wrap (solderless)					
	Point to point	5.04	0.35	0.70	0.64	0.87
	Lay in U channel		0.41	0.87	0.69	1.10
	Lace harness		0.69	1.38	0.98	1.61

SUMMARY DATA FOR SECS. 17 & 18. *(Continued)*

Reference Section	Operation (Run Time Values in Minutes)	Machine Prep. Set-Up Min.	Machine Prep. Length 15"	Machine Prep. Length 60"	Hand Prep. Length 15"	Hand Prep. Length 60"
18.2	Bus-wire-cut, crimp, solder (2) ends			0.43	0.50	
18.3	Resistor, etc. (2) ends					
	Crimp and solder to terminals	0.75		0.70	0.78	
	Crimp to PC board and dip solder	0.75		0.42	0.53	
18.4	Sleeving-C/L and thread to lead			0.35	0.24	
18.5	Shielded cable-prepare and install (2) ends					
	Single conductor				4.82	5.29
	Double conductor				6.16	6.47
18.6	Coax cable-prepare and install (2) ends					
	Ground lead termination				4.82	5.29
	Connector termination	2.52			9.67	10.47
	Connect 1st end, ground lead 2nd end	2.52			7.25	7.88
17.10	Connector-mechanical assembly to harness					
	Small 5–10 pin				3.11	
	Medium 10–25 pin				4.89	
	Large 25–40 pin				7.08	
17.12	Pull tubing over cable					
	Regular = 0.978 per foot	1.38				
	Heat Shrink = 1.782 per foot					
17.13	Twist cable wire per foot				0.09	
17.14	Spot tie harness per foot			0.69	1.15	
19.4	Seam solder per inch				0.14	
18.2	Develop wire list from schematic: Per wire				2.50	2.50
17.16	Fabricate harness nail board	4.86 × number of wires				
18.13	Set up components at work stations	0.027				

SUMMARY DATA

SUMMARY DATA FOR SECS. 21 & 22. Inspection and Test

Reference Section	Labor Category	Percent of Fabrication and Assembly Hours		
		Minimum	Average	Maximum
21.0	Inspection			
21.1	Receiving & Source	2	5	7
21.1	Production	5	10	15
21.1	Composite	7%	15%	22%
22.0	Test			
22.1	Receiving	1	2	4
22.1	Production	9	18	36
22.1	Composite	10%	20%	40%

Note: See section 21 & 22 for elemental inspection and test times.

SUMMARY DATA FOR SEC. 24. Packing and Packaging

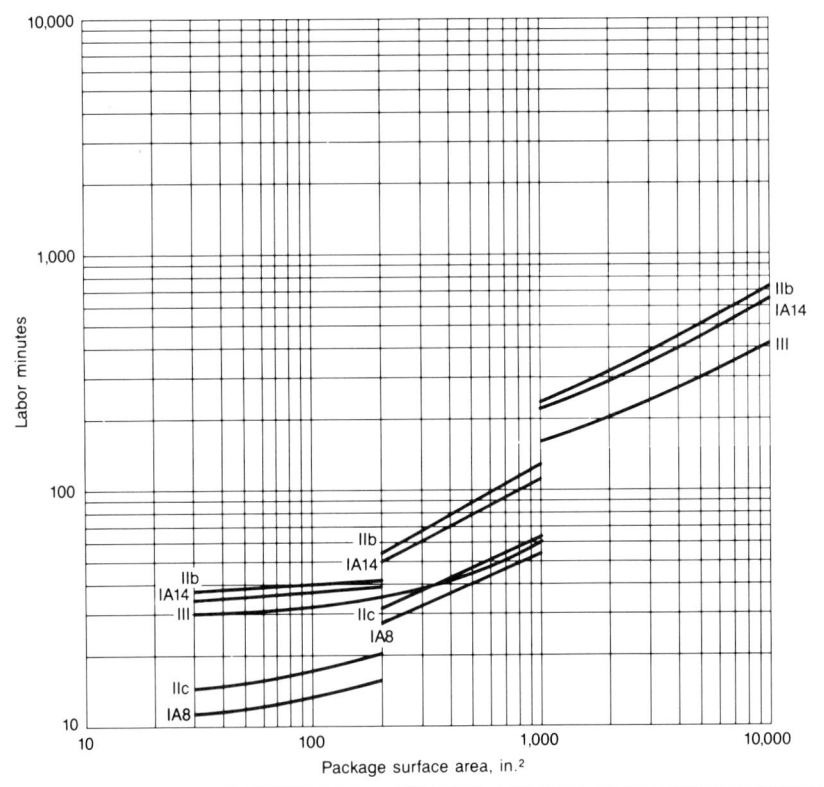

Note: Code identified with each curve refers to level of protection from MIL − P − 116 (see Table 24.1, Section 24).

SUMMARY DATA FOR SEC. 24. *(continued)*

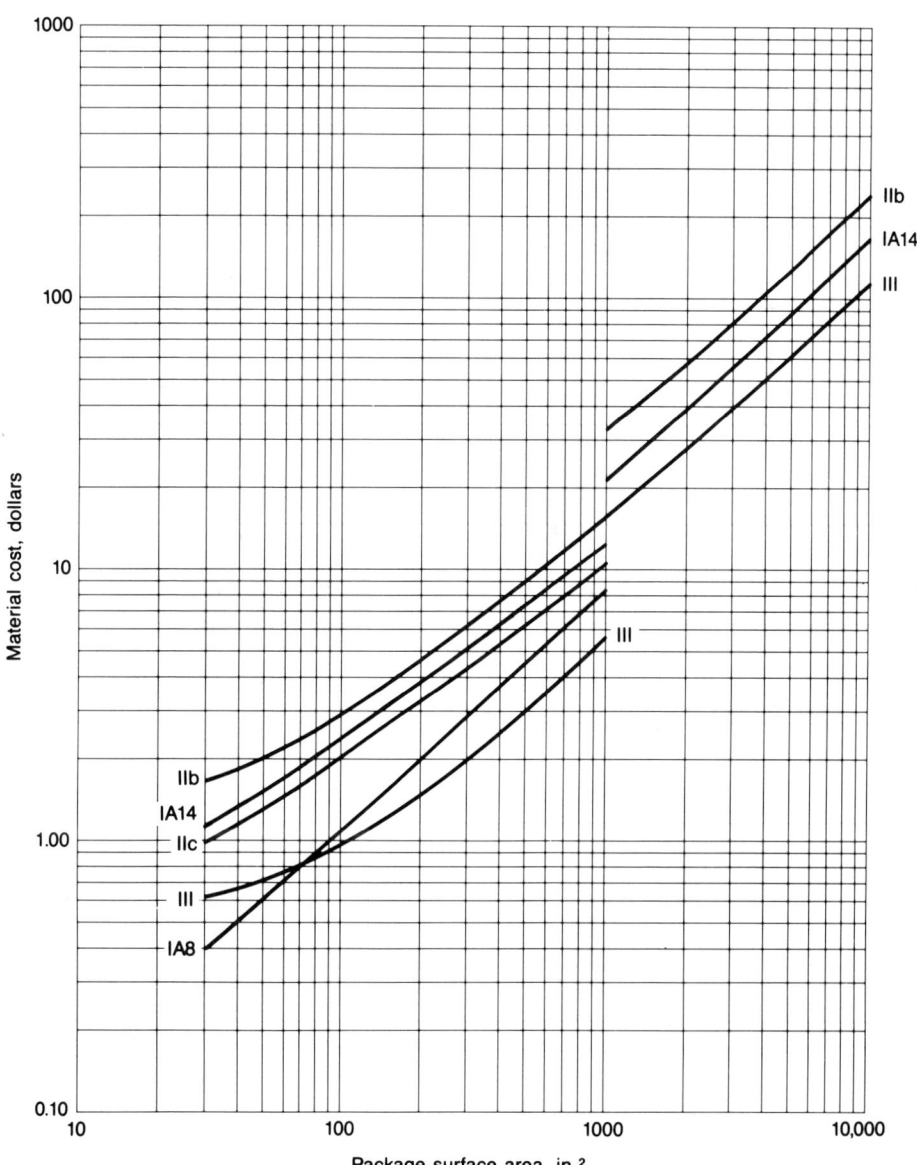

Note: Code identified with each curve refers to level of protection from MIL – P – 116 (see Table 24.1, Section 24).

34 SUMMARY DATA

SUMMARY DATA FOR SEC. 25. Learning Curve

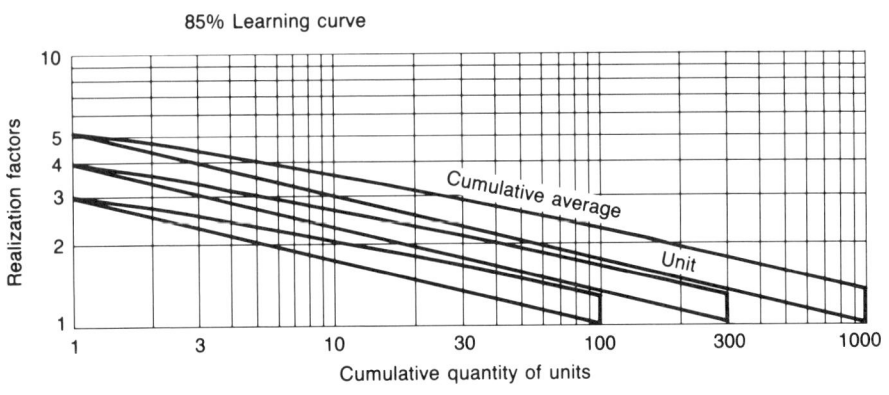

SUMMARY DATA FOR SEC. 26. Labor Allowances and Multipliers

Reference Section	Description	Factor
	Labor	
26.2	Standard hour multipliers;/minutes-to-hour converter	0.0167
25.0	Learning curve (realization factor) (See Figs. 25.3–25.4)	between 1 and 10
26.1	Total bid allowance	% of hours
	Variance from measured labor	5–20%
	Normal rework and repair	10–20%
	Engineering change allowance	0–15%
	Design growth allowance	0–30%
	Engineering prototype allowance	15–25%
	Misc. & Unobserved	4–9%
	Average value	55%

Labor Hours = (Std. time values in minutes from tables) × Learning curve losses (factors) × $\dfrac{100}{(100 - \text{Sum of Allowances from Sec. 26.1})} \times 0.0167$

SUMMARY DATA

SUMMARY DATA FOR SEC. 27. Material Cost and Allowances

Reference Section	Description	Factor
	Material	
27.1	Raw material costs (Boston, Ma., 1982)	
27.1.1	Steel	$0.37/pound
27.1.1	Stainless Steel	1.50/pound
27.1.1	Aluminum	1.84/pound
29.0	Quantity price reductions	
	Method 1: Price material per required buy quantity	
	Method 2: Adjust an existing price to a different buy quantity by using the material discount curve (Summary Data for Section 29)	
27.3	Total bid estimating ratios	% of MTL cost
	Line stock (hardware, chemicals, aluminum, steel, etc.)	¼–4%
	Attrition (normal mix of electronic components)	1–5%
	Engineering change allowance	0–10%
	Design growth allowance	0–30%
	Inflationary price increases	3%/year

SUMMARY DATA FOR SEC. 29. Material Discount Curves

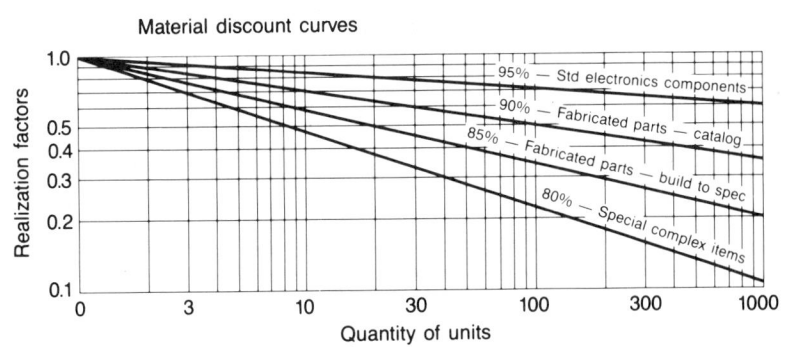

SUMMARY DATA FOR SEC. 30. Special Tooling and Test Equipment

Cost Element	Manufacturing Start-up Costs — Degree of Implementation	Percentage of Recurring Manufacturing Costs — Lot Quantity			
		10	100	1,000	10,000
Production Planning	High	20%	6 %	1.7%	0.50%
	Medium	10	3	0.8	0.25
	Low	5	1.5	0.4	0.12
Special Tooling	High	10	6	3.5	2
	Medium	5	3	2	1
	Low	3	1.5	1	0.5
Special Test Equipment	High	10	6	3.5	2
	Medium	6	3	2	1
	Low	3	1.5	1.0	0.5
Composite Total	High	40%	18 %	8.7%	4.50%
	Medium	21	9	4.8	2.25
	Low	11	4.5	2.4	1.12

Typical Proportions of Test Equipment Cost Elements

Cost Elements	Percentage of Hours	Percentage of Cost
Direct Labor		
Test engineer	25	
Drafting and design	20	
Chassis fabrication	10	
Assembly and wire	30	
Inspect and check out	15	
Total direct labor	100%	
Burden		
Burdened labor		60
Material		40
Total test equipment cost		100%

SUMMARY DATA FOR SEC. 30. (Continued)

Test Equipment Design & Drafting

Type Design	Description		Manual Operation		Computer Aided Design (CAD)	
	Standard Drawing Size	Ft.^{2} Drawing	Hours/ Ft.^{2}	Hours/ Drawing	Hours Ft.^{2}	Hours/ Drawing
Original	C	2.5	15	38	7.5	19
Concept	D	5.0		75	(PR = 2.0)[a]	38
	H	9.0		135		68
	J	11.0		165		82
Layout	B	1.0	10	10	5.0	5
	C	2.5		25	(PR = 2.0)[a]	12
	D	5.0		50		25
	H	9.0		90		45
	J	11.0		110		55
Detail	A	0.7	3	2.1	0.8	1.7
or Copy	B	1.0		3.0	(PR = 3.7)[a]	2.4
	C	2.5		7.5		6.0
	D	5.0		15.0		12.0
	H	9.0		27.0		21.6
	J	11.0		33.0		26.4

[a] Productivity ratio—improvement due to use of CAD. (See also Section 32.4, CAD/CAM Benefits & Costs.)

SUMMARY DATA FOR SEC. 31. Manufacturing Engineering

		Number of Units in Lot					
31.0	Manufacturing Engineering	1–20	25	50	100	500	1000
31.2	Minimum planning Percent of fabrication and assembly hours	15	—	—	—	—	—
31.2	100 percent planning Start-up hour/lot: ME hour/standard hour of fabrication/unit	—	13	13	13	13	13
	ME hour/standard hour at assembly/unit	—	27	27	27	27	27
	Sustaining Percent of fabrication and assembly hours		7	7	7	3	2
31.4	Composite Percent of fabrication and assembly hours	15	24	21	16	8	4

SUMMARY DATA FOR SEC. 33. Hourly Labor and Burden Rates, Earnings

Government Sales		Product Earnings-Markup		
CPFF	Fixed Price	Commercial Sales		
Manufacturing ↓ User	Manufacturing ↓ User	Manufacturing ↓ 20% Manufacturing sales ↓ 10% User	Manufacturing ↓ 20% Manufacturing sales ↓ 10% Distributor ↓ 20% User	Manufacturing ↓ 20–25% Manufacturing sales ↓ 10–15% Distributor ↓ 20–30% Retailer ↓ 30–50% User
		Typical Total Markup		
5–8%	9–12%	30%	50%	100%

PART TWO
PROGRAM PLANNING

SECTION FOUR

SCHEDULE DETERMINATION

4.1. Manufacturing Phasing Charts	41
4.2. Delivery Schedule Estimating	45

Three basic questions need to be addressed when responding to prospective customer's order:

Will the proposed equipment satisfy the customer's requirement?
How much will it cost?
When can it be delivered?

This section provides guidance for establishing production program delivery schedules. A list of the essential elements of program scheduling is provided along with a general example of how a delivery schedule can be developed using basic learning curve considerations.

4.1. MANUFACTURING PHASING CHARTS

A master phasing chart, which develops into the master production schedule, is an important part of any major production proposal, whether for in-house or bid use. Figure 4.1 is an example of a phasing chart for the production program of a typical electronic system. The same phasing elements generally apply to any program, regardless of the complexity of the equipment.

42 SCHEDULE DETERMINATION

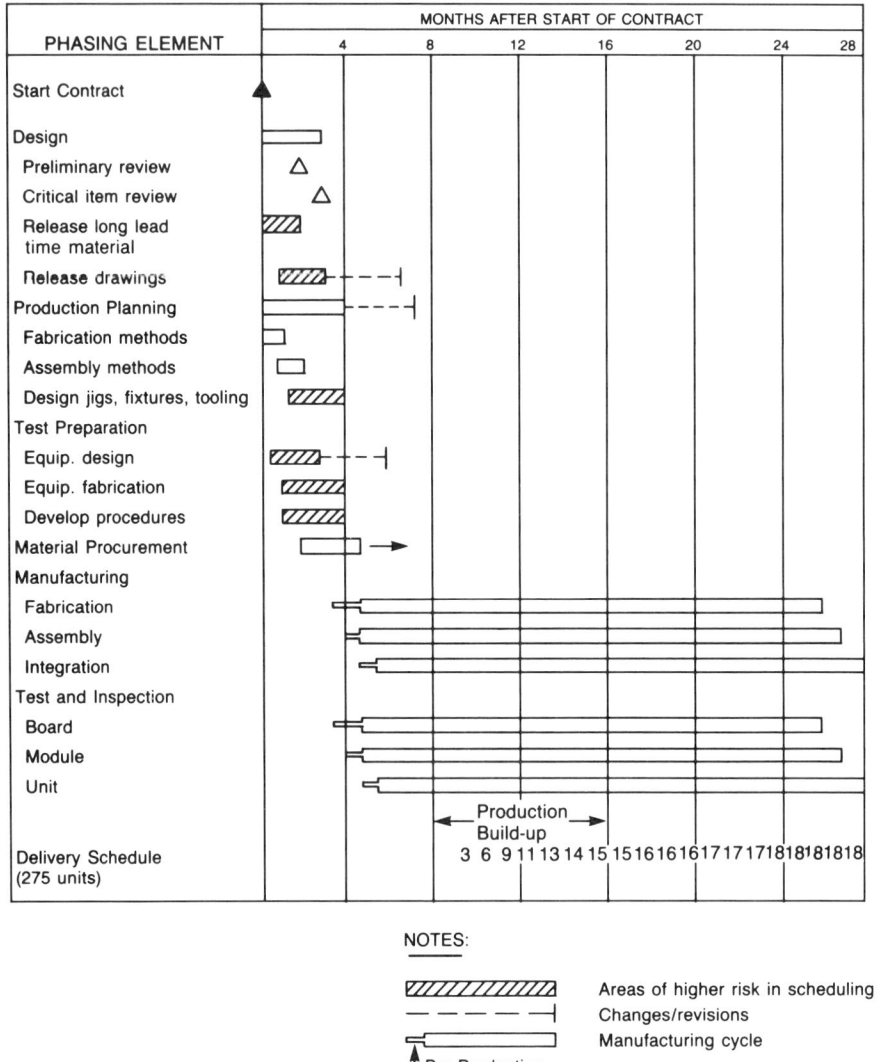

FIGURE 4.1. Manufacturing phasing chart.

The time required for a program depends both on the complexity (work content) of the equipment and on the skill and quantity of the available work force. As discussed below, other factors such as procurement of material, design information, or production resources other than labor may also lead to constraints on the production schedule. It is important to recognize the difference between the ideal and the real world in addressing these factors, and to allow sufficient margin for those areas in the scheduling which most often cause problems and present higher risk of missing milestones (e.g., long lead material purchases, drawing releases, etc.).

Although the phasing of a major program may be broken down into endless detail, the following elements constitute the basics for nearly all programs. (Refer to Figure 4.1.)

Design. Even though a conceptual design and a specification has been prepared in the course of bidding the job, detailed production design is part of the contract effort. Normally a preliminary design has been conducted early in the program in order to identify problems and to plan the detailed design approach.

Technical reviews are conducted during the course of design to ascertain progress towards meeting the technical requirements of the product. A *preliminary review* at the halfway point evaluates technical adequacy and risk on a schedule-cost basis, and confirms compatibility of the product with other equipments it may interface. A *critical review* is usually held at sometime near design completion to ensure that the engineering and performance requirements are met, that interface compatibility requirements are fully spelled out, and that the design permits producibility within cost and schedule constraints.

Design, planning, and production are sequential operations, but production of one part of the system can proceed while another is still in the design or planning stages. Consequently, the phasing of these events is an important aspect of the schedule.

Because procurement lead times of some materials are very long, it is usually necessary to identify early the material with long lead time and to expedite the scheduling of specifications and purchase releases. Under normal procurement lead times, material specifications would be prepared and released on a schedule dictated by the production manufacturing plan. Long lead time material, on the other hand, must be handled differently in order to allow sufficient margin for uncertainties in delivery.

A major problem confronting the designer and management is the timely recognition of those materials that indeed are long-lead items and the allocation of sufficient resources for generation of purchase specifications early enough in the program.

Besides identifying material requirements, the design team is also responsible for the preparation and release of drawings for production. Scheduling problems in this area result in late releases as well as errors in the drawings themselves. A late release leads to delays in manufacture. Drawing errors and omissions result in time lost rectifying problems through the revision and Engineering-Change-Notice (ECN) cycle. Drawings are normally released by engineering over a period of time in a logical sequence of packages, which permits production control and manufacturing to begin work on those parts of the product where the design is complete. A complete "clean release" of all drawings as a single package is a rare event, usually occurring only in those situations involving repeat orders on equipment previously built.

Production Planning. Planning is normally done at three or four levels. The top level is "planning the planning," that is, organizing the work so that the plans required first by production are available first. If time were no object, then the order of planning would not matter, as each phase would be completed prior to the start of the next. However, in the real world this is not possible.

The planning effort is conducted in accordance with the order of production. Fabrication is planned first, followed by assembly, and then by systems integration. Each of these is made as complex and as detailed as necessary to properly manage the production effort, which includes specification of manufacturing methods and processes, labor standards, tooling, and plant facilities. In concert with these basic manufacturing operations, supporting functions of producibility and manufacturing engineering, product assurance and quality control, and production control are also defined and invoked.

Manufacturing Test Engineering. Testing is an integral part of production. It is never enough to simply build something to a theoretical design; the equipment must be demonstrated. Preparation for testing includes design and fabrication of the test equipment and the development of in-process and final acceptance test procedures. As a practical matter, construction of adequate test equipment and preparation of test procedures is often delayed to the extent that production is affected. As with design, testing is a complicated but essential business.

Material Procurement. Part of the production process is to allow sufficient time to acquire parts and materials before the build cycle begins. Material lists and specifications must be generated, sources identified, and purchase orders placed. This may require as little as 30 days for repeat orders on small systems and as much as four months for complex systems and all new programs. Consequently, material specification is an integral part of the design phase. Once the orders have been placed, delivery cycles for standard stock parts and materials generally run from 30 to 60 days. Long-lead and special order items, of course, will take longer.

Manufacturing. The production effort itself is divided into three main parts: fabrication of parts and boards, assembly of these into modules, and integration of the modules into complete systems. Additional levels may be useful, depending on the physical or logical breakdown of the system.

The various levels are, of necessity, sequential in nature for any individual system (parts and boards must be fabricated before they can be assembled, etc.) but overlapped for a production program, so that at the height of production, all levels are occurring simultaneously. Start-up and shut down are, of course, in the sequential order.

Test and Inspection. Testing and inspection take place in conjunction with manufacturing, and are divided into the same levels.

Figure 4.1 represents a logical, idealized production plan. There are several points at which a real program can be expected to depart from such a plan. For instance, the release of drawings for manufacture is shown as a simple bar, with a well-defined end. In reality, the release of drawings for production is hardly ever well defined. Customer-requested changes, engineering delays, coordination problems, and design bugs that are discovered during the later stages of design conspire towards missed milestones.

In addition, it is unusual for manufacturing to receive a clean drawing package. The discovery and rectification of drawing errors in the form of Engineering Change Notices (ECNs) and revisions inevitably lead to delays in the production process. These tend to cost both time and money, because such problems are usually found during the production phase, after the workforce is in place.

As shown in Figure 4.1, delays in drawing release have a direct impact on production. There are also costs that are difficult to show on a schedule, but which can become significant. These include design and dimensional tolerance errors that do not surface until the product reaches the manufacturing floor, resulting in delay and overtime costs in both engineering and manufacturing.

As a practical matter, there are degrees of delay to be expected. The more complex the system, the greater the likelihood of error and chance of delay. For small systems, where roughly 200 or fewer drawings are involved, it is reasonable to allow a cushion of 10% of the calendar duration of the design effort. Note that it is not necessary, nor perhaps advisable, to simply extend the completion date. It is, however, appropriate to make sure that there will be no catastrophic effects if the drawings are somewhat late.

For medium-size systems, on the order of 1000 drawings, the schedule ought to contain a 20% cushion in the design and fabrication time. For larger systems, of more than 5000 drawings, 25% to 30% is more in order.

These margins vary considerably from plant to plant and program to program. They are given here only as representative values and for use in the absence of any historical information.

One of the major benefits reported for the use of extensive computer-aided design (CAD) systems is that of error reduction, which leads to the cleaner release of drawings and to fewer ECNs and revisions. In many instances the saving from the improvement in the quality of the information far outweigh any other benefits of the CAD system.

4.2. DELIVERY SCHEDULE ESTIMATING

The rate of delivery usually increases with time, because the accumulation of experience leads to reduced labor per unit, as described by the learning curve. The balance between workforce size and experience and the work content of the production units determines the shape of the delivery curve. As illustrated

46 SCHEDULE DETERMINATION

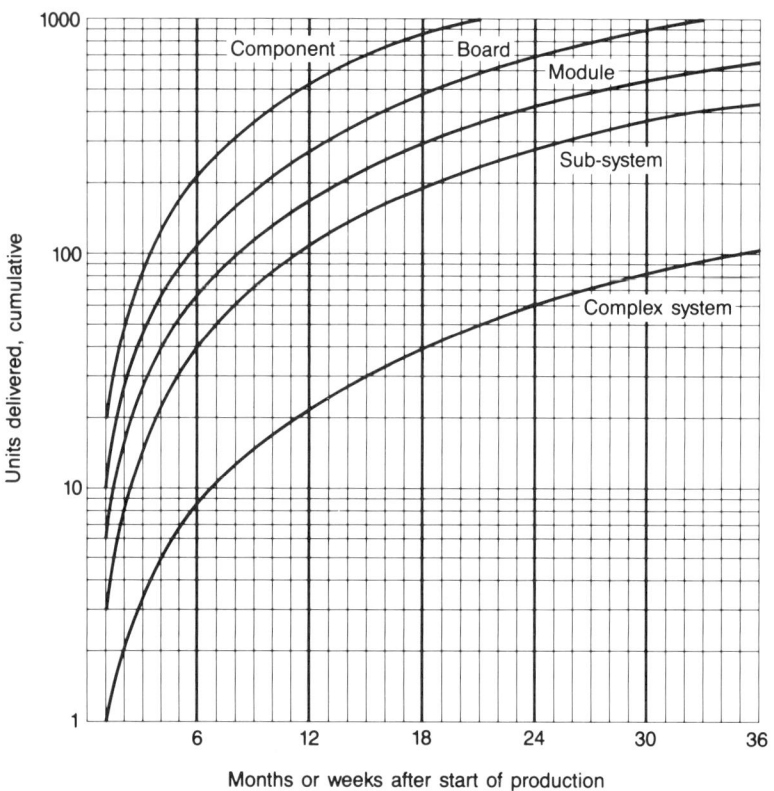

FIGURE 4.2. Manufacturing delivery for different degrees of unit size and complexity.

by Figure 4.2, simple systems achieve higher rates of delivery more quickly than do complex ones, for a given workforce size and experience level.

The most common request is for delivery at the most economic rate. This type of delivery schedule is calculated by using the learning curve to predict the decrease in labor per unit with experience, and estimating the number of units that can be delivered in each production period (week or month) by the workforce. A short computer program will relieve the tedium of such a calculation, as well as make it considerably faster and more accurate. Note that this method is based on the learning curve; thus proper application of the learning curve is essential, taking into account the size and skill level of the workforce and their experience on the type of work at hand. (See Section 25 for learning curve applications.)

Workforce Factor. The simplest and most economical situation is to maintain a constant workforce throughout the job. In this situation, the benefits of learning are retained without loss of skills during part of the production run. More often this is not possible, with frequent changes of the production workforce over the course of the program being the rule. Increases, decreases, or high turnover rates caused by breaks in production often result.

Increased production rates with corresponding increases in new workforce often reduce the benefits of the experience gained by the original workforce and incur additional expense to train and break in replacements. Their learning will generally not be as rapid as that of the original workforce, because they are learning at an individual level whereas the original workforce was learning both as individuals and as a production team. The original learning included experience in debugging various production systems, which the replacement workers will not have the benefit of unless the addition to the workforce is substantial enough to cause major changes to the production management and methods. The effects of adding workers to an existing program can be estimated by including their learning in the overall cumulative average labor content per unit.

If the workforce decreases, those retained presumably still have the benefits of their earlier experience. Estimating the production rate for a decreasing workforce is mostly a matter of allowing for the reduced production rate due to the reduced workforce and the slower learning curve over time, as fewer units will be produced per unit of time.

There is usually a certain amount of turnover in the course of any program. If turnover is unusually high, then projections of work content per unit will be unrealistically low, due to the continual need for training new people. In this case, the delivery rate calculations, and all other estimates based on the learning curve, must be adjusted. It is possible to factor the turnover percentage into the workforce at different points on the learning curve. As a practical matter, it is adequate to simply use a higher learning curve percentage (lower rate of learning) than is otherwise recommended. See Table 25.2 of Section 25 for guidance in learning curve selection.

Production Discontinuities. Program breaks are not desired but often cannot be avoided. Funding and approval delays are the most frequent causes of program slippage along with material shortage or missed deliveries. When restarted, production rates and unit costs will be higher than when the program was halted, due to the relearning required. It is difficult to precisely quantify the effects of a break in production. Primarily, they are:

Loss of familiarity; a need for at least partial retraining.
Loss of experienced personnel; a need for complete training of some individuals.

Note that the loss of certain key individuals may be much more significant than the loss of others.

The worst case would be a complete break in production and a restart by a new workforce, so that complete relearning was required. As an approxima-

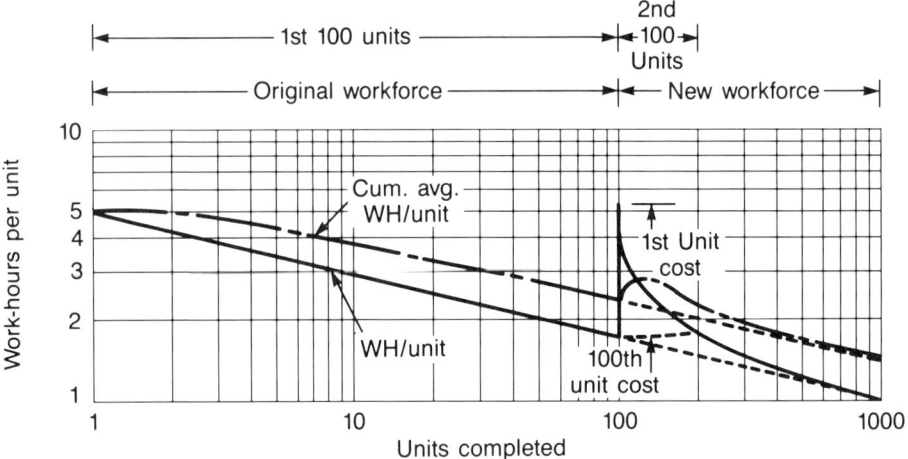

FIGURE 4.3. Worst case effect of a break in production on labor requirements.

tion, the effects of a real break lie somewhere between this worst case and the best case—complete retention of learning. Figure 4.3 illustrates these boundaries. Assume a production of 1000 units, with a break after 100 units. The solid line depicts the hours-per-unit learning curve, and the broken line is the cumulative-average learning curve.

After the break, the hours-per-unit curve begins again at the first unit, no-learning rate. In the course of the second hundred units, it will drop again to the 100th unit level. Since the distance from unit 101 to unit 200 is much shorter than from unit 1 to unit 100 on a log log scale, the curve seems much steeper. The curve continues to be somewhat steeper than the original curve but becomes asymptotic to it at high numbers of units, where the hundred unit difference becomes less significant.

The cumulative-average curve rises then approaches the original curve. The cumulative average for the second hundred units would be the same as that for the first hundred units, so the combined cumulative-average curve would be at the 100-unit level after 200 units.

Note that these increased production hours do not include any of the other costs of a break in production, such as loss of income, cost of carrying inventory, costs of recruiting new workforce, and so forth.

From the foregoing, the importance of repeat orders and benefits gained from having produced the product before are clearly obvious. This has the lowest risk and best learning-curve experience.

SECTION FIVE

PERSONNEL AND FACILITY PLANNING RATIOS

5.1.	Direct Labor Proportions	49
5.2.	Burden Expense Ratios	50
5.3.	Space Planning Ratios	51
5.4.	Sales and Facility Ratios of Selected Electronic Companies	51

Planning ratios are an extension of the cost estimating process, which allow the approximation of unknown elements from known or predetermined data. Such ratios aid in arriving at costs where rough estimates are acceptable or information is insufficient for detailed estimates. For example, if it is determined that 100 direct labor personnel are required for a project, it is reasonable to expect that 60 to 80 indirect persons will be required as well. Moreover, the combined total of 160 people will also require approximately 20,000 square feet of floor space. The ratios and percentages presented in this section pertain to electronic equipment manufacturing. They are a useful tool, but should be applied with good judgment and modified wherever possible to fit the particular application or situation.

5.1. DIRECT LABOR PROPORTIONS

The following proportions illustrate typical divisioning of labor among the primary categories of the manufacturing process for electronic equipment and systems. These proportions are contingent upon all work (including machining and sheet metal) being performed in-house. Two situations are presented: one where the majority of the manufacturing operations are performed manually, as in the case of prototype builds or limited production runs, and the second case, which favors a moderate degree of automation where techniques such as automatic component insertion and assembly of printed circuit boards and automatic test equipment (ATE) for in-process inspection and final test are standard practice.

Compared to manual operations, automation reduces the labor requirements in all areas except for manufacturing engineering. Labor increases for this last category because of the necessary added engineering for automation applications.

Process Operation	Manual		Partial Automation	
	hrs	% hrs	hrs	% hrs
Fabrication	40	27	30	27
Assembly	60	40	35	32
Inspection	15	10	11	10
Test	20	13	12	11
Manufacturing, engineering and production control	15	10	22	20
Total	150	100	110	100

5.2. BURDEN EXPENSE RATIOS

There are two major classifications of burden expenses: variable and fixed. Variable burden are those costs that are directly associated with the making of a product. It is determined by the size of the job and its labor and material requirements. It includes such cost items as hourly personnel fringe benefits (vacation, sick leave, health insurance, and pension plans), perishable tools, and operating utility costs attributed solely to the job. Variable burden is a recurring cost element that must be calculated for each new project.

Fixed costs are those that accrue independently of the amount of work on hand and exist whether products are being manufactured or not. They are related to the size of the facility. Included in this category are indirect labor (supervision other than D/L, engineering support, quality control, production control, accounting, personnel, plant and industrial engineering, etc.), operating supplies, state and local taxes, heat and light (not chargeable to a specific job), maintenance, equipment and building depreciation, leases, and so on.

Distribution of the burden expense as a function of the Direct Labor (DL) base is presented in Table 5.1. The number of indirect to direct labor personnel ranges from 50 to 100% depending upon manufacturing activities and the nature of products produced. Sixty percent is a reasonable average. Variable burden costs represent about 37% of the DL costs while indirect labor costs add

TABLE 5.1. Burden Expense Ratios (Typical Proportions)

Item	Personnel	Dollars
Direct Labor Base (DL)	100%	100%
Burden Expenses		
Variable (Fringes)	a	37
Fixed		
Indirect labor (IL)	60	71
Equipment (Depreciation, leases, etc.)	a	19
Building (Amortization, taxes)	a	23
Total Burden	60%	150%

[a] Expenses not related to personnel.

another 71%. The balance represents expenses not related to personnel, such as equipment, building, taxes, and depreciation. The combined variable and fixed burden costs add up to 150% of the DL costs.

5.3. SPACE PLANNING RATIOS

Floor space requirements can be estimated based on the type of manufacturing operation or plant activity involved. A convenient estimating ratio is the amount of floor space required per worker. Total space needs can be calculated by multiplying each of the plant function planning ratios by the total number of employees assigned to the job and then summing the totals. Table 5.2 provides rough estimating ratios for the principal plant functions. Some of the functions are more product related than others and appropriate adjustments should be made based on the type of product being produced. For example, bonded storage areas for high-value materials and work-in-process may require more space than conventional stockrooms. Also, the space requirements for assembling and testing small electronic parts and assemblies would differ significantly from those for a large system. Large systems also often require an intermediate step called integration where equipments and assemblies are mated prior to system testing. For instance, in order to estimate the area required to integrate and test a large, multi-rack system, one must consider the area of the racks, cable trays, support and test equipment, and other services, plus make suitable allowance for access and walkways. This area estimate, therefore, would be more accurate than an estimate that is based on the number of persons assigned to the job.

5.4. SALES AND FACILITY RATIOS OF SELECTED ELECTRONIC COMPANIES

A number of ratios can be used to determine the dollar sales volume and cost of a facility based on the number of employees. Table 5.3 is a representative list of

TABLE 5.2. Space Planning Ratios

Function	Actual Ft.^{2} per Person	Net^a Ft.^{2} per Person	Percent of Total Area
Production			
Machine shop and sheet metal fabrication	160	215	12%
Process and etched circuitry	210	280	13
Assembly	50	70	25
Inspection (in process)	20	30	3
Test	90	120	11
Production subtotal			64%
Production control			
Including stockrooms	320	430	28
Manufacturing engineering	55	75	6
Plant administration	125	170	2
Per person average	Total Plant 129	174 Total Plant	100%

Miscellaneous
  Cafeteria—25 ft.^{2} per employee
  Parking Lot—144 ft.^{2} per vehicle (excluding access to public artery)

^{a}Add 35% for aisles, stairs, rest rooms, boiler room, telephone switching room, and so forth.

companies covering a broad range of electronic products. The companies have been divided into three groups based on sales volume. Within each of these categories, specific products are identified for each entry. The significant cost elements are specified by the column headings. The table is structured to provide a wide variety of applications depending upon the planning function that may be of interest. For example, if representative facility costs and typical employment levels are desired, but only the type of product is known, a scan of the product listing would identify those companies that are representative of that product along with the cost data under the appropriate column.

Although the names of the companies are not identified, they are established firms within their fields. The cost data was taken from current financial reports. It is interesting to note that the ratio of sales per employee remains relatively constant among companies within the same size class. Moreover, large companies of basic materials such as plastics and coatings have higher sales per employee than do suppliers of components and equipment. This is due to the lower labor requirements characteristic of the former kind of industry.

TABLE 5.3. Sales and Facility Ratios of Selected Electronic Companies

Product categories	(A) No. of Employees	(B) Sales ($1000)	Sales per Employee (B ÷ A)	(C) Facility Cost ($1000)[1] Land	Bldgs.	Equip.	Other	Total	Facility Cost per Employee (C ÷ A)	(D) R&D Investment ($1000)	R&D as Percentage of Sales	Notes
Materials				*Small-Sized Companies (<$100 Million in Sales)*								
Sealants, laminates, adhesives	570	28,819	50,560	305	5,023	5,693	4,420	15,441	27,089	2,782	9.6	
Components												
Switches	711	36,124	50,807	424	5,887	11,382		17,693	24,885	2,568	7.1	
Components				*Medium-Sized Companies ($100 Million–$2 Billion in Sales)*								
Linear circuits	2,735	156,236	57,865	13,069	36,428	4,406	5,451	59,354	21,983	8,600	5.5	
Connectors	4,698	256,764	54,631					97,968	20,844	9,420	3.7	
Integrated circuits	9,915	275,593	27,838	3,347	43,223	96,451	5,527	148,548	15,005	28,309	10.3	
Connectors, switches	19,650	1,234,295	62,974	15,062	176,750	435,674		627,486	32,015	111,000	8.9	
Terminals, connectors	4,698	256,764	54,631					97,968	20,844	9,420	3.7	
Semiconductors	35,725	1,110,053	31,094	5,714	94,228	327,929	82,433	510,304	14,284	96,043	8.6	

(continued)

TABLE 5.3. (Continued)

Product categories	(A) No. of Employees	(B) Sales ($1000)	Sales per Employee (B ÷ A)	Facility Cost ($1000)[1] Total				Facility Cost per Employee (C ÷ A)	(D) R&D Investment ($1000)	R&D as Percentage of Sales	Notes
				Land	Bldgs.	Equip.	Other				
Equipment											
Data communications terminals	24,079	1,530,699	63,514	13,924	143,641	373,094 215,529		15,495	54,500	3.6	
ATE	2,735	156,236	57,865	13,069	36,428	59,354 4,406	5,451	21,983	8,600	5.5	
										① Facilities at Cost	
Systems											
ATE	3,050	159,942	52,440	5,583	33,597	103,021 55,309	15,532	35,417	20,377	12.7	
Militarized minicomputer	4,414	294,576	66,949	12,152	123,746	89,169 132,938	20,333	20,201	20,056	6.8	
Small- to medium-sized computers	14,370	588,525	40,870	17,511	72,736	236,396 82,213	63,936	16,416	74,573	12.7	
Test equipment	17,000	704,161	41,421	1,943	21,079	67,499 44,477		3,971	10,500	1.5	
Components					*Large-Sized Companies (>$2 Billion in Sales)*						
Chips	64,000	3,578,000	55,906	78,000	789,000	1,448,000 581,000		22,625	347,000	9.7	No employee, facility cost & R&D investment breakdown by division.
Semiconductors, gate arrays, FET's	76,500	5,636,000 (2,644 Elec. Div.)	73,673	23,356	341,843	1,601,594 962,313	274,082	20,936	166,100	2.9	
Circuit boards	75,400	4,942,835 (2,954,573 Elec. Div.)	65,555	42,404	419,257	1,380,725 919,064		18,312	251,000	5.1	As above.

Numerous products & services	133,000	8,004,800 (4,410,300 Elec. Div.)	60,186	662,300	2,741,400 2,079,000		41,224	461,000	5.7	As above.
All products & services	402,000	27,854,000 (8,850,000 Elec. Div.)	69,289	164,000	12,705,000 2,581,000 7,121,000	2,839,000	21,604	1,700,000	6.1	As above.
									① Facilities at Cost	
Equipment										
Test equipment	64,000	3,578,000	55,906	78,000	1,448,000 789,000 581,000		22,625	347,000	9.7	
Computer peripherals	90,513	4,239,849 (2,707,398 Elec. Div.)	46,849	33,507	1,886,557 375,996 1,477,054		20,843	336,470	7.9	As above.
Systems										
ATE, scientific computers	64,000	3,578,000	55,906	78,000	1,448,000 789,000 581,000		22,625	347,000	9.7	
General computers	76,500	5,636,000 (2,644 Elec. Div.)	73,673	23,356	1,601,594 341,843 962,313	274,082	20,936	166,100	2.9	
EDP systems	90,513	4,239,849 (2,707,398 Elec. Div.)	46,849	33,507	1,886,557 375,996 1,477,054		20,843	336,470	7.9	As above.
Numerous products & services	133,000	8,004,800 (4,410,300 Elec. Div.)	60,186	662,300	2,741,400 2,079,100		41,224	461,000	5.7	As above.
All products & services	402,000	27,854,000 (8,850,000 Elec. Div.)	69,289	164,000	12,705,000 2,581,000 7,121,000	2,839,000	31,604	1,700,000	6.1	As above.
Materials										
Plastics, Encapsulants, Coatings, adhesives	63,800	11,873,000	186,097	532,000	10,984,000 866,000 7,933,000	653,000	172,163	404,000	3.4	
									① Facilities at Cost	

SECTION SIX

CONCEPT ESTIMATING

6.1. Concept Estimating Parameters for Hardware   57
6.2. Concept Cost Estimating Parameters   63

Concept estimating is the estimating of the cost of a unit or system that has not yet been designed. The result is a tentative cost based on limited information, such as preliminary design studies, a verbal description, or whatever other data may be available at the time the concept estimate is required. The term "concept estimating" is descriptive, because what is being estimated is only a concept in the absence of engineering drawings, parts lists, and production plans and schedules.

Estimates of this nature are of value in spite of their lack of precision. The initial estimate is a useful management tool for judging whether a product can be priced competitively. Concept estimating methods, combined with detailed analyses of critical parts of the design, are also used to determine the relative costs of each of several design alternatives.

Another important aspect of the concept estimate is its value as a tool for comparing estimated versus actual cost as the design progresses through the development cycle. It often happens that significant cost elements are not identified in the initial design concept, due either to the fact that the design has not reached maturity (omission of essential functions) or that insufficient allowance was made for costs attributed to such non-design-related areas as manufacturing support, production, and management. As the product is developed, more and more problems appear and are solved. The cumulative effect of the solutions on cost can be disconcerting—on the order of doubling the initial estimates. If the costs of changes are tracked, or if the product costs are revised at intervals during development, anomalies can be detected early enough for corrective action.

Finally, the concept estimate is the only source of information about the size and duration of the program. It is used to plan and schedule each phase of the program from prototype development to production. As better information becomes available, these plans and schedules are refined in an iterative process using the concept estimate as a basis.

The material in the following sections is useful in two ways. First, a procedure is given for concept estimating. The procedure is independent of the sources of data used (if more precise in-house data is available, it may be substituted for data provided here). Second, the data is in the form of cost estimating ratios.

It must be understood that this data is respresentative only to the extent described. The determined estimator can certainly extrapolate or interpolate to find "ballpark" costs, which will increase the level of confidence over that developed here.

6.1. CONCEPT ESTIMATING PARAMETERS FOR HARDWARE

Whereas a complete hardware (and software) specification is provided to the detailed cost estimator, this is lacking when concept estimates are required. The first part of the concept estimator's job is to develop a reasonable definition of the product. Only then can its cost be estimated.

One method of assessing the hardware required for a given function is to focus on the number of active components, such as individually packaged transistors, or integrated circuits (IC). Once this is established, the number of passive components can be determined through the use of suitable experience ratios. Based on the number of components and the degree of packaging required, it is then possible to estimate the number and size of enclosures or boxes.

Note that the term "active component" as used here refers to a physical part containing one or more active elements, usually assembled on a printed circuit board. Technological advances may increase the functional capability of these components, but without causing a corresponding increase in the size of the component package or the enclosure of the electronic system. Likewise, the cost to assemble boards, chassis, and cabinets will not change proportionately either with increasing functional capability. The cost estimating methods and parameters presented by the following are concerned only with physical attributes of the parts, once the technology for implementing the function has been selected.

The first step in the concept estimating process, then, is to estimate the number of components, especially the number of active components. This in turn depends on the number of circuits required.

58　CONCEPT ESTIMATING

Given the functional block diagram of a system, it is not uncommon for an experienced professional to estimate circuit requirements from a library of basic circuit functions. From this, component counts can then be derived using quantities from the selected standard circuits. Table 6.1 lists typical component counts for a number of such basic circuits. It should be remembered that circuit requirements are also dependent on functional requirements and operating conditions, such as power levels and frequency ranges. A high-power radar oscillator is clearly quite different from a computer clock generator or an audio frequency oscillator. The data in Table 6.1 is typical for a normal range of operations.

The number of components also depends on technology assumptions. Not all circuit functions can employ high-density integrated circuit packaging; applications are typically limited to lower power levels and frequency ranges. However, where possible, the use of integrated circuits allows enormous reductions in system size or corresponding increases in function for a given system package.

Medium scale integration (MSI), which can place up to several dozen transistors plus supporting circuit elements on a single chip, in a single compo-

TABLE 6.1. Circuit Part Counts for Basic Circuit Types, Discrete Components

Function	Total	Transistor	Diode	Resistor	Capacitor	Inductor
NAND gate	8	1	4	3		
NOR gate	6	1	2	3		
Schmidt trigger	8	2		5	1	
Astable multivibrator	15	2		8	5	
Monostable multivibrator	10	2		5	3	
Bistable multivibrator	12	2		6	4	
Amplifier (small signal)	3	1		2		
RF amplifier	8	1			5	2
Differential amplifier	18	5	2	11		
Operational amplifier	26	8	2	16		
Oscillator	8	1		1	4	2
Crystal oscillator	6	1		2	1	1+ (crystal)
Interstage circuits						
Single-tuned		1			1	1
Double-tuned		1			2	3

CONCEPT ESTIMATING PARAMETERS FOR HARDWARE

nent package is presented in Table 6.2. The first entry, for instance, is a packaged chip with four NAND logic gates. If implemented with individual (discrete) circuit elements, per Table 6.1, eight components would be required for each gate. Thus, a total of 32 components could be replaced by this single chip, with an attendent savings in component and assembly cost, as well as system size.

The device prices given in Table 6.2 illustrates the cost competitiveness of complex circuit packages over discrete component circuits.

Large scale integrated circuit technology (LSI) (thousands of transistors) offers even greater circuit densities, with corresponding increases in capacity per circuit chip. Table 6.3 is a list of typical LSI devices; the range of capabilities is enormous compared with their implementation in discrete components. For instance, the circuity required to perform the functions of the 16-channel, 12-bit data acquisition system, which includes a multiplexer, differential amplifier, sample-and-hold amplifier, precision reference voltage source, comparator, and analog-to-digital converter, are packaged in two IC chips.

TABLE 6.2. Typical MSI Integrated Circuit Chips

(Quantity) Function Per Chip (Comparable to Table 6.1)	Typical Prices[a]
(4) NAND gates	$ 1.56
(4) NOR gates	1.56
(6) Schmidt triggers	2.56
(2) Multivibrator	1.89
(2) Low-noise preamplifier	1.65
(1) Differential amplifier	3.95
(1) Operational amplifier	4.25[b]
Other Functions	
Digital	
4-Bit arithmetic logic unit	9.80
Dual 4-input multiplexer	3.76
16-input multiplexer	3.76
4-Bit binary counter	3.69
Variable module counter	4.40
8-bit serial-to-parallel converter	5.86
Analog	
Voltage regulator, 5 volt, 5 amp	35.00
Precision instrumentation amplifier	14.30
High-impedance comparator	19.00

[a]Quantity 100, qualified per MIL-883/B.
[b]Requires typically 2–4 external resistors and 0–2 external capacitors as circuit additions.
Note: Represented are single chip devices in ceramic DIP packages, meeting military temperature ranges, under MIL-883/B.

TABLE 6.3. Typical LSI Integrated Circuit Devices

Device	Typical Prices[a]
Analog computational circuits	
High performance multiplier/divider	$ 75.60
Multi-function device: $V = Y(Z/X)M$	107.00
Digital-to-analog converters	
Complete 12-bit DAC	81.25
Ultra high speed (100 MHz) 10-bit DAC	367.00
Analog-to-digital converters	
Complete 10-bit ADC	77.00
Very fast, complete ADC	310.50
8-Channel, 8-bit combined ADC and multiplexer	
(not MIL-883/B)	24.60
Voltage-to-frequency converter (100kHz)	138.00
Frequency-to-voltage converter (1kHz–200kHz)	70.00
Sample-and-hold amplifier (6 microsecond)	32.10
Precision sample-and-hold amplifier with 16-channel multiplexer	280.00
Complete 16-channel data acquisition system	439.00
(Multiplexer, differential amplifier, sample-and-hold	
amplifier, precision voltage reference source, comparator,	
A/D converter on two IC chips)	
16-channel multiplexer	42.50
Microprocessors	
8-Bit, 2.5 MHz, MIL-883/B	75.65
Military grade, not MIL-883/B	60.65
Industrial grade	26.10
16-Bit, 4.0 MHz, MIL-883/B	860.00

[a]Quantity 100, qualified per MIL-883/B.

Complete single-chip computers are not only possible, but relatively inexpensive. The microprocessors described here typically require support circuitry, such as interface and memory chips to enhance design flexibility. True single-chip microcomputers are readily available, for special purposes.

The prices for several quality levels of microprocessors are provided to show the additional costs typically associated with the high quality required for military applications. Similar ratios apply to the other devices in this and the other tables. The differences in costs are due to the additional testing and auditing required to assure the high quality levels; the industrial grade, military grade, and military grade qualified to the requirements of MIL-883/B devices may all have been produced in the same lot.

Clearly, estimating the component count requires the services of a professional who is familiar with the current state-of-the-art technology and has some knowledge of the functional attributes of the system concept. It is not possible to provide general guidelines for this complex task beyond the foregoing.

CONCEPT ESTIMATING PARAMETERS FOR HARDWARE

If only the active components have been estimated, Table 6.4 may be used to determine the count of support components. The ratios of support components to active components vary with the function of the equipment, particularly regarding operating power and frequency ranges, which influence the choice of available technology options (discrete, MSI, LSI). Table 6.4 is based on ground-based, forward deployment military radar signal processing equipment designed in the mid-1970s. Two sets of estimating ratios are given: one for RF/video signal generating-and-processing circuits, and one for low-power logic circuits. In these examples, the number of active components to the total number of all components in the circuit is 1 to 10.9 for medium-power circuits and 1 to 5.6 for low-power logic circuits.

Once the component count has been established, the physical requirements can be estimated. In this context, packaging refers of course to the circuit boards, chassis, and other mechanical enclosures rather than to shipping protection. Equipment environment influences the packaging selection. Some considerations are:

Commercial vs. military
Airborne vs. ground
External exposure

Unusual environmental conditions, such as heat, moisture, vibration, shock, high altitude, and so on.

TABLE 6.4. Electronics Components Estimating Ratios

	Components per Active Circuit[a]	
	Medium Power (RF and Video Circuits)	Low Power (Logic Circuits)
Active Components		
Transistor (discrete)	0.85	0.25
Integrated circuit	0.15	0.75
Support Components		
Diode	0.55	0.10
Resistor, fixed	4.15	2.50
variable	0.15	0.10
Capacitor, fixed	4.10	1.70
variable	0.02	—
Inductor, fixed	0.30	—
variable	0.40	—
Crystal	0.05	—
Plug and connector	0.07	0.10
Other—switch, lamp, relay, etc.	0.10	0.10
Total circuit	10.9	5.60

[a]These ratios were derived from non-airborne military radar equipment of mid-1970s design.

These affect choices such as:

Printed circuit boards vs. wire-wrapped or terminal boards
Chassis, rack, or box configuration of enclosure
Tolerance of machined and sheet metal work

Estimating the number of packaging units required is best based on ratios developed from past experience on similar work. Key ratios are:

Components per board or per in.^{2} of board.
In the absence of other values, 3.2–3.8 comp./in.^{2} is a reasonable range (see Table 6.5).
Boards per chassis or per cubic foot of chassis.
In the absence of other values, 5 to 10 boards per chassis of 19 in. width is reasonable for medium- to high-power applications using primarily discrete

TABLE 6.5. Electronics Components Estimating Ratios[a]

	Components per Board[b]	
	Medium Power (RF and Video Circuits)	Low Power (Logic Circuits)
Active Components		
Transistor (discrete)	15	9
Integrated circuit	3	30
Support Components		
Diode	10	6
Resistor, fixed	73	95
variable	3	6
Capacitor, fixed	72	66
variable	0.3	—
Inductor, fixed	5	—
variable	6	—
Crystal	1	—
Plug and connector	6	9
Other—switch, lamp, relay, etc.	1	6
Total	195	227
Average components per in.^{2} =	3.2	3.8

[a]These ratios were derived fom non-airborne military radar equipment of mid-1970s design.
[b]Typical 6 × 10 in.

CONCEPT COST ESTIMATING PARAMETERS 63

transistors, and up to 14 boards per chassis in low-power, integrated circuit equipment.

Chassis per rack or per cabinet.
In the absence of other values, eight chassis per 7 ft. rack is reasonable.

Component costs and enclosure costs estimated above may need additional allowances for special requirements such as environmental control, RFI/EMI/EMP shielding, and so on.

The concept estimator must be both imaginative and cautious. He or she is without the exhaustive material lists that will later be available, and so must use experience and creativity to define the concept in equipment terms. As many aspects of the system as possible must be considered to avoid omissions which could later be costly.

6.2. CONCEPT COST ESTIMATING PARAMETERS

Given as complete a system description as can be prepared, the final step is to estimate its cost. In general, the information presented above is used to develop an estimate of material cost, fabrication labor, and assembly labor. Once these are established, the various ratios, percentages, and other rules-of-thumb described in the program planning sections of this book can be used to develop total program costs, plans, and schedules.

The following work-hour requirements are based on system estimates and include all labor from the handling of sheet metal and the machining of parts through wiring and final assembly. Fabrication labor includes printed circuit fabrication as well as chassis fabrication.

Although these are useful numbers, it cannot be overemphasized that numbers developed from an estimator's own experience or in-house sources are far valuable than general values.

Labor Concept Estimating Ratios	Standard Hours		
	Fab	Assy	Total
Per active element (transistor or I.C.), all labor for system	.30	.35	.65
Per circuit board 6" × 10" with 200 components, labor for circuit board assembly only:			
each of 50	1.4	1.7	3.1
each of 100	0.8	1.7	2.5

(continued)

Labor Concept Estimating Ratios	Standard Hours		
	Fab	Assy	Total
Per 19" rack chassis, all labor for assembly of the chassis			
High	50	50	100
Average	20	20	40
Low	10	10	20
Per 7' rack (average 8 chassis/rack)			
High	20	90	110
Average	10	60	70
Low	5	40	45
Fabrication labor to modify a purchased rack only. Labor to assembly and wire already completed chassis into the rack.			

Note that the values here are standard hours. The learning curve (see Section 25) as well as Labor Allowances (Section 26) must be applied to determine the labor content of multi-unit production runs.

PART THREE
COST ESTIMATING ANALYSIS

SECTION SEVEN

MACHINE SHOP OPERATIONS

7.1.	Cut Raw Material	68
7.2.	Turret Lathe—1-in. Diameter Stock	70
7.3.	Engine Lathe—6-in. Diameter Stock	73
7.4.	Milling	77
7.5.	Drilling	79
7.6.	Broach	81
7.7.	Grind, Centerless	82
7.8.	Grind, External Cylindrical	84
7.9.	Grind, Internal Cylindrical	87
7.10.	Grind, Surface	88
7.11.	Gear Hobbing	89
7.12.	Speeds and Feeds	94
7.13.	Tolerances and Surface Finishes	94

The standards of this section are highly summarized. The usual detailed machine-time tables have been replaced by a few benchmark table values for each type of machine shop operation. The values are simplified to make it practical to use them on systems containing a great number of individual and diverse piece parts. The loss of accuracy inherent to summarized and averaged time values is usually acceptable when the total time per system is the desired end product. However, it must be strongly stressed that the pricing of individual parts should be done using detailed methods analysis and data. The accuracy of the results of using summary tables is downgraded dramatically if the correct manufacturing method is not selected.

A word of caution—estimating standards of the benchmark nature are aimed at the person already familiar with machine shop operations. The descriptive information in this section merely touches on the high points of machining techniques, customs, and practices. The benchmark values, tempered by judgment or interpolated for intermediate stock sizes, will give reasonable answers for the system-type estimate. When individual parts are being priced to stand or fall on their own merits, there is only one way to do it and that is with detail standard data for the specific feed, speed, depth cut, and so on, required by the part.

As a complement to the data in this section it is strongly suggested that a complete set of machine shop standard data be available for references. McGraw-Hill publishes the *American Machinist's Manufacturing Cost Estimating Guide*. The *Machining Data Handbook* published by the Machinability Data Center of Cincinnati, Ohio contains optimum practical speeds and feeds.

Machine operating times are given for 6061 aluminum, 1040 mild steel, and 302 stainless steel. The times for aluminum may be adjusted for magnesium and plastics. The time for mild steel may be adjusted for bronze and brass. The times for stainless steel may be adjusted for drill rod and hard steels. Figure 7.1 may be used as a guide. Adjust the metal cutting times only. Set-up times do not change. Speeds and feeds have been taken from *Machining Data Handbook*. Aluminum values are for 6061 T-6 with a Brinnel Hardness range of 70 to 150 (500 kg). The mild steel values are for 1040, BHN 125–175. The stainless steel was Austenitic Type 302, BHN 225–275, cold drawn. Times were based on disposable carbide tooling for turning and milling. Other times are based on high-speed steel tooling.

7.1. CUT RAW MATERIAL

Set-Up Analysis:

Very little set-up time is required for changing jobs on either the band saw or the power hack saw. Resetting the stops, vise adjustments, and blade changes are all that is necessary for a normal set-up.

Set-Up Time: 8.08 minutes per set-up
Run Time Analysis:

7.1.1. Handle Part

The element includes pick up part from input tote, position to the saw, aside cut part to output tote. The time value covers input stock and weighing up to 60 pounds.

Handle Part 0.30 minutes per output piece

CUT RAW MATERIAL

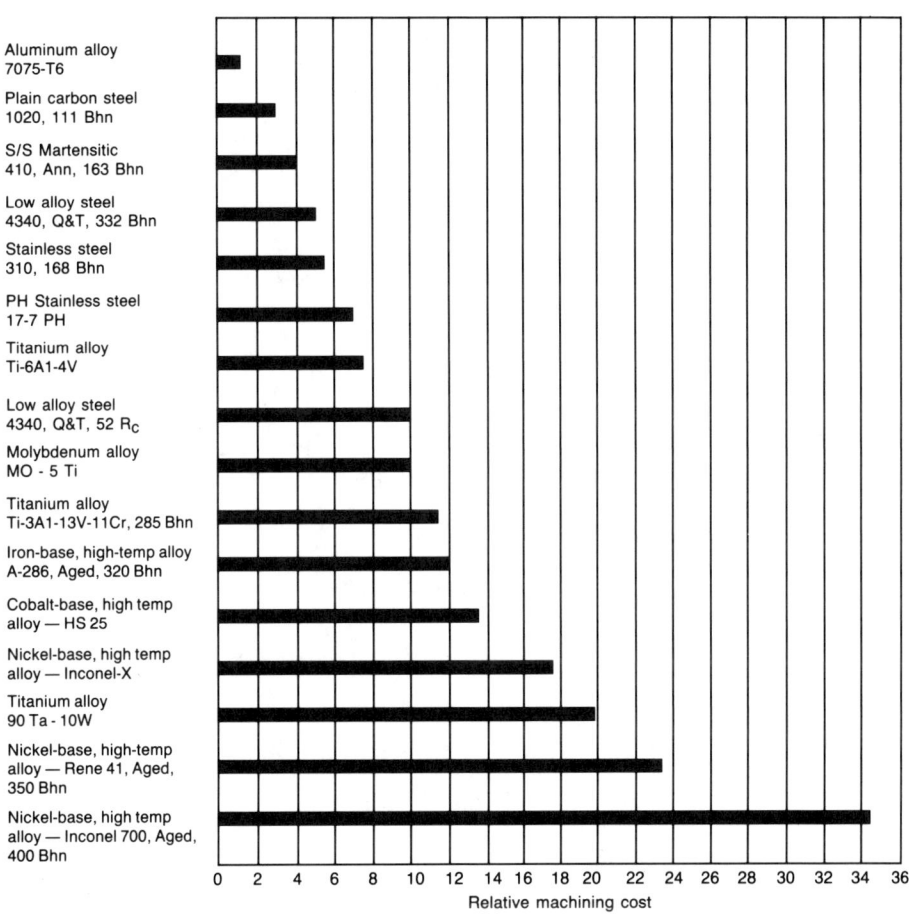

FIGURE 7.1. Factors which may be used to adjust metal cutting times. (Adapted from Relative machining costs for hard and soft materials, The Machining Data Handbook, Third Edition)

7.1.2. Band Saw, Do-All Type

Band sawing time depends largely upon the feed rate, which varies as the operator pushes the stock past the blade. A weight and cable arrangement sometimes is used to augment the pressure exerted by the operator. A few shops use powered positive feed devices for straight cuts. Times reflect blades with 65Rc teeth.

Machine time is given in minutes to make a cut 1 in. long in metal of the given thickness. Thus, a cut 10 in. long would be 10 times the 1-in. value, or 0.20 minutes for 10 in. of ⅛-in. aluminum. Apply values to straight cuts in flat stock. Add extra handling time for irregular or curved cuts.

	Minutes per Inch of Cut		
Thickness	Aluminum	Mild Steel	Stainless
⅛-in. stock	0.02	0.04	0.12
½-in. stock	0.06	0.16	0.22

7.1.3. Weld Blade for Internal Cut

Process steps:

Open saw guard, remove slide, cut saw blade and remove, pass saw through work, grind blade ends, clamp blade in welding fixture, weld, anneal, unclamp, grind to gauge, install on pulleys, replace side, adjust saw for tension, close guard. Remove blade from work. Store or discard at end of job.

8.03 minutes per weld

7.1.4. Power Hack Saw

Machine time to cut each in. of metal thickness. Apply to rod, bar, and shapes.

Minutes per inch of metal thickness		
Aluminum	Mild Steel	Stainless
0.60	1.60	3.03

7.2. TURRET LATHE—1-in. DIAMETER STOCK

Set-Up Analysis:
Warner Swasey No. 3 type lathe.
A summary of the basic set-up element is:

Job assignment, obtain work authorization and information packet, read instructions, complete time card and production report.	2.50
Trip to tool crib to acquire necessary gauges, and so on, and to return tools upon completion of job.	11.64
Set-up three measuring instruments, clean, and tear down at job completion.	5.12

TURRET LATHE—1-IN. DIAMETER STOCK

Install collet or chuck bar stock.	8.65
Install and square off bar stock.	1.89
Gauge and inspect first piece.	6.09
Subtotal—constant set-up and tear down.	35.89 minutes

Add per Cutting Tool:

Install hex turret tools (average) 6 at 4.36 minutes	26.16
Install cross slide tools (average) 2 at 3.60 minutes	7.20
Tear down, clean, put away (average) 8 at 2.00 minutes	16.00
Subtotal—per lot of (8) Tools	49.36 minutes
Total—including (8) tools and set-up	85.25 minutes
Install cutting tools—average per tool	6.17 minutes

Run Time Analysis:

7.2.1. Handling Time per Part

Release collet chuck, advance bar to stop, tighten collet chuck.	0.82
Stop and start machine.	0.08
Position coolant.	0.13
Change spindle speed.	0.28
After cut off, remove part from machine and set aside to tote pan.	0.34
Check part (1 in 15).	0.09
Total—per part produced	1.74 minutes

7.2.2. Turn and Bore—1-in. Diameter Stock

The time to turn or bore along 1 linear in. of 1-in. diameter stock may be used as a basic time unit in estimating small machined parts. When used with discretion, it serves as an average time per cut to turn, bore, drill, ream, knurl, form, or cut off.

Turn or Bore:

Remove up to 0.150 in. of stock from the radius per rough cut and 0.025 in. per finish cut. Allow one roughing pass per 0.150 in. (or fraction thereof) of stock removal and add time for one finish pass. For tolerances below ±0.002 in. or finish finer than 63 micro inches allow an additional finish cut. Grinding is recommended for tolerances below ±0.001 in. or finishes finer than 32 micro inches.

	Minutes per Cut		
Operation	Aluminum	Mild Steel	Stainless
Rough Turn	0.21	0.26	0.30
Finish Turn	0.23	0.32	0.37

7.2.3. Tap

Prepare to operate.

	Minutes per Hole
Stop and start machine.	0.08
Change to slower spindle speed.	0.28
Index turret.	0.22
Apply tapping fluid.	0.10
Blow tap clean.	1.13
Reverse spindle direction to remove tap from work.	0.08
Total	1.89

7.2.4. Thread

Prepare to operate.

Stop and start machine.	0.08
Change to slower spindle speed.	0.28
Index turret.	0.22
Blow die head clean, allow once per hole.	0.16
Total	0.74

Machine Time

Based on automatic or self-opening dies. ¼" diameter × NF 32 threads per inch of thread. Allow once for each inch or less of thread length. (The following values are not applicable to material harder than R_c-36 or BHN 332.)

Minutes per Inch of Length		
Aluminum	Mild Steel	Stainless
0.289	0.435	0.514

7.2.5. Taper

Prepare to operate.

	Minutes per Taper Angle or per Piece
Release compound, or taper attachment, adjust to proper angle, return to normal position.	0.95
Start and stop lathe.	0.08
Change spindle speed and return.	0.28
Index turret.	0.22
Total	1.53

Machine time—up to 7° taper.
Turn taper.

Minutes per Linear Inch of Taper		
Aluminum	Mild Steel	Stainless
0.244	0.370	0.488

7.3. ENGINE LATHE—6-in. DIAMETER STOCK

Set-Up Analysis:
Monarch 10 × 20-in. lathe or equivalent.

The basic set-up time for chucked work is about the same as for the turret lathe. The number of tools is normally fewer for an engine lathe operation, but the adjustments done during the operation are more time consuming. Thus, set-up time is decreased, but run time is proportionately more than on a turret lathe. Each is installed by hand.

Obtain job assignment.	2.50
Obtain gauges and tools and return them to tool crib.	11.64
Set-up measuring instruments.	5.12
Tear down and clean.	14.18
Install and remove tools to compound (1.89 minutes each—average of 2 tools).	3.78

Set speeds and feeds.		0.74
Mike diameter.		0.69
Total set-up time		38.65 minutes

Allow once per job set-up.
Add for trial cut:

Minutes		
Aluminum	Mild Steel	Stainless
0.47	0.69	0.92

7.3.1. Handling Time per Part Produced

Bar and tube stock approximately 10 lbs. per part.

Pick up piece, install in universal chuck, align by hand, and secure for work.
Remove after machining and place in tote.   1.41

7.3.2. Align by Eye   1.54

7.3.3. Turn and Bore

The time to turn or bore 1 linear inch on a 6-in. diameter piece of stock may be used as a basic time in estimating machining of parts. When used with discretion, it serves as an average time for a cut along 1 in. of surface for turning, boring, drilling, reaming, knurling, spin forming, or cut-off.

	Minutes per Cut		
Operation	Aluminum	Mild Steel	Stainless
0.175 Rough cut	0.09	0.31	0.34
0.025 Finish cut	0.23	0.32	0.37

Time per piece produced on 6" diameter.

ENGINE LATHE—6-IN. DIAMETER STOCK

7.3.4. Thread

Prepare to operate.

Release compound, stop spindle, swing into position, and secure.	0.55
Change tool in tool holder.	1.89
Position tool, set for proper micrometer depth, start machine, engage feed at the proper lead.	1.06
Cut (see below).	
Blow threads clean.	0.10
Check with "go" gauge.	0.55
Check with "no go" gauge.	0.20
Clean up threads (remove burrs) with emery cloth.	0.44
Total	4.79

Machine Time

External or internal V thread, 12 threads/in., single-point tool.

Time Factor	Aluminum	Mild Steel	Stainless
Number of passes	4	6	8
Time per pass	0.204	0.257	0.564
Time to thread	0.820	1.540	4.520

7.3.5. Taper

Prepare to operate.

Stop spindle, change gears, start spindle.	0.20
Position compound.	0.55
Set taper attachment.	1.21
Install and remove tool.	0.72
Mike dimensions.	0.69
Total	3.37

Machine Time

Remove up to 0.150 in. of stock from radius per rough cut and 0.025 in. per finish cut. Allow one roughing pass per 0.150 in. (or fraction thereof) of stock removal and add time for one finish pass.

MACHINE SHOP OPERATIONS

| | Minutes per Cut | | |
Operation	Aluminum	Mild Steel	Stainless
Rough cut	0.28	0.52	0.77
Finish cut	0.36	0.88	1.17

7.3.6. Sample Calculation for Taper Turning

To calculate time for taper turning:

1. Determine maximum depth of cut from part dimensions.
2. Allow one roughing pass for each 0.150 in. of stock removal.
3. Allow one finish pass if tolerances are ±0.002 and finish is 63. Allow extra finish pass if tolerances are ±0.001 or finish is 32.

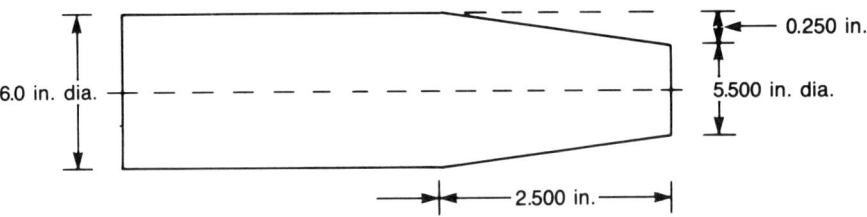

Material is 1040 steel. Depth of cut is 0.250 in.

Complete necessary 2 roughing passes at 0.52	1.04
1 finish pass at 0.88	0.88
Time for 2.5 in. of taper. (Time for 1 in. is 1.92—1.92 × 2½.)	4.80
Set-up of lathe (previously allowed for other turning operations on this part).	0.00
Prepare to operate for taper.	3.37
Set aside completed part (not previously allowed for other turning operations on this part).	1.41
Total	11.50 minutes

7.4. MILLING

Set-Up Analysis:
Milwaukee No. 2 or equivalent.
A summary of the basic set-up elements is:

Obtain job assignment, work authorization, and information packet, read instructions, complete time card and production report.	2.50
Trip to tool crib to acquire necessary tools, gauges, measuring instruments, return at completion of job.	11.64
Set up three measuring instruments, clean and tear down at completion of job.	5.12
Gauge and inspect first piece.	6.09
Install and remove part holding device and align.	11.27
Install and remove cutter.	7.49
Adjust table to initial start position.	1.70
Total set-up and tear down	45.81 minutes

7.4.1. Handling Time per Part Produced

Pick up part and clamp in vice.	0.71
Check part with micrometer and so forth, 1.00/20.	0.05
Set aside part to tote pan.	0.71
Clean vice to receive next part.	0.05
Total	1.52 minutes

Note that for securing parts in a more complex mill fixture, or aligning part with a dial indicator, you must double the above handling time.

7.4.2. Advance and Back-Off, or Adjust Table

	Minutes per Occurrence
Start machining and advance work to cutter.	0.09
Back work from cutter and stop machine.	0.09
Approach and overrun allowance for cutter diameter (for each roughing and finished pass).	0.09
Set table at proper position for work by moving up, down, or in saddle.	0.20
Index dividing head.	0.15

7.4.3. Profile or End Mill

½ in. deep × ¾ in. wide.

Maximum rough cut depth equals 0.050 in. Maximum finish cut depth equals 0.015 in. Ten roughing cuts and one finished cut are needed for ½ in. deep profile.

	Minutes per Linear Inch		
Material	Rough	Finish	Total
Aluminum (10 × 0.091)	0.91	0.12	1.03
Carbon steel (10 × 0.257)	2.57	0.23	2.80
Stainless steel (10 × 0.655)	6.55	0.80	7.35

7.4.4. Surface or Face Mill

½ in. deep × 6 in. wide cut.

Maximum roughing cut depth is 0.300 in. The maximum finish cut depth is 0.040 in. Allow adequate roughing passes and one finish pass.

Type Cutters:

Plain, helical, slab, or shell end mills.
Cutter diameters, 2½ in.−4½ in.
Width of cutter face, 2 in.−6 in.
   Total, ½ in. depth × 6 in. width

	Minutes per Linear Inch		
Material	Rough	Finish	Total
Aluminum (2 × 0.03)	0.06	0.03	0.09
Carbon steel (2 × 0.08)	0.16	0.07	0.23
Stainless steel (2 × 0.09)	0.18	0.11	0.29

7.4.5. Side Mills, Straddle Mill, Slotting

⅕ in. deep × 1 in. wide cut.

Maximum roughing cut depth is 0.250 in. Maximum finish cut depth is 0.025 in. Allow adequate roughing passes and one finish pass.

DRILLING

Type Cutters:

Side, staggered tooth, half-side milling cutters
Cutter diameters, 4–8 in.
Width of cutter face, ¼–1 in.
  Total, ½ in. depth × 1 in. cutter face

	Minutes per Linear Inch		
Material	Rough	Finish	Total
Aluminum (2 × 0.02)	0.04	0.02	0.06
Carbon steel (2 × 0.07)	0.14	0.06	0.20
Stainless steel (2 × 0.125)	0.25	0.10	0.35

7.4.6. Corners, Grooves, Slots

All of these cuts are in approximately the same time range. Maximum roughing at depth is 0.250 in. Maximum finish cut depth is 0.025 in. Allow adequate roughing passes and one finish pass.

Corner rounding, ½ in. radius
V groove or chamfer, ½ in. depth × ½ in. width
Key slot, ½ in. depth × ⅜ in. width

	Minutes per Linear Inch		
Material	Rough	Finish	Total
Aluminum (2 × 0.02)	0.04	0.02	0.06
Carbon steel (2 × 0.07)	0.14	0.06	0.20
Stainless steel (2 × 0.125)	0.25	0.10	0.35

7.5. DRILLING

Set-Up Analysis:
Medium and heavy duty drill presses
Typical set-up includes the following elements:

Obtain job assignment, work authorization and information packet. Read instructions. Complete time card and production report.	2.50
Trip to tool crib to acquire necessary tools, gauges, and so forth, and to return them upon completion of job.	11.64
Set up three measuring instruments, and clean and tear down at completion of job.	1.23
Gauge and inspect first hole.	1.57
Constant time in handling jigs, fixtures and vises.	2.31
Make necessary machine adjustments, and change speeds and feeds.	1.39
Make necessary feed stop adjustment.	0.79
Install drill in spindle.	2.39
Total	23.82 minutes

7.5.1. Drill

Handling Time per Hole.

This element includes: move the part from hole to hole, and lower the drill to the surface of the part. Add the machine cutting time per inch to depth to the handling time per hole.

	Handling Time (Minutes per Hole)	Material	Minutes per Inch Depth
⅛-in. Diameter hole, medium, general purpose press (spindle 500–2000 RPM)	0.55	Aluminum Carbon Steel Stainless Steel	0.09 0.25 0.55
2-in. Diameter hole, heavy duty drill press (spindle 1–1000 RPM)	1.10	Aluminum Carbon Steel Stainless Steel	0.33 0.68 1.34

7.5.2. Drill, Tap, Countersink

	Handling Time (Minutes per Hole)	Hole Drill	Machine Tap	Countersink
⅛-in. Diameter hole Aluminum	0.55	⅛ in. ×1 in. 0.09	⅛ in. ×NS40 0.06	⅛ in. × 1/16 in. 0.05

	Handling Time (Minutes per Hole)	Hole Drill	Machine Tap	Countersink
Carbon steel	0.55	0.25	0.10	0.15
Stainless steel	0.55	0.55	0.18	0.30
2-in. Diameter hole		2 in.×1 in.	2 in.×4½ TPI	2 in.×¼ in.
Aluminum	1.10	0.33	0.06	0.26
Carbon steel	1.10	0.68	0.10	0.37
Stainless steel	1.10	1.34	0.18	0.69

7.6. BROACH

Simple broaching machine for cutting internal keyways.
Set-Up Analysis:
Milwaukee No. 2 or Equivalent
A summary of the basic set-up elements is:

Obtain job assignment, work authorization, and information packet. Read instructions and complete time card and production report.	2.50
Trip to tool crib to acquire necessary tools, gauges, measuring instruments, and return them at completion of job.	7.14
Set-up two measuring instruments, clean and tear down at completion of job.	0.82
Install and remove tooling—simple clamping fixture.	4.68
Gauge and inspect first piece.	1.21
Total	16.35 minutes

7.6.1. Handling Time per Part

Pick up part, position in fixture, fasten, remove, and set aside at job completion.

Weight in Pounds	0–20	21–40	41–80	Over 80
Minutes per piece	1.32	1.64	2.04	5.36

Position head in start position, install broach in work and in head, remove broach, start and stop machine. Clean chips away from work, tool and machine   2.97 minutes

7.6.2. Machine Time

The machine time is for broaching internal keyways (single broach for single keyway) and includes deburring.

Size of Keyway (in.)	Depth	1/16	1/4	1/2
	Width	1/8	1/2	1
Broad Length		32 in.	50 in.	56 in.
Number of cuts		1	2	2
Minutes per Keyway				
Aluminum		0.16	0.42	0.52
Carbon steel		0.31	0.92	1.02
Stainless steel		0.61	1.82	2.02

7.7. GRIND, CENTERLESS

Set-Up Analysis:
A summary of the basic set-up elements is:

Obtain job assignment, work authorization, and information packet. Read instructions and complete time card and production report.   2.50

Trip to tool crib to acquire necessary tools, gauges, measuring instruments, and return them at completion of job.   11.64

Set up one measuring instrument, clean and tear down at completion of job.   0.82

GRIND, CENTERLESS

Install and remove tooling (adjust work rest and blade).	10.46
Change and dress wheel.	10.82
Gauge and inspect first piece.	6.09
Total	42.33 minutes

Handling parts to and from grinder usually takes place during the grinding cycle.

The amount of grinding stock left on the part to be ground should not exceed 0.015 in. although where rough material is ground, stock may exceed this considerably. The number of cuts will depend on the amount of stock to be removed. The roughing cuts wheel in-feed is 0.005 in. per pass, while the finishing cuts wheel in-feed is only 0.0015 in. per pass.

7.7.1. Through-Feed Method

This method is used for straight cylindrical surfaces without interfering shoulders. The parts are positioned either manually or by gravity chute to the grinder. They then pass through the machine. The machine time is used as the unit time since handling can usually be accomplished in equal or less time. Allow one roughing pass for each .010″ of stock removed from a diameter plus allow one finish pass of .003″ or less stock removed from diameter. Add an extra pass if the tolerance is less than ±.0005″.

Material	Minutes per Pass per Inch Length Ground
Aluminum	0.04
Mild steel	0.04
Stainless steel	0.04

7.7.2. In-Feed Method

This method is used for work that has a shoulder. The parts are individually positioned to the grinder, fed into the depth of the shoulder, and either manually or mechanically removed from the feed side of the grinder.

Handling Time

Pick up part and position between feed rollers. Remove ground part, and set aside to tote. Also includes time to feed stock into the depth of the shoulder (gravity chute not used).

Weight in pounds	0–20	21–40
Minutes per piece	1.40	1.70

Run Time Analysis

Grinding time—maximum stock removal per roughing pass is 0.010 in. and 0.003 in. per finish pass. Add an extra pass if the tolerance is less than ±0.0005 in.

Material	Minutes per Pass per Inch Length Ground
Aluminum	0.05
Mild Steel	0.05
Stainless steel	0.05

7.8. GRIND, EXTERNAL CYLINDRICAL

Set-Up Analysis:

A summary of the basic set-up elements is:

Obtain job assignment, work authorization, and information packet. Read instructions, and complete time card and production report.	2.50
Trip to tool crib to acquire necessary tools, gauges, measuring instruments, and return them at completion of job.	8.64
Set up two measuring instruments, clean and tear down at completion of job.	0.82
Install and remove tooling, set dogs.	1.12
Change and dress wheel.	10.82
Gauge and inspect first piece.	1.19
Total set-up and tear down	25.09 minutes

7.8.1. Handle Parts to Chuck, Centers, or Taper Arbor and Set Aside. Start, Position and Stop Grinder.

Prepare to operate:

GRIND, EXTERNAL CYLINDRICAL

Start work spindle, position grinder to work by hand, and feed in proper depth for cut, clear grinder from work, stop work spindle.

Pick up part, install in universal chuck, support other end in steady rest or footstock center, secure for work. Release part and set aside. 1.09 minutes per part installed

Note that the handling time to secure work piece between centers or on a taper arbor is approximately the same.

Weight in pounds	0–20	21–40	41–80	Over 80
Minutes per piece	1.78	2.08	2.48	5.21

7.8.2. Dial Indicate for Concentricity

When there is a concentricity note, allow this additional time to true up, using a dial indicator. 2.81 minutes per occurrence

7.8.3. Machine Time, Rough and Finish

The following formula will give approximate grinding times:

Rough Cut: $T = \dfrac{(L)(T_s)(D)}{2(w)(F)(CS)}$

Finish Cut: $T = \dfrac{(L)(T_s)(D)}{(w)(F)(CS)}$

T = Time, minutes
L = Length of ground surface, inches
T_s = Total amount of stock to be removed, inches
D = Diameter of part, inches
w = Width of face of grinding wheel, inches
F = Feed (depth of cut), inches
CS = Cutting speed, surface feet per minute

The required data are derived as follows:

L = Length of grind surface—from print
T_s = Total stock removed per pass—see table below

Part Diameter	Aluminum		Mild Steel		Stainless Steel	
	Rough	Finish	Rough	Finish	Rough	Finish
½ in.	0.010 in.	0.002 in.	0.010 in.	0.001 in.	0.010 in.	0.001 in.
1 in.	0.010 in.	0.002 in.	0.010 in.	0.001 in.	0.010 in.	0.001 in.
4 in.	0.020 in.	0.002 in.	0.020 in.	0.001 in.	0.020 in.	0.001 in.

D = Diameter of part—from print
w = Width of grinding wheel—average, 2 in.
F = Feed (in inches)—from table below

Part Diameter	Aluminum		Mild Steel		Stainless Steel	
	Rough	Finish	Rough	Finish	Rough	Finish
½ in.	0.004 in.	0.002 in.	0.001 in.	0.00025 in.	0.002 in.	0.0005 in.
1 in.	0.004 in.	0.002 in.	0.002 in.	0.00025 in.	0.002 in.	0.0005 in.
4 in.	0.003 in.	0.002 in.	0.002 in.	0.00025 in.	0.002 in.	0.0005 in.

CS = Cutting speed in surface feet/minute

Part Diameter	Aluminum	Mild Steel	Stainless Steel
½ in.	25	30	25
1 in.	55	30	25
4 in.	75	30	25

Machine time per above formula:

			Minutes Per In.		
			Aluminum	Mild Steel	Stainless Steel
½ Diameter × 0.010 in. Stock Removed		Rough	0.013	0.042	0.098
		Finish	0.010	0.033	0.002
		Total	0.023	0.075	0.100
1 Diameter × 0.010 in. Stock Removed		Rough	0.111	0.083	0.159
		Finish	0.009	0.067	0.003
		Total	0.120	0.150	0.162

GRIND, INTERNAL CYLINDRICAL

		Minutes Per In.		
		Aluminum	Mild Steel	Stainless Steel
4 Diameter × 0.020 in. Stock	Rough	0.089	0.667	0.340
Removed	Finish	0.027	0.267	0.012
	Total	0.116	0.934	0.352

7.9. GRIND, INTERNAL CYLINDRICAL

Set-Up Analysis:
Milwaukee No. 2 or equivalent.
A summary of the basic set-up elements is:

Obtain job assignment, work authorization, and information packet, read instructions, complete time card and production report.	2.50
Trip to tool crib to acquire necessary tools, gauges, measuring instruments, return them at completion of job.	11.64
Set up three measuring instruments, clean and tear down at completion of job.	5.12
Change and dress grinding wheel.	10.82
Gauge and inspect first piece.	2.84
Set speeds and feeds.	1.09
Total set-up and tear down	34.01 minutes

7.9.1. Handle to Chuck

Pick up part, install in pneumatic or hydraulic chuck and aside to tote.

Weight in pounds	0–20	21–40	41–80	Over 80
Minutes per piece	1.20	1.50	1.90	4.73

Advance grinding wheel to work and back off, 0.33 minutes per occurrence.

7.9.2. Dial Indicate for Concentricity

First Part   5.86
Successive Part 0.58

7.9.3. Machine Time

Minutes to grind 0.001 in. of stock from inside of the hole.

Diameter of Hole	Depth of Hole	Stock to Remove	Aluminum (Minutes/ 0.001 in.)		Mild Steel (Minutes/ 0.001 in.)		Stainless Steel (Minutes/ 0.001 in.)	
			Stock	Total	Stock	Total	Stock	Total
½ in.	1 in.	0.006 in.	0.042	0.252	0.130	0.78	0.130	0.78
1 in.	2 in.	0.010 in.	0.420	0.042	0.130	1.30	0.130	1.30
4 in.	3 in.	0.018 in.	0.072	1.300	0.220	3.95	0.220	3.95

7.10. GRIND, SURFACE

Set-Up Analysis:

Planer type such as Pratt & Whitney 14 in., Hydraulic Horizontal.

Obtain job assignment, work authorization, and information packet. Read instructions, complete time card and production report.	2.50
Trip to tool crib to acquire necessary tools, gauges, and so on, and to return them upon completion of job.	8.64
Set up one measuring instrument, clean and tear down at completion of job.	0.82
Gauge and inspect first piece.	1.58
Set speeds, feeds, coolant, and so forth.	1.09
Change grinding wheel and dress.	4.45
Total	19.08 minutes

Run Time Analysis:

Handle parts to and from magnetic chuck

Weight in pounds	0–10	10–20	21–40	41–80	Over 80
Minutes per piece	0.15	0.48	1.27	1.67	4.50

Machine Time

Remove 0.001 in. of stock per square inch from multiple small parts. Position 150 parts on magnetic chuck and block for work. Set aside to tote after grind.	22.50
Advance to work, and so on.	0.26
Check parts.	0.05
Total	22.81 minutes
Divide by total in.^{2} (⅞ in. × 3 in. × 150 parts).	400 in.^{2}
Average handling time/in.^{2}	0.057 minutes
Divide by 0.001 in. of stock removed.	8
Average handling time/in.^{2} per 0.001 in. of stock removed.	0.007 minutes

Machine time

Remove 0.001 in. of stock from part per in.^{2} of *table area occupied by the part* (not just the area to be ground).	0.001 minutes
Total handling and machine time per in.^{2}	0.008 minutes

7.11. GEAR HOBBING

Gears can be made by molding, stamping, sintering, cutting room bar stock, milling, and hobbing. Stamping and drawn bar stock are the cheapest and produce the least-precise gear. Milling is a low-tooling, fair-precision method usually used on only small quantities. Hobbing requires specialized machines and high-cost cutters, but the accuracy and speed of production make them the most practical for production gear cutting.

Set-Up Analysis:

The average time to set up a hobbing machine of the Barber–Colman type is shown below. The work included is job assignment, select and install correct hob, spacers, arbors, fixtures, set speeds and feeds, adjust coolant, obtain and set-up dial indicator, adjust machine, and measure the first piece produced.

The tear down and clean up is included. Includes required reference to production instructions, handbooks, tables, and so forth. Set-up gauges and dial indictor if required.

Gear Type	Minutes
Spur	35.0
Helical	65.0

Run Time Analysis:

7.11.1. Handling Time per Gear

Pick up *first* blank, install on arbor, install spacers, and tighten lock nut. Remove and set aside.	1.00
Pick up *successive* blanks in each set-up for cutting, install on arbor. Remove and set aside.	0.50
Advance work to cutting position.	0.06
Move overarm bracket up and tighten to overarm. Adjust center to work.	0.60
If concentricity is called for, dial indicator part for alignment	0.40
Start/stop machine.	0.06
Back away work from cutting position.	0.06
Release overarm bracket from overarm and position into clear.	0.40
Move hob clear.	0.05
Remove gear from arbor.	0.35
Visually inspect gears (per gear).	0.15*
Total per one gear set-up	3.63 minutes

7.11.2. Machine Time (MT)

1. Example:
   The summary time values are used in the following manner.
   From print:
   Gear, 64 pitch
   Class, commercial
   Material, steel
   100 teeth
   0.125 in. face width
   0.250 in. width at hub

   .125 in.
   .250 in.

*Add 0.65 minutes for each additional gear on arbor when multiple gears are cut at one time.

GEAR HOBBING

From summary table:

	Minutes per Inch		
	Aluminum	Mild Steel	Stainless Steel (S/S)
Machine Time (MT) per:			
Tooth/in.	0.37	0.42	0.48
Face width	0.37	0.42	0.48

Calculate:

Distance to be traveled by hob (width at hob)	0.250 in.
Number gears per in. 1.00/0.25	4
MT/tooth/gear 0.42/4	0.105 minutes
Number of teeth	100
MT/Gear 0.105 × 100	10.50 minutes

2. Time Value Summaries:

	Aluminum	Mild Steel	S/S
MT/tooth/inch of face width; commercial, rough cut only	0.37	0.42	0.48
Precision 1			
Rough cut	0.37	0.42	0.48
Finish cut (at 2 × rough)	0.74	0.84	0.96
Total	1.11	1.26	1.44

Precision 2
The net labor cost of P2 gears is increased by the following factors:

a. More elaborate handling techniques
b. More frequent checking in process

	Aluminum	Mild Steel	S/S
c. More finish and/or rework operations to bring dimensions within tolerance			
d. Sorting total lot into classes			
e. Lower yield rate pr log			
Total (at 2 × P1 value)	2.22	2.52	2.88

MACHINE SHOP OPERATIONS

	Aluminum	Mild Steel	S/S
Precision 3 The labor cost of P3 gears runs about three times that of the P1 class.			
Total (at 3 × P1 value)	3.33	3.78	4.32

3. Time Value Analysis
   The basic formula for estimating hobbing time is:

$$T = \frac{N(W + A)}{(F)(HR)}$$

Symbol	Description	Source
T	Machine time, minutes	Print
N	Number of teeth	Print
W	Width of face, in.	Print
A	Approach distance, In. per pass $A = \sqrt{d(D - d)}$	Calculate
D	Hob diameter, in.	
d	Whole depth of gear tooth. (Approach via the above calculation usually averages 20% of hob diameter.)	
F	Hob Feed, In./Revolution	Table 7.1
HR	Hob rpm $HR = \dfrac{4\,CS}{D}$	Calculate
CS	Cutting speed, rpm	

Example of Hobbing Formula:
Given:

$$\text{24 pitch gear} \quad T = \frac{N(W + A)}{(F)(HR)}$$

Mild steel
15 teeth
0.187 in. width of face
0.406 in. width at hub

Determine from Table 7.1:

        Hob diameter   2.5 in.

GEAR HOBBING

Whole depth of tooth 0.0937 in.
Feed 0.038 in.
Cutting speed 50 rpm

Calculate approach dimensions:

$$A = \sqrt{0.0937\,(2.5000 - 0.0937)} = 0.48 \text{ in./pass}$$

Assume machine blanks (10) per pass = 0.048 in./gear

Calculate hob rpm:

$$HR = \frac{4(50)}{2.5} = 80 \text{ rpm}$$

Calculate time to hob:

$$T = \frac{15\,(0.406 + 0.048)}{(0.038)\,(80)} = \frac{6.81}{3.04} = 2.24$$

$T = 2.24$ minutes

MT/tooth/in. of face width

The same formula is used to develop the "minutes per inch" time values. In the above example,

$$15 \text{ teeth} \times (\text{width} + \text{approach})$$

gives the equivalent linear inches of hob travel. The feed × hob rpm gives feed in terms of minutes per inch.

$$\frac{6.81 \text{ in.}}{3.04 \text{ minutes/in.}} = 2.24 \text{ minutes}$$

TABLE 7.1. Gear Hobbing Feeds and Speeds (in.)

Pitch	Whole Depth Tooth (in.)	Hob Diameter (in.)	Feed/Hob (in.)			Cutting Speed (RPM)		
			Al.	M. Steel	S/S	Al.	M. Steel	S/S
24	0.0937	2.5	0.057	0.038	0.031	150	50	40
64	0.0363	1.5	0.033	0.022	0.016	150	50	40
96	0.0249	1.15	0.026	0.017	0.011	150	50	40

For minutes/tooth assume one tooth with a 1-in. face. The feed and speed values using a steel gear come from Table 7.1.

$$24 \text{ Pitch} = \frac{1\,(100)}{0.038[4(50)/2.5]} = \frac{1.00}{3.04} = 0.33 \text{ minute}$$

$$64 \text{ Pitch} = \frac{1\,(1.00)}{0.022[4(50)/1.5]} = \frac{1.00}{2.93} = 0.34 \text{ minute}$$

$$96 \text{ Pitch} = \frac{1\,(1.00)}{0.017[4(50)/1.25]} = \frac{1.00}{2.72} = 0.37 \text{ minute}$$

The pitch of the gear (tooth size) has very little influence on the machining time because of the inverse relationship between pitch and tool feed. Lower, coarse pitch gears (large teeth) use high rates of feed whereas higher, fine pitch gears (small teeth) require relatively low rates of feed. The net result is nearly the same machining time per tooth for all pitches from 24 to 96. It is also evident from Table 7.1 that the feed for cutting aluminum gears is theoretically faster than steel, thus requiring less machining time. In actual practice, however, precision class gears do not reflect these savings. A practical machining time differential for cost estimating aluminum gears is to use 10% less than for steel gears. Machine time per tooth per inch of face width is summarized as follows from the above examples and detail calculations.

                Steel   0.36 minute
                Aluminum 0.32 minute

7.12. SPEEDS AND FEEDS

As discussed in the beginning of this section, the estimating standards in Sections 7.1–7.11 take into account suitable speeds and feeds, as well as set-up time, handling, and so on. The resulting standards are of sufficient accuracy for estimating the cost of machined parts as a part of the cost of the complete system. If greater accuracy is required, the necessary level of machining time data is beyond the scope of this book. A primary reference should be consulted, such as *The Machining Data Handbook*, published by the Machinability Data Center, Cincinnati, Ohio.

7.13. TOLERANCES AND SURFACE FINISHES

The following tables list the tolerance and surface finishes normally held on production machines. The data can act as a guide in determining when an additional grinding operation or reaming operation is required. Figures 7.2 and 7.3 give tolerance data by machine type and cut type. Figure 7.4 provides the

Machine	Operation	Working Tolerance[a]		Minimum Tolerance[b]	
		Angular	Linear	Angular	Linear
Automatic Screw	Turning Internal	±15'	+0.003 −0.000	±5'	±0.001
	External	±15'	+0.000 −0.003	±5'	±0.001
	Shoulder Location	±15'	±0.002	±5'	±0.001
Turret Lathe	Turning Internal	±15'	+0.002 −0.000	±5'	±0.0005
	External	±15'	+0.000 −0.002	±5'	±0.0005
	Shoulder Location	±15'	±0.002	±5'	±0.001
Engine Lathe	Turning Internal Through 2.000 dia.	±15'	+0.002 −0.000	±5'	±0.0005
	2.001 to 5.000 dia.	±15'	+0.005 −0.000	±5'	+0.002 −0.000
	Over 5.000 dia.	±15'	+0.010 −0.000	±5'	+0.004 −0.000
	External Through 2.000 dia.	±15'	+0.000 −0.002	±5'	±0.0005
	2.001 to 5.000 dia.	±15'	+0.000 −0.005	±5'	+0.000 −0.002
	Over 5.000 dia.	±15'	+0.000 −0.010	±5'	+0.000 −0.004
Grinding	Grinding Clyindrical	±1'	±0.0005	±15"	±0.0002
	Surface (Thickness)	±5'	±0.001	±30"	±0.0005
	Disc	±1°	±0.020	±30'	±0.010
	Threads through 1.000 dia.	±5'	±0.0005	±1'	±0.0002
	Threads over 1.000 dia.	±5'	±0.001	±1'	±0.0005
Planer	Planing (Thickness)	±15'	±0.010	±5'	±0.005
Duplicator	Duplicating	±15'[c]	±0.005[c]	[c]	[c]
Profiler	Profiling	±15'[c]	±0.005[c]	[c]	[c]
Hone	Honing Through 6.000	±5'	±0.005	±1'	±0.0002
	Over 6.000	±5'	±0.001	±1'	±0.0005

Notes:

[a] The column of working tolerances should be considered the minimum for production drawings.

[b] The column of minimum tolerances should be restricted to tooling work.

[c] These tolerances are governed by the accuracy of the master or die.

FIGURE 7.2. Recommended operational tolerances for machined metal parts. (Reprinted through the courtesy of Thompson Ramo Wooldridge Inc.)

Machine	Operation	Working Tolerance (1)		Minimum Tolerance (2)	
		Angular	Linear	Angular	Linear
Shaper	Shaping	±15'	±.005	±5'	±.003

Broaching	Broaching
	Surface (Thickness)
	  Through 1.000
	  1.001 through 4.000
	  Over 4.000
			±.001
			±.0015
			±.002		

Milling	Milling
	  Flat Work
	  Small Surfaces
	  Face & Side
	  2 or more surfaces
	  Slot
	
	  Contour
	End
	  Through .250 dia.
	
	  Over .250
	
	  Straddle
	  Spline
	  Thread
	Hollow
	  Through .250 dia.
	  Over .250 dia.
		±15'
		±15'
		±15'
		±15'
		±15'
		
		±15'
		
		±15'
		
		±15'
		
		±15'
		±15'
		±15'
		
		±15'
		±15'
			±.004
			±.002
			±.002
			±.005
			±.005
			−.001
			±.005
			
			+.004
			−.000
			+.008
			−.000
			±.005
			±.002
			±.001
			
			±.006
			±.010
				±5'
				±5'
				±5'
				±5'
				±5'
				
				±5'
				
				±5'
				
				±5'
				
				±5'
				±5'
				±5'
				
				±5'
				±5'
					±.002
					±.001
					±.001
					±.003
					±.002
					
					±.002
					
					+.002
					−.000
					+.004
					−.000
					±.003
					±.001
					±.0005
					
					±.002
					±.004

Boring	Boring
	  General
	  Mill
	  Jig
	  Shoulder Location
		±15'
		±15'
		±5'
		±15'
			±.0015
			±.002
			±.0005
			±.002
				±5'
				±5'
				±1'
				±5'
					±.0005
					±.001
					±.0002
					±.001

Notes:
(1) The column of working tolerances should be considered the minimum of production drawings.
(2) The column of minimum tolerances should be restricted to tooling work.
(3) These tolerances are governed by the accuracy of the master or die.

FIGURE 7.3. Recommended operational tolerances for machined metal parts (contd.). (Reprinted through the courtesy of Thompson Ramo Wooldridge Inc.)

Surface Classification	Operation	1000	500	250	125	63	32	16	8	4	2	1
Very Rough	Sand Casting	■	■									
	Hot Rolled Surface	■	■									
	Torch Cut	■	■									
	Weld	■	■									
	Saw	■	■									
	Forging		■	■								
	Turning/Boring		■	■								
	Milling/Shaping		■	■								
Coarse	Disc Grinding			■								
	Cold Rolled Surface			■	■							
	Turning/Boring			■	■							
	Milling/Shaping			■	■							
	Drilling			■	■							
	Surface Grinding			■	■							
Medium	Cold Rolled Surface				■							
	Permanent Mold Casting				■	■						
	Investment Casting				■	■						
	Turning/Boring				■	■						
	Milling/Shaping				■	■						
	Surface Grinding				■	■						
Fine	Die Casting						■					
	Extrusion						■					
	Turning/Boring						■	■				
	Milling/Shaping						■	■				
	Surface/Cylindrical Grinding						■	■				
	Plaster Mold Casting						■	■				
	Cold Draw						■	■				
	Reaming						■	■				
Smooth	Cylindrical Grinding								■	■		
	Microhoning									■	■	
	Lapping									■	■	■

*The roughness height rating range indicates the most practical values associated with economy of production.

FIGURE 7.4. Surface finish data. (Reprinted through the courtesy of Thompson Ramo Wooldrige, Inc.)

MACHINE SHOP OPERATIONS

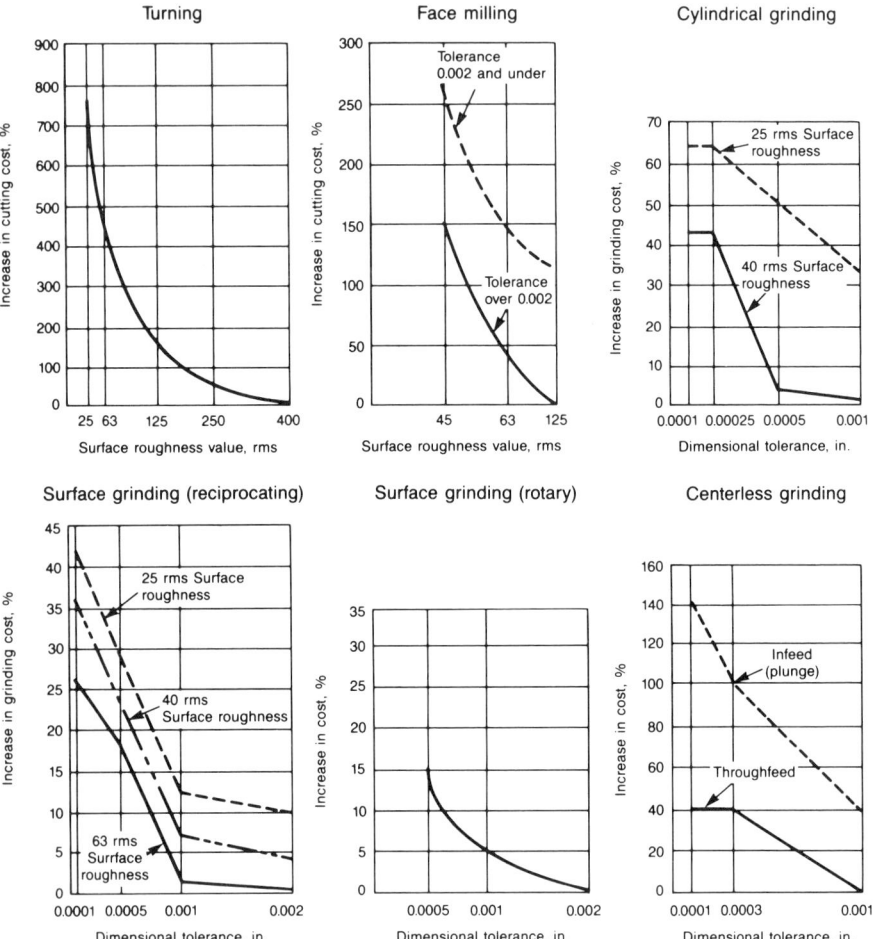

FIGURE 7.5. Effect of tolerance and surface roughness on machining costs for several operations. (L. J. Bayer, Analysis of manufacturing costs relative to product design, Paper No. 56-A-9, American Society of Mechanical Engineers, New York, NY—The Machining Data Handbook, Third Edition.)

normal range of surface finishes produced by various machining and casting methods. Both tolerances and finish tables give the preferred "working" tolerances and the minimum tolerances that can be held for special cases. It is best if the minimum tolerances be used only for tool room work, but it seems that they still keep cropping up on production drawings. Figure 7.5 shows the effect of tolerances and surface roughness on machining costs for several operations. Figure 7.6 shows the relative machining costs and surface roughness for steel parts.

TOLERANCES AND SURFACE FINISHES

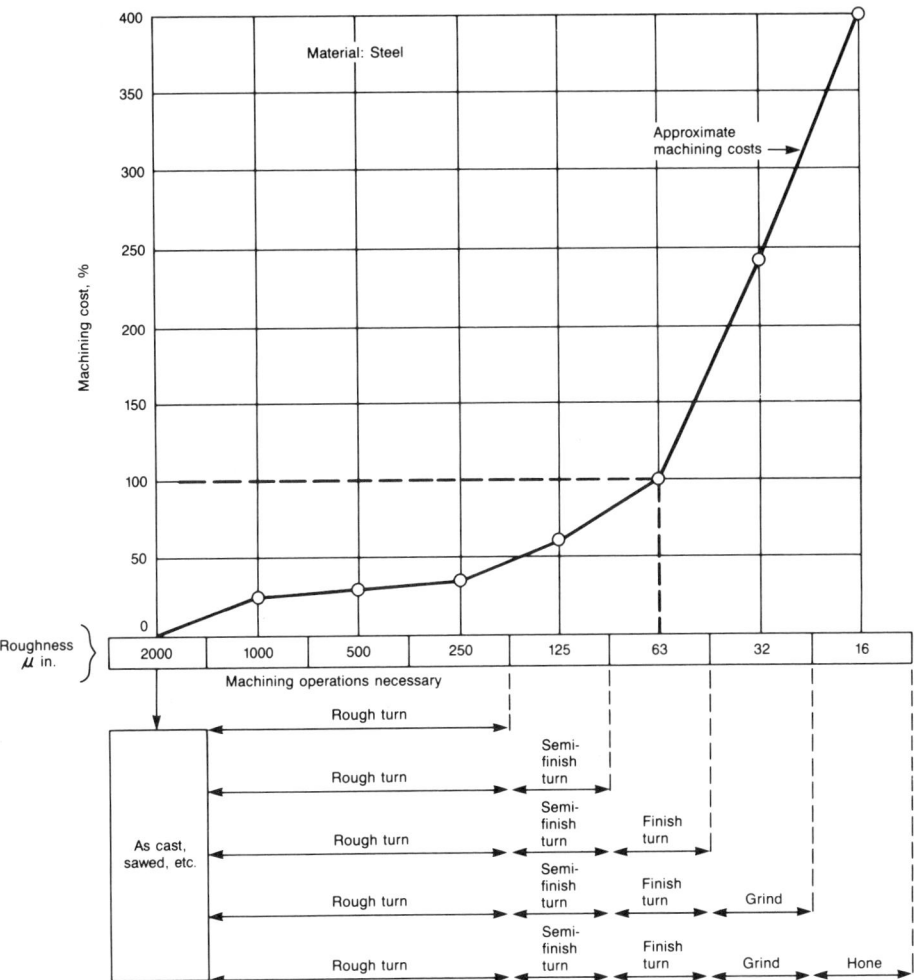

FIGURE 7.6. Relative machining costs and surface roughness for steel parts. (Courtesy of General Electric Company—The Machining Data Handbook, Third Edition.)

The shop doing the work (and the estimator) has the option of making regular lathe and milling cuts but slowing down the machining process to attain the borderline tolerances, or to do only the rough cuts on the standard equipment and to add grinding, reaming, and boring operations for the finish dimensions. The latter method is preferable for production operations. In any event, close tolerances and finishes mean more machining time, more precise and costly tooling, and more inspection time. The final cost estimate must include all these factors.

The limiting tolerance for the lathe work is usually 0.0005 in. When tolerances are closer than this, grinding is the most economical finishing method. In many cases grinding will be the most economical finishing method for tolerances as loose as 0.001 in.

The following rules of thumb can be used as rough approximations where the detail tables are not required:

Description	Number of Cuts	Maximum Tolerance (in.)	Surface Finish	Approximate Machine Cost Index
Medium rough cut	1	0.002"	To 63	1.00
Rough turn, finish turn	2	0.001"	To 32	2.50
Rough turn, finish grind	2	0.0002"	To 16	4.00
Rough turn, grind, lap	3	0.0002"	To 8	6.00

When machining a thin disc to close tolerances (4 in. in diameter × ½ in. thick and up), it is often necessary to stress relieve and age the part in an oven after the rough turning operation. This relieves the distortions built up during the heavy cutting.

SECTION EIGHT

SHEET METAL OPERATIONS

8.1.	Cut Blank	102
8.2.	Notching	103
8.3.	Holes	103
8.4.	Press Operations	106
8.5.	Trim, Profile, Rout	113
8.6.	Burr Removal—Bench, Belt, Vibrator	115
8.7.	Arc Welding	116
8.8.	Spot Weld	118
8.9.	Rivet	120
8.10.	Surface Preparation	121
8.11.	Paint	121
8.12.	Silk Screen and Engraving	123

Since the invention of the first radio, the sheet metal chassis has been the major packaging choice for electronic components. Printed circuit boards are the only alternative technique that has found widespread use in the industry, and even then, the case, cabinet, or box that holds the printed circuit boards is usually a sheet metal structure.

It is not unusual for the labor required to fabricate the cabinets, chassis, terminal boards, and other packaging hardware to be a significant cost factor. A cost item of this magnitude is not to be glossed over in making an estimate.

Steel is the most economical chassis and cabinet-making metal. The basic sheet stock costs less per square foot than aluminum and, also, the cleaning and

SHEET METAL OPERATIONS

finishing operations require fewer steps. Although the overall fabrication costs of aluminum are slightly more than for steel, the light weight of aluminum makes it preferred for airborne equipment and for any type of portable equipment.

High-quality "engineering" plastics are replacing metal in some applications. The range of available physical properties has increased considerably, and RFI/EMI shielding problems can be solved. However, relatively large production runs are required to amortize the tooling costs. (See also Section 10.)

The operations included in this section are listed in the general sequence by which they would be done in the shop. The time values are given for three representative part sizes. They are given as reference points, and it is expected that parts of intermediate size or special configuration will require a deviation from the given values. The parts handling time is included in the operation time unless it is listed separately.

8.1. CUT BLANK

Set-Up Analysis:
Straight blade or guillotine shear with 8-in. blade with clutch remotely activated by foot control or treadle.

The set-up time includes setting a stop on the squaring arm or adjusting the position of the back gauge. Incoming stock and outgoing completed blanks are moved to shear by a material handler but stock and blanks are handled by the shear crew for all shearing operations. A crew size of 1 is allowed on the 3×3 in. blanks and crew size of 2 is allowed on the 18×18 in. and 30×30 in. blanks.

Obtain job assignment.	2.50
Set stop or gauge.	2.58
Measure first cut for accuracy and squareness, and visually check quality of sheared edge.	1.12
Total set-up time	6.20 minutes

8.1.1. Run Time Analysis

The maximum number of cuts required for a rectangular blank is four cuts for four sides. In practice, the number of cuts is usually closer to one cut per blank because each cut that is made frees at least one other blank. And, on the initial cuts one cut is actually cutting the side of several blanks.

The power shear average time per piece produced, in minutes, is given below.

HOLES

Size of Finished Piece

3 × 3	18 × 18	30 × 30
0.36	1.16	1.81

8.2. NOTCHING

Set-Up Analysis:
Power notcher with mechanical or hydraulic drive.

Set-up time includes setting the stop for notch size. Sheared blanks are delivered to the notcher by a material handler. Operator notches an average of four places and puts blank aside for material handler to remove. Scrap falls into scrap bin to be removed by material handler.

Notching time may be applied to punching or blanking of holes from flat blanks, but these times are not applicable to press operations that start with coil stock rather than a blank.

Obtain job assignment.	2.50
Set stops.	3.32
Measure first cut for accuracy, and visually inspect quality of sheared edge.	1.12
Total set-up time	6.94 minutes

8.2.1. Run Time Analysis

Pick up and position a part in the notcher. Depress the foot pedal to activate the machine. Reposition for the next notch. Set aside the notched part.

	Size of Incoming Part		
	Small 3 × 3 in.	Medium 18 × 18 in.	Large 30 × 30 in.
Average time per notch (in minutes)	0.16	0.27	0.49

8.3. HOLES

Holes may be produced by a variety of methods. Punching, drilling, fly cutters, circle shears, circular hole saw, profilers, routers, milling machines, and blank pierce dies are all utilized. The turret punch, drilling, and the drilling fly cutter attachment are the most common techniques for close tolerance and low-quantity chassis work. Numerical Control (N/C) punching is increasingly common.

8.3.1. Turret Punch

Set-Up Analysis:
Weideman type RA 41P press with 28 × 40 in. work table and 3 in. maximum diameter punch.

This machine utilizes a Pantograph type stylus, template, and work piece holder.

Obtain job assignment.	2.50
Obtain template from tool crib, position and secure position, adjust work holding device.	15.28
Change punches (average of 3).	14.99
Total set-up time	32.77 minutes

8.3.2. Run Time Analysis

Handle and Clamp Part:
Pick up blank, position tooling holes to locating pins on table. Clamp two to four places. Disassemble part from holding fixture after piercing and lay aside.

Punch Hole:
Move stylus to enter template hole. (Pantograph action moves blank to position under punch. When stylus enters hole, punch is tripped.) Remove stylus from hole and move to next hole. Rotate turret to proper punch size as required.

	Size of Incoming Part		
	Small 3 × 3	Medium 18 × 18	Large 30 × 30
Handle part (minutes per part)	0.29	0.47	0.84
Punch hole (minutes per hole)	0.06	0.06	0.06

8.3.3. Drill Hole with Drill Press

Set-Up Analysis:
Secure drills, jigs, and so on. Install to drill press. Set stops, speeds, and so forth. See detail analysis in Section 7.5.

HOLES

Average set-up time   23.82 minutes

Run Time Analysis:
Handle and Jig Part

	Size of Incoming Part		
	Small 3 × 3	Medium 18 × 18	Large 30 × 30
Handle part to jig. Lock in place, unlock after drilling, and lay aside. Minutes per part.	0.29	0.47	0.84

Drill Holes:
See Table 8.1 and the following explanations.

1. Machine time for the ⅛ to 1 in. diameters remain fairly constant. Holes less than ⅛ in. in diameter require twice the machine time because of the additional care to prevent drill breakage. Holes of the 1 or even 2 in. sizes can be drilled in the same time because the larger holes require heavy-duty drill presses allowing higher rates of metal removal.

2. The extra time allowed to drill a hole with no template is for the operator to visually line up the part to the point laid out and punched. In this case, there is no template to guide the drill into the final position.

3. The move time from hole to hole also includes lowering and raising the drill spindle to the work piece before and after actual cutting takes place. The

TABLE 8.1. Hole Drilling Time Values[a]

Hole Diameter (in.)	Hole Depth (in.)	Location Guide	Move Time (hole to hole)		Machine Time		
			Up to 40 lbs.	Over 40 lbs.	Mild Steel	S/S	Aluminum
⅛	⅛	Scribed lines	0.078	0.110	0.05	0.15	0.02
⅛	⅛	Template	0.061	0.087	0.05	0.15	0.02
¼	¼	Template	0.061	0.087	0.13	0.109	0.11
¾	⅜	Template	0.061	0.087	0.23	0.42	0.05
1	½	Template	0.061	0.087	0.35	0.61	0.07
2 to 4 in. Flycutter	⅛	Scribed lines	0.078	0.110	2.16	3.76	0.25

[a] Scribed lines are previously laid out, and punched. Template includes drill bushings. Hole must be deburred after drilling.

time will cover moving the part from hole to hole at a single-station drill press, or moving the part from spindle to spindle at a multiple set-up.

4. The Flycutter is a means of cutting holes larger than the normal drill sizes. These are usually lightning holes in electronic chassis. It is essentially a boring bar attached to a drill shank. The boring bar may be adjusted to any diameter by loosening the set screw and tightening at the desired radius.

If the part weighs over 40 pounds, use 0.110 minutes for each hole-to-hole move with no template and 0.087 minutes for each hole-to-hole move with a template. Typical hole location accuracy in a drill press using a bushing in a drill jig is ±0.002 in.; for a drill press without bushing or drill jig, it is ±0.015 in.

8.4. PRESS OPERATIONS

Forming is the normal operation that follows blanking, notching, and hole punching. Blanking and piercing via die sets are included in this group because they can be done on a press brake as well as regular punch press equipment. Blanking and piercing also bring the part to the same stage of completion as those that were individually blanked, notched, and drilled.

The statement that forming operations normally follow hole punching or filling must be qualified. Making holes prior to forming is the most economical method. For a flat sheet the holes can be punched in rapid succession, or drilled in stacks. Also, tooling cost for a flat part usually amounts to a simple template, whereas tooling for a formed part may be an expensive box jig or similar three-dimensional device.

If drilling of deep-drawn or severely formed parts is required, the drilling operation follows the forming operation because of the metal flow during deep drawings or severe forming.

Dimpled, joggled, or brake-formed parts can usually be punched prior to forming, provided the hole location in reference to the form line is not too critical. A rule of thumb for brake forming is that hole locations can be held to plus or minus 0.010 in. as referenced to the brake line. The desired tolerance is 0.030 in. but 0.010 in. can be held for special runs. Hole grouping within the flat surface of a chassis can be held to 0.005-in. tolerance in relation to each other, but when referenced to the form line, both the distortion of the metal and the physical holding of the blank to the brake make the 0.010-in. tolerance a minimum. Parts with tighter tolerances than this, or with multiple form lines, must be individually drilled after forming.

8.4.1. Blank and Pierce

This is a punch press operation utilizing individual blanking die sets, piercing sets, or combination blank-and-pierce sets. Twenty- to 75-ton presses are sufficient for light electronic chassis work. Press brakes may also be set up to use small blank-and-pierce die sets. The die sets consist of one or more punch-and-die elements mounted in a common die shoe-guide post-punch holder combination. The punch-and-die elements are permanently aligned in the die set. Guide posts must be rigid enough to maintain precise die alignment.

Set-Up Analysis:

Because of the single-unit die set construction, the actual set-up labor is limited to fastening the die shoe to the press bed and adjusting the press bed height or stroke length so that the proper punching depth is achieved. These basic elements along with miscellaneous adjustments, trips to the tool crib, and so on, can usually be accomplished by one or two workers with a hand or battery-operated lift cart.

Job assignment is included in die set-up.

	Install and Remove Die Set Size of Blank		
	Small 3×3 in.	Medium 18×18 in.	Large 30×30 in.
Minutes per set-up	21.15	24.85	30.85

Run Time Analysis:

The blank-and-pierce operation includes handling individual blanks to the die set, actuating the press with double-hand safety buttons, machine cycle, removing finished part from die, setting aside to tote, and clearing scrap for the next blank.

Minutes per Blank Produced Size of blank		
Small 3×3 in.	Medium 18×18 in.	Large 30×30 in.
0.36	0.42	1.53

Note that individual handling of parts to and from the die is the slowest type of blank-and-pierce operation. When the raw material can be fed from strip stock, or coil stock, the output of the press is proportionate to its steady running RPM. Press cycles vary from 10 to 200 or more strokes per minute. Therefore, the potential output is 10 to 200 pieces per minute. Naturally, this full potential is reduced by frequent stops to set up new stock, remove skeleton material from the die, and other maintenance adjustments. As a result, typical punch press operations range from 1.0 hour per 1000 pieces for hand-fed strip stock blanking to 0.1 hours per 1000 pieces for coil-fed automatic blanking.

8.4.2. Dimple or Joggle

Dimple and joggle operations are performed on the same type of press as blank and piercing. The only difference is that dimples and joggles utilize a forming die rather than the blanking and piercing type. Dimples are added to the edges of large (2- to 6-in. diameter) lightening holes to add rigidity. Joggles are made for the purpose of lap joints. One of the two overlapping pieces of metal is offset so as to create a flush surface where the metals are joined.

Set-Up Analysis:

(Same as blank and pierce operation.)

Set-up minutes		
Size of blank		
Small 3 × 3 in.	Medium 18 × 18 in.	Large 30 × 30 in.
21.15	24.85	30.85

Run Time Analysis:

Includes individual parts handling to and from die in same manner as blank and pierce operation.

Size of blank		
Small 3 × 3 in.	Medium 18 × 18 in.	Large 30 × 30 in.
0.36	0.42	1.53

8.4.3. Brake Form

The press brake and the shear are the backbone of the sheet metal shop. A wide variety of bending and forming can be done with standard dies. Special blanking, piercing, and forming dies can also be utilized on the press brake.

Set-Up Analysis:

Most jobs are done on the press brake without changing dies. Positioning stops must be set, and space between the ram and bed must be adjusted for material thickness and/or die sizes.

Obtain job assignment.	2.50
Adjust stops and test measure first part.	6.39
Total set-up time	8.89 minutes

Note that if dies must be changed, add: 4.87 minutes for top die, and 7.65 minutes for bottom die.

Run Time Analysis:

The handling and machine time for press bending is the same as for the previously discussed blanking and dimpling operations. The time values include handling the individual part to and from the forming die.

Minutes per Blank		
Size of Blank		
Small 3 × 3 in.	Medium 18 × 18 in.	Large 30 × 30 in.
0.23	0.38	0.72

8.4.4. Roll Form

The roll form machine shapes flat sheet stock into round cylindrical shapes. Full cylinders, half cylinders, and a wide variety of radii can be formed on this machine. The machine consists of three properly spaced rolls, two of which feed the stock into the machine, and the third, which bends the sheet into the desired radius. By adjusting the location of its longitudinal axis relative to the fixed axis of the other two rolls, different radii may be generated. Some roll formers have only one roll the position of which is adjustable.

The set-up required is to adjust the front two rolls to the thickness of the metal, and the back roll to the desired radius.

Set-Up Analysis:

Job assignment	2.50
Roll adjustment	2.32
Total	4.82 minutes

Run Time Analysis:

The blank is inserted between the front two rollers, and the crank handle is turned by hand to feed it through. A radius of approximately the same diameter as the forming rolls (½ to 1 in.) can be turned in one pass. Separate corner radii can be turned in two passes.

	Size of Blank		
	Small	Medium	Large
Time per pass	0.18	0.43	0.77
Time per inch on blank length	0.02	0.03	0.06

8.4.5. Deep Draw

Drawing is a specialized method for forming round or rectangular cans, covers, boxes, and so on. It replaces the notch, brake, weld, and grind operations that are used to produce the same type of part when quantities do not warrant making up expensive draw dies. Deep drawing is done on high-tonnage, hydraulic presses. The die sets range in cost from $1000 to $4000 for typical electronic chassis parts.

The operation sequence for a drawn part will run as follows:

Blank
Draw
Anneal
Redraw
Trim
Bead or roll edges

The blanking and at least one drawing operation are always required. The annealing operation is required prior to each additional draw. The trimming operation is usually required only for special flanged edges or close tolerances. The edge beading or hemming is performed as a separate operation when called for by the part design.

The number of draws is always a hard point to determine. The three controlling factors are easily identified:

Height-to-diameter ratio
Ductility of the stock metal
Corner radius

With this the simplicity ends. In practical application, these overall guides, plus formulas, tables, experience, and the seasoning of a little trial and error are what determine the number of draws needed to produce a completed part. In addition to the drawing operation, the part must usually be annealed for each subsequent draw because the drawing process work hardens the metal. Annealing restores the metal to its original ductility and thus enables it to be drawn farther on each subsequent operation.

A guide for estimating new designs is as follows:

1 Draw—depth ⅓ of punch diameter
2 Draws—depth ½ of punch diameter
3 Draws—depth ¾ of punch diameter
4 Draws—depth equal to punch diameter

These ratios are for a corner radius of four times the stock thickness. If the radius is large enough, even a one-to-one reduction can be made in one draw. Selecting the proper lubricant is very important.

Set-Up Analysis:

Setting up form dies for deep drawing is very similar to setting up blanking and piercing dies on a punch press. In addition to alignment and material thickness adjustments, deep drawing requires precise control of the press pressure and speed.

Job Assignment Included In Die Change	Size of Blank		
	Small 3 × 3 in.	Medium 18 × 18 in.	Large 30 × 30 in.
Install And Remove Die Minutes Per Set-Up	21.05	24.85	30.85

Run Time Analysis:

Pick up and position blank to die, trip the two hand actuator buttons, machine cycle, remove finished part from die, stack in tote, and blow die clear for next load.

	Size of Blank		
	Small 3 × 3 in.	Medium 18 × 18 in.	Large 30 × 30 in.
Minutes Per Blank Feed	0.25	0.40	0.74

8.4.6. Annealing

If a drawing operation involves a reduction of more than 40%, intermediate annealing is usually required. Percentage of reduction is defined as:

$$\left(1 - \frac{\text{New diameter}}{\text{Old diameter}}\right) \times 100$$

SHEET METAL OPERATIONS

Reduction		Number of Annealing Operations
First draw	40%	1
Redraw	25%	2
Redraw	15%	3
Redraw	11%	4

A preferred method of annealing is to heat the work piece to the required temperature in an exothermic atmosphere. The subsequent cooling cycle is also in a controlled atmosphere chamber. If the proper atmospheres are not maintained in the heating and cooling cycles, extra cleaning operations are required before drawing operations are possible.

Set-Up Analysis:

This is usually a continuous operation utilizing conveyor, walk-in, or reach-in furnace. A job changeover may require a change in temperature setting with a delay before the furnace can be used. Controlled atmosphere furnaces are increasingly common because their use minimizes the need for cleaning operations.

Obtain job assignment	2.50
Set furnace to required temperature. Set conveyor speed, if required. Install load couple, if required. Turn fans on/off, adjust dampers, open/close doors.	5.38
Total time	7.88 minutes

Run Time Analysis:

The time values are for the manual parts handling only. Machine time is normally figured as part of the overhead cost on the direct work-hours.

	Minutes per Blank		
	Size of Blank		
	Small 3 × 3 in.	Medium 18 × 18 in.	Large 30 × 30 in.
Annealed in batch furnace	0.15	0.60	1.88
Annealed in conveyor furnace	0.13	0.50	1.00

8.4.7. Hydroform

The hydroform process utilizes a hydraulic press in the same manner as deep drawing, but only the male half of the die is required. A caged rubber mat is attached to the upper platen of the press and, on coming in contact with the blank, the rubber forces the blank to take the contour of the form block on the lower platen. Aluminum up to 0.125-in. stock can be formed and even sheared by the hydroform method.

The chief advantage of hydroforming is the low tooling cost as compared to the conventional punch-and-die sets. Moreover, more severe reductions are possible—as high as 60 to 70% on the first draw—because the metal is held tightly against the sides as well as the end of the punch. A disadvantage is the relatively low production output, and the hand finishing that is often required when metal shrinking occurs. Application is limited to stock gauge in the 0.010- to 0.065-in. range for low carbon steel, stainless steel, and aluminum.

Presses are usually equipped with mechanical feeding devices that locate the blank over the punch. Formed parts are removed mechanically also. Blanks up to 25 in. in diameter can be formed to a maximum depth of 12 in. Most of the parts shaped by hydroforming are so irregular or complex in shape that mating dies for conventional forming presses would be very expensive. In addition, marks on the finished blank from a conventional die set are significantly reduced by hydroforming, since a rubber pad forms one of the die surfaces over which the displaced metal must flow. In some cases greater reductions are possible in the hydroformer than in conventional die sets because metal flow is easier to achieve.

Set-Up Analysis:

Obtain job assignment.	2.50
Adjust press parameters.	3.42
Install and remove die.	22.95
Total	28.87 minutes

Run Time Analysis:

Pick up and position blank to form block, position form block under press, trip actuator buttons, machine cycle, remove finished part from die, stack in tote, and blow die clear for next load.

Minutes per Blank		
Size of Blank		
Small	Medium	Large
3 × 3 in.	18 × 18 in.	30 × 30 in.
0.25	0.40	0.74

8.5. TRIM, PROFILE, ROUT

Trimming is sometimes required after a deep draw or hydroforming operation. Profiling or routing will be required on some contoured parts. A router,

SHEET METAL OPERATIONS

pantograph, nibbler, circle shear, or profile milling machine may be used. The router, pantograph, and profiler make the trimming cut with a rotating cutter similar to a drill or end mill. The router bit is operated at 15,000 to 20,000 RPM, and produces a smooth machined surface. The nibbler and circle shear make the trimming cut by a reciprocating action of the cutting tool. The edge finish is not as smooth as that obtained by the router bit.

A template or N/C tape (or computer control) is required for all of the above trimming operations, unless there are only one or two parts which are easier to lay out by hand. A flat template is used for flat parts, and a cage-like jig must be used for formed parts. A template may be used to merely scribe the pattern, or it may be clamped to the part and used to guide the cutter around the trim line. On a pantograph or profile milling machine, the template is positioned separately from the work, and the stylus and pantograph arm transmit the pattern to the work. Templates are not required on N/C type machines, which continuously reposition either the cutter or the part that is clamped securely in a holding fixture.

Set-Up Analysis:

Router
Pantograph
Profile miller

The maximum set-up time for any of these machines should not exceed 0.5 hours. This includes a trip to the tool crib for template and cutters, and set-up and alignment of same.

Average set-up time   28.50 minutes

Run Time Analysis:

8.5.1. Handle and Jig Part

Pick up part and position to template. Clamp in place. Disassemble and lay aside after trimming.

Handle Parts
(Minutes per Part Produced)

Size of Part		
Small 3 × 3 in.	Medium 18 × 18 in.	Large 30 × 30 in.
0.26	0.54	1.14

8.5.2. Trim, Profile, Rout (and Burr Edge)

Aluminum, brass, phenolic, or epoxy. Laminate to ¼ in. thick.

	Minutes per Part		
	Size of Part		
	Small 3 × 3 in.	Medium 18 × 18 in.	Large 30 × 30 in.
Machine Time Per In.	0.05	0.05	0.05
Burr Edge After Cutting	0.02	0.02	0.02
Total	0.07	0.07	0.07

8.6. BURR REMOVAL—BENCH, BELT, VIBRATOR

A note included on practically all blue prints is "break sharp edges." A belt sander, portable power sander, hand file, or emery cloth are used for straight edges or large holes. A "Flapper" is often used on contoured shapes. For some applications, tumbling or washing with an abrasive media is used. A hole drilled in a flat surface can usually be deburred by a portable power vibrator sander. The vibrator will not deburr the inside of a hole. This requires a scraper or end file, but the vibrator is much faster and gives an acceptable surface for most applications.

Set-Up Analysis:
Belt Sander or Portable Power Vibrator

The only set-up required is a trip to the tool crib, and occasional change of the emery belt or pad.

Set-up time   6.23 minutes

Run time Analysis:

8.6.1. Handle Part

Pick up and position to bench or sander. Lay aside to tote.

	Size of Part		
	Small 3 × 3 in.	Medium 18 × 18 in.	Large 30 × 30 in.
Minutes per Part	0.13	0.28	0.62

8.6.2. Burr Edge

Aluminum, epoxy laminate, and so on. Belt sander, hand file, and emery cloth.

	Size of Part		
	Small 3 × 3 in.	Medium 18 × 18 in.	Large 30 × 30 in.
Minutes machine time per in.	0.02	0.02	0.02

8.6.3. Burr Flat Surface

Portable power vibrator sander.

	Size of Part		
	Small 3 × 3 in.	Medium 18 × 18 in.	Large 30 × 30 in.
Minutes machine time per ft.^{2}	0.20	0.50	0.50

8.6.4. Burr Hole

Hand scraper or end file.

	Size of Part		
	Small 3 × 3 in.	Medium 18 × 18 in.	Large 30 × 30 in.
Minutes per hole	0.03	0.03	0.03

8.7. ARC WELDING

Welding is one of the most practical fabricating methods for low-production items where casting patterns or form dies are not economical. Gas welding is utilized less and less because of the high labor cost due to low deposition rates and because of higher skill requirements than for some of the electric welding

ARC WELDING

practices. MIG (metal arc inert gas), TIG (tungsten arc inert gas), and Resistance (spot-and-seam) welding are the most common methods. Some induction welding is done, especially in connection with roll-forming lines. All aluminum alloy groups can be welded, but superior joining is achieved by using alloys specifically formulated for welding (especially the 5xxx alloy group).

8.7.1. Degrease and Caustic Etch Aluminum

See Section 9.

8.7.2. Welding

Set-Up Analysis:

Set up an MIG welder, obtain wire and gas as required. Install correct size electrode, set speed, current, and gas flow. Turn on/off gas, power, and control.

12.91 minutes

Set up a weld fixture, obtain and release fixture and parts. Assemble and verify set-up. Test on first set-up and adjust as required.

31.82 minutes

8.7.3. Handle Parts to Weld Fixture

The standard handling time is allowed for ample positioning and alignment of parts.

	Size of Part		
	Small 3 × 3 in.	Medium 18 × 18 in.	Large 30 × 30 in.
Minutes	0.29	0.47	0.84

Note that where an intricate fixture is used, time must be allowed to lock holding devices and/or assemble and disassemble fixture to parts. Add 0.06 minutes per locking handle.

Run Time Analysis:

The time values include the average. They will hold for aluminum, steel, or magnesium, provided the proper alloy and surface preparation has been used.

SHEET METAL OPERATIONS

	Mild Steel Small	Stainless Medium	Aluminum Large
Weld 0.062″ stock per in.	0.25	0.25	0.25
Weld 0.125″ stock per in.	0.40	0.40	0.40
Weld 0.250″ stock per in.	0.75	0.75	0.75

8.7.4. Stress Relief

A furnace heat-treating operation must sometimes be used to relieve the stresses built up by the rapid heating and cooling of a localized area during the welding operation. Preheating, restraining holding fixtures, and planned welding sequences help to avoid warping, but the stress relieving and straightening operation is still required on some parts due to close dimensional tolerances or highly stressed working parts. A minimum set-up time is required. The run time is merely the handling of parts to and from the heat treating furnace, or furnace racks.

Set-up time   6.0 minutes

	Small	Medium	Large
Minutes per unit	0.29	0.47	0.84

8.7.5. Grind or Burr Fillet

Welds made on exterior corners of chassis and cabinets must be ground smooth prior to the final paint operation. Portable power sanders are used for large cabinets. Floor model belt sanders or grinders are used for smaller parts.

Set-up time   6.0 minutes

	Small	Medium	Large
Machine time, grind or burr per in.	0.03	0.04	0.05

Handling time must be added to the above values.

8.8. SPOT WELD

Spot welding is one of the most economical sheet metal fastening methods. The spot welding operation itself takes about the same time as riveting, providing the rivets are hopper fed, but spot welding does not require drilling holes in the mating parts as rivets do. The disadvantage, which prevents spot welding from being used in many applications, is its lack of structural strength (especially peel strength) as compared to rivets. This is inherent to the process. The surge of electric current momentarily melts the two pieces of metal to fuse them together, and in so doing reduces the metal to the "as cast" state. As a result, the weld has less strength than the surrounding metal, the strength of which has been enhanced by cold working and often heat treating. Thus, in comparing a rivet to a spot weld of the same diameter, the rivet is considered a stronger joint. On the other hand, a number of spot welds will soon equal the strength of one rivet. A spot weld has a lower resistance to tensile stress than shear stress so the assemblies "peel" apart more readily when spot welded as compared to mechanical fastening with rivets and bolts with nuts.

Set-Up Analysis:
100 amp output, single-head spot welder

The set-up of a spot welder involves installing and adjusting the contact points, current, timing, and holding fixtures as required by the part.

Set-up time   12.50 minutes

Run Time Analysis:
Handle parts:
The appropriate size ranges must be chosen when handling a large base part and one, two, or more small parts to be attached. For example, the handling time for three small parts being welded to one medium part is 0.65 minute. (0.16 + 0.16 + 0.16 + 0.17).

	Minutes/Operation	
Small 3×3 in.	Medium 18×18 in.	Large 30×30 in.
0.16	0.17	0.49

Spot Weld:
This value includes moving the chassis from spot to spot, depressing the foot pedal, and the machine cycle.

	Minutes/Operation	
Small 3×3 in.	Medium 18×18 in.	Large 30×30 in.
0.07	0.07	0.10

SHEET METAL OPERATIONS

8.9. RIVET

This section covers rough time values as would apply to a sheet metal job shop rather than a high-production mechanical assembly section. See Section 15 for highly mechanized riveting.

Set-Up Analysis:

The set-up includes installing and aligning the anvil, hammer, and raceway for the proper size rivet.

Set-up time   19.77 minutes

Run Time Analysis:

8.9.1. Handle Parts

The appropriate size ranges must be chosen when handling a large base part, and one, two, or more small parts to be attached. For example the handling time for three small parts being riveted to one medium part is 0.67 minute.

	Minutes per Part		
	Small 3 × 3 in.	Medium 18 × 18 in.	Large 30 × 30 in.
	0.13	0.28	0.62
Rivet-Hand Feed Rivet (Machine Driven)			
Pick up and place rivet through joining parts.	0.13	0.13	0.13
Press foot pedal, machine cycle time, and shift to next hole.	0.05	0.05	0.05
Total	0.31	0.46	0.80

8.9.2. Rivet-Automatic Feed Rivet (Machine Driven)

	Minutes		
	Small	Medium	Large
Press foot pedal, machine cycle time, and shift to next hole.	0.05	0.05	0.05

8.10. SURFACE PREPARATION

Surface preparations are treated separately under Section 9. A summary of these operations are presented here (Table 8.2) for each base metal. Although common, they are by all means not the only ones that can be used. The "Minutes to Process One Batch" column in the table constitute the set-up time described in the preceding subsections. Section 9, however, should be consulted for details and more precise applications.

8.11. PAINT

Typical applications are summarized in Table 8.3. The times represent set-up and operation cycles once surface preparations have been performed. The

TABLE 8.2. **Electroplating and Chemical Surface Treatment of Metals (Summary of Section 9) (Labor Times only)**

	Minutes to Process One Batch	
Process	First	Successive
Anodize Aluminum, clear	28	19
Anodize Aluminum, dye	31	22
Anodize Aluminum, hard, clear	25	16
Anodize, Aluminum, hard, dye	28	19
Black oxide coat iron and steel[a]	26	17
Cadmium plate steel[b]	23	14
Chemical conv. coat aluminum	22	13
Chrome plate copper[a]	23	14
Chrome plate steel[b]	25	16
Copper plate copper[a]	22	13
Gold plate aluminum	31	22
Gold plate copper[a]	23	14
Nickel plate[a]	21	12
Nickel, electroless[a]	27	18
Passivate stainless steel[a]	19	10
Phosphate treat steel[a]	23	14
Rhodium[a]	25	16
Silver plate aluminum	31	22
Silver plate copper[a]	23	14
Tin plate aluminum, hot oil fuse	35	26
Tin plate copper[b]	22	13
Zinc	24	15

[a]Heat treated parts require abrasive cleaning.
[b]Heat treated parts require abrasive cleaning and baking.

TABLE 8.3. Painting Operations (Summary of Section 13)

Reference Section	Operation	Set-Up or Drying (Min.)	Minutes/Unit/Spray Coat Part Size			
			To 3"	To 8"	To 20"	To 30"
13.1	Surface preparation (typical)					
9.6	Steel or Aluminum-phosphate treat					
9.6	Aluminum-chemical film		See Section 3, Table 9			
9.6	Magnesium-dichromate					
13.2	Mask and de-mask part for painting (See also 13.11, below)					
	Minor		0.50	1.70	3.40	6.00
	Average		1.00	3.40	6.40	10.00
	Major		2.50	8.50	19.00	25.00
13.3	Set-up time per paint type	Set-up 22.10				
13.4	Primer					
	Wash primer	Dry 12	0.50	0.77	1.04	1.27
	Zinc chromate	Dry 12	0.80	1.22	1.74	2.17
13.5	Surfacer (includes power buff)	Dry 12	2.40	3.52	4.14	6.02
13.6	Lacquer					
	Flat	Dry 18	0.80	1.22	1.74	2.17
	Gloss	Dry 18	1.10	1.82	2.54	3.22
13.7	Enamel or Modified Vinyl					
	Flat	Dry 30	0.80	1.22	1.74	2.17
	Gloss	Dry 60	1.10	1.82	2.54	3.22
13.9	Plastic protective film, strippable	Dry 60	0.70	1.12	1.59	1.92
13.10	Fungicide (spray application)	Dry 30	0.90	1.42	2.04	2.52
13.11	Miscellaneous detail values					
	Handling time	/part	0.20	0.26	0.32	0.42
	File or burr edge	/in.	0.02	0.02	0.02	0.02
	Sand or grind by power	/sq. ft.	1.09	1.09	1.09	1.09
	Sand by hand	/sq. ft.	2.18	2.18	2.18	2.18
	Blow off surface	/sq. ft.	0.05	0.05	0.05	0.05
	Wash surface with solvent	/sq. ft.	0.05	0.05	0.05	0.05
	Spray paint	/sq. ft.	0.06	0.06	0.06	0.06
	Brush paint, flat	/sq. ft.	0.24	0.24	0.24	0.24
	edges, lips	/sq. ft.	1.00	1.00	1.00	1.00
	Apply & remove masking tape	per in.	0.02	0.02	0.02	0.02
	Trim tape	per in.	0.11	0.11	0.11	0.11
	Assemble & remove masking plugs and stencils	per item	0.10	0.10	0.10	0.10

TABLE 8.4. Silk Screen Printing and Engraving (Summary of Section 14)

Reference Section	Operation	Time (Minutes)
14.2	Photographic operations to prepare silk screen stencil	
14.2.1,3,4	From line copy	48.8
14.2.2,3,4	From continuous tone copy	57.1
14.3	Silk screen printing, including decals	
14.3.1	Set-up for production, per color	14.5
14.3.2	Print, per color, per print	0.73
14.3.3	Add two cycles (14.2.1 and 14.2.2) for decals (to apply clear lacquer layers)	
14.4	Engraving	
14.4.1	Set-up	10.2
	per letter	0.3
14.4.2	Engrave and lacquer...	
	per piece	1.1
	per letter, per piece	
	Letter size	
	to ⅛ in.	0.17
	to ¼ in.	0.21
	to ⅜ in.	0.24
	to ½ in.	0.28
	to ⅝ in.	0.32

most common paint types are given, but Section 13, "Painting Operation," should be consulted for detailed estimates.

8.12. SILK SCREEN AND ENGRAVING

Silk screening and/or engraving is the final identification operation performed on electronic chassis and panel work. The set up values are summarized in Table 8.4. The analysis and detailed estimates for silk screening are covered in Section 14.

SECTION NINE

ELECTROPLATING AND CHEMICAL SURFACE TREATMENT OF METALS

9.1.	Plating Method	125
9.1.1.	Time Variables	126
9.1.2.	Limitations	127
9.2.	Analysis of Minutes Per Lot, Batch, and Part Values	127
9.2.1.	Basic Formula	127
9.2.2.	Element Description	128
9.2.3.	Calculation Method and Example	135
9.3.	Process Descriptions	138
9.3.1.	Anodize Aluminum—Clear	138
9.3.2.	Anodize Aluminum with Dye	138
9.3.3.	Anodize Aluminum, Hard, Clear	139
9.3.4.	Anodize Aluminum, Hard, with Dye	139
9.3.5.	Black Oxide Coat Iron and Steel	139
9.3.6.	Cadmium Plate Steel	140
9.3.7.	Chemical Conversion Coat Aluminum (Chromate)	140
9.3.8.	Chrome Plate Copper	141
9.3.9.	Chrome Plate Steel	141
9.3.10.	Copper Plate Copper	142
9.3.11.	Gold Plate Aluminum	142
9.3.12.	Gold Plate Copper	143
9.3.13.	Nickel Plate Copper	143
9.3.14.	Electroless Nickel Plating	143
9.3.15.	Passivate Stainless Steel	144

9.3.16.	Phosphate Treat Steel	144
9.3.17.	Rhodium Plate Copper	145
9.3.18.	Silver Plate Aluminum	145
9.3.19.	Silver Plate Copper	146
9.3.20.	Tin Plate Aluminum and Hot Oil Fuse	146
9.3.21.	Tin Plate Copper	147
9.3.22.	Zinc Plating	148

There are two basic types of facilities used in production plating: the manual, individual tank types used by job shops, and the automatic or semi-automatic plating machines employed by large-order production houses. The automatic type of plating machines have a conveyor mechanism that transports the parts through the complete process of cleaning and plating, including timing the cycle for each bath. The labor cost in this case is reduced to handling parts to and from the conveyor. The need for specialized estimating standards is significantly reduced; hence the costs of plating with such equipment are not covered here.

This section does address the costing of plating small parts in bulk and in rotating barrels, which is easily done on both manual and automatic lines.

9.1. PLATING METHOD

The time values given are for a job shop plating facility. The shop will typically have rows of open top tanks 1 to 3 ft. deep and 2 to 6 ft. square. Each line of tanks is supplied with electric current, air agitation, tap water, and drainage. Bench space and storage are provided for racking and storing parts. The general method of plating is as follows:

1. The plater determines the type process, plating thickness, and racking method from the work order, blueprint, and so forth.
2. The parts are then hung on racks, hooks, or wired in strings so as to facilitate dipping in the plating baths.
3. The third step consists of cleaning the parts in preparation for plating. Vapor degreasing, an alkaline bath, or an electrolytic bath may be employed.
4. The fourth step requires soaking parts in the particular plating baths. The plater must time each batch of parts in each bath for from 0.5 minute to 3 hours.

5. The last step consists of drying the parts, removing them from the racks, and repacking.

Several other steps may be necessary.

1. If the parts are heavily corroded, they must first be mechanically cleaned by a grit blast, buff wheel, or similar abrasive action.
2. If some surfaces of the parts are not to be plated, they must be masked and then demasked. The masking details depend on the type and extent of surface to be masked and the chemical solutions involved in the plating or treatment process. For instance, if only a small area is to be plated, it may be more economical to first plate the entire part then mask the desired areas and strip the plating from the remainder of the part. Where significant masking is required, the cost of masking can be *10 or 20 times* that of the plating.
3. Some parts will not plate uniformly if simply connected to the current source and immersed. Because plating thickness is related to the local electrical field potential, internal areas such as holes and cavities tend to plate less than exterior areas, or even not at all. If plating on these interior areas is important, it is necessary to arrange specially shaped anodes that project into the hole or cavity to assure adequate field distribution and plating deposition. Construction of these special anodes is called fixturing. Fixturing is seldom required for components of electronic equipment, but if required may be many more times as expensive as the plating itself, and should be separately considered.
4. A final baking operation may be necessary for some processes and basic metals, to eliminate hydrogen embrittlement, or to harden the plating.

9.1.1. Time Variables

The tables of time values at the end of the section reflect the following variables:

1. Parts handling time—varies by part size and number of parts to be handled.
2. Number of baths—varies per process.
3. Soaking time per bath—varies per bath.
4. Number of parts dipped per batch—this is the number of parts that can be placed in a tank at one time. The batch size is determined by the size of tank or area of part vs. current capacity of the tank. (Also termed the amps/ft.^{2} limitation of the tank.) The anodize bath and chrome plate bath both fall in the category where the maximum amps/ft.^{2} will be reached before the physical size of the parts fill the tank to capacity.

9.1.2. Limitations

The greatest single limitation of the tabular time values is that they are based on averages. This means parts with an average amount of corrosion, average plating thickness, average time per bath, and so on. It is to be expected that individual cases will vary from the table values, but still, over a period of time the table values and actual times should be about equal. Where high-quantity runs are made to a particular specification, the detailed "Analysis of Minutes per Part" time values should be used to develop a refined time standard to match the specification and the facility.

9.2. ANALYSIS OF MINUTES PER LOT, BATCH, AND PART VALUES

This section addresses the time values presented in Summary Table 9 of Section 3 and repeated here as Table 9.1A. Because it would be impractical to publish all the detail elements of each process, a sample analysis is given that incorporates the basic elements of most all processes. With this data the user can make variations in the present tables or use them to develop tables for new processes.

9.2.1. Basic Formula

The time for processing a given lot of parts is given by:

$$t = \underset{\text{lot}}{[s]} + \underset{\text{batch}}{[nb_t \times \Sigma b_t]} + \underset{\text{part}}{[np \times (h + c + m + b_k + i)]}$$

- t = Total time, minutes per lot.
- s = Sign in and out time, including all time that applies to the entire lot, such as blueprint reading.
- nb_t = Number of batches required to process the entire lot.
- b_t = Labor time for processing a batch through the various baths.
- np = Number of parts.
- h = Handling time.
- c = Abrasive cleaning time, if required.
- m = Masking time, if required.
- b_k = Baking time, if required.
- i = Inspection time.

The formula is merely a statement of the lot-batch-and-part-related time elements that make up the overall processing time. A description of these elements is presented below followed by a calculation worksheet and a worked out example.

TABLE 9.1A. Electroplating and Chemical Surface Treatment of Metals (Summary)

Process	Minutes to Process one Batch[c]	
	First	Successive
Anodize Aluminum, clear	28	19
Anodize Aluminum, dye	31	22
Anodize Aluminum, hard, clear	25	16
Anodize Aluminum, hard, dye	28	19
Black oxide coat iron and steel[a]	26	17
Cadmium plate steel[b]	23	14
Chemical conv. coat aluminum	22	13
Chrome plate copper[a]	23	14
Chrome plate steel[b]	25	16
Copper plate copper[a]	22	13
Gold plate aluminum	31	22
Gold plate copper[a]	23	14
Nickel plate[a]	21	12
Nickel, electroless[a]	27	18
Passivate stainless steel[a]	19	10
Phosphate treat steel[a]	23	14
Rhodium[a]	25	16
Silver plate aluminum	31	22
Silver plate copper[a]	23	14
Tin plate aluminum, hot oil fuse	35	26
Tin plate copper[b]	22	13
Zinc	24	15

[a]Heat treated parts require abrasive cleaning
[b]Heat treated parts require abrasive cleaning and baking
[c]Labor times only.

9.2.2. Element Description

1. *Sign-in time for a new job(s)*

    (a) Receive material, work order, shop order, and blueprint — 2.50

    (b) Carry lot to bench. — 0.70

    (c) Read work order and blueprint instructions. — 2.00

    (d) Determine racking methods and get racks, hooks, wire and so on. Return racks when done. — 3.00

    (e) Carry lot to storage after processing. — 0.70

(f) Miscellaneous—sign off job, return paper work, and so forth.

Total   8.90 minutes

This time is incorporated into the "first batch" column of Table 9.1A.

TABLE 9.1B. Number of Batches per Lot Quantities

			Process			
			Chromium Gold Nickel Passivate Stainless Rhodium Silver Tin		Anodize (All) Black Oxide Cadmium Chem. Conv. Coat Aluminum Copper Nickel Electroless Phosphate Treat Zinc	
			Small Tank 4 × 12 × 15 in.		Medium Tank 14 × 26 × 32 in.	
Type Part	Size (in.)	Lot Quantity	Parts Capacity	Number of Batches	Parts Capacity	Number of Batches
I	½ × ½	1	180	1	300	1
		20	180	1	300	1
		100	180	1	300	1
II	2 × 2	1	16	1	200	1
		20	16	2	200	1
		100	16	7	200	1
III	4 × 4	1	9	1	80	1
		20	9	2	80	1
		100	9	11	80	2
IV	4 × 12	1	3	1	30	1
		20	3	7	30	1
		100	3	33	30	4
V	4 × 4 × 12	1	2	1	20	1
		20	2	10	20	1
		100	2	50	20	5
VI	19 × 12	1			13	1
		20			13	2
		100			13	8
VII	19 × 12 × 4	1			4	5
		20			4	1
		100			4	25

130 ELECTROPLATING AND CHEMICAL SURFACE TREATMENT OF METALS

2. *Number of Batches (nb_t).* The number of batches required to process a given lot is determined by the size of the part and the capacity of the tank. Two tank sizes have been used: a small tank with usable dimensions of $12 \times 4 \times 15$ in. (commonly used for precious metal processes), and a medium tank with $26 \times 14 \times 32$ in. These dimensions allow for the tank space occupied by air pipes, anodes, heating elements, and so forth. Table 9.1B shows the maximum number of parts of each of the seven types per batch in the small and medium tanks.

3. *Dipping Time per Bath (b_t).* The labor time required per bath depends on the length of time the parts stay in the bath. If the time in the bath is long, the operator places the parts in the tank, starts and adjusts the process; returns to check the parts while in process; and returns again to remove the parts. If the bath time is short, the operator stays at the tank throughout the process time. For a long bath, the operator's time is essentially independent of the bath time, being 2.5 minutes as shown in Table 9.2. For a short bath, the operator's time is the bath time plus time for those tasks that cannot be done during the bath, which add up to 0.5 minute.

TABLE 9.2. Dipping Time Per Bath

	Labor Time in Minutes	
	Long Bath	Short Bath
Chemical Baths		
Place parts in tank		
1. Pick up two loaded racks of parts from bench or previous tank	0.10	0.10
2. Carry to tank (ave. 25 ft.)	0.10	0.10
3. Hang or clamp to bar	0.20	0.20
4. Set current, air, water level, and so on	0.10	0.10
5. Adjust and inspect at start of process	0.20	During bath cycle
Check parts while in process		
6. Walk to tank (ave. 25 ft.)	0.10	Not required
7. Inspect parts	0.30	Not required
Remove parts from tank		
8. Walk to tank (ave. 25 ft.)	0.10	Not required
9. Inspect to determine if finished	0.30	During bath cycle
10. Pick up two racks from tank (paid in element (1), above)		
11. Pick up and lay aside 3rd and 4th rack if required by quantity parts in batch	1.00	Part of bath time
Total	2.50	0.50
Rinses		
Rinse-dip in tank, agitate, and remove	0.50	0.50

ANALYSIS OF MINUTES PER LOT, BATCH, AND PART VALUES

Exceptions to this are the rinse cycles. Here the standard 0.5 minute allows for the minimal traveling and handling required, in addition to the nominal 15–30 second rinse.

4. *Dipping time per batch (b).* The total labor time per batch is the sum of the labor times required for the individual baths plus an allowance for lot time.

Table 9.3 shows the times for the chemical conversion coat (Chromate) on aluminum process. The first column lists the baths, and the second lists the bath times. The third column contains either the standard 2.5 minutes for long baths, or the bath time plus 0.5 minutes, whichever is less. This is the obviously required time.

The fourth column contains the excess bath time, the time when the parts were in process and under the responsibility of the operator even though he or she was not actively attending to them. If this excess or waiting time could be 100% utilized on other work, there would be no lost time. In practice, only 75% of this time can be utilized, so 25% is lost. The 25% lost time is inherent to a mixed process, job shop type of operation.

TABLE 9.3. Dipping Work-Time Per Batch (Example)

1. Process Times
   Chemical Conv. Coat on Aluminum

Bath	Bath Time	Work-Time	Excess Bath Time
1. Alkaline cleaner	4.0	2.5	1.5
2. Rinse	0.5	0.5	—
3. Deoxidize	7.0	2.5	4.5
4. Rinse	0.5	0.5	—
5. Chromate (Iridite)	2.0	2.5	—
6. Rinse	0.5	0.5	—
7. Water displacer	0.5	0.5	—
8. Degreaser	2.0	2.5	—

2. Totals   17.0   12.0   6.0
3. Lost time = 25% of excess bath time = 1.5 minutes
4. Time on other jobs = 75% of excess = 4.5 minutes
5. Factor to distribute lost time =
$$\frac{\text{Work-time} + \text{Lost time} + \text{Time on other jobs}}{\text{Work-time} + \text{Time on other jobs}} = \frac{18}{16.5} = 1.091$$
6. Standard time = Work-time × Factor
   = 12.0 × 1.091 = 13 minutes

132 ELECTROPLATING AND CHEMICAL SURFACE TREATMENT OF METALS

Operations that can be done with the waiting time and the limitations that lead to the 25% lost time are:

(a) Operate other processes. Limitation—on the average, one plater can operate 3 processes at one time. Even though more processes could be handled physically, the mental task of controlling, starting, soaking, and finishing times becomes too great.
(b) Racking, deracking, and cleaning parts that have been, or will be processed. Limitation—Availability of racks and parts.
(c) Maintenance of tanks and racks. Limitations—Only a nominal portion of the working day is required for maintenance.

Of the lost time, some belongs with the job being estimated, and the remainder belongs with the other jobs that are being worked during the 75% of the excess time that is on those jobs. The "Factor to Distribute Lost Time" shown in Table 9.3, when multiplied by the work-time, will give a total work-time that includes a proportionate share of the lost time.

5. *Handling Time per part (h).* Handling time per part depends on the size and general shape of the part. Seven types of parts have been established as shown in Table 9.4. Handling time includes racking, unracking, miscellaneous packing, and minor cleaning. Where protective wrapping or bagging of each part is involved, additional time is required, as shown.

6. *Abrasive Cleaning, if required (c).* Heat treated parts require mechanical cleaning to remove the heat treat scale prior to plating. The operation elements and time values are given in Table 9.5.

7. *Masking (m).* When portions of a part are not to be plated or treated, it is necessary to mask off the areas to be left unchanged. There are three common types of masking.

Corks may be used to plug holes.

Chemical-resistant plater's tape may be used to mask off portions of relatively smooth surfaces.

Wax or heavy plastic compound may be applied by painting or dipping. In some cases a primer is required to assure adequate bond between the wax and base metal.

Removal of corks and tape is straightforward. Wax is cut and peeled. Primer, if applied, is removed in a degreaser.

Masking time must be applied to each part. The times shown in Table 9.1. include elements for handling the part to and from the masking workbench as well as for applying and removing the corks, tape, or wax masking. Sometimes small areas are to be plated. It may be more economical to plate the entire part, mask the area to be plated, and then strip the remainder of the plating than to mask very large areas. For estimating purposes, the stripping bath may be assumed to be of 10 minute duration (work-time = 2.5 minutes).

TABLE 9.4. Parts Handling Time (Minutes)

	_____ Type Part (in.) _____						
Element	I[a] ½ × ½	I[b] ½ × ½	II 2 × 2	III 4 × 4	IV 4 × 12	V 4 × 4 × 12	VI 19 × 12	VII 19 × 12 × 4
Assemble part to rack (hand, spring clamp, or thread wire and bend)	0.03	0.13	0.13	0.13	0.21	0.21	0.28	0.30
Disassemble from rack, lay aside to tote	0.02	0.05	0.10	0.15	0.20	0.30	0.40	0.50
Miscellaneous, pack, clean, or brush some parts	0.01	0.05	0.20	0.20	0.30	0.40	0.40	0.40
Subtotal	0.20	0.05	0.40	0.50	0.70	1.00	1.20	1.40
Total/lot quantity								
Quantity 1	0.2	0.05	0.4	0.5	0.7	1.0	1.2	1.4
Quantity 20	4	1	8	10	14	20	24	28
Quantity 100	20	5	40	50	70	100	120	140

[a]Bulk handling to basket or barrel.
[b]Individual handling.

TABLE 9.5. Cleaning and Baking Parts

		Clean Heat Treat Scale	Bake
Equipment		Pressurized Grit blast, or motor driven wire brush wheel.	Oven
Operations		Pick up part and position. Clean all surfaces. Lay aside part.	Transport parts to oven. Place in oven. Bake approximately three hours at 375°F. Remove parts and lay aside.
Type Part	Size (in.)	Minutes per part	Minutes per part
I	½ × ½	0.20	0.20
II	2 × 2	0.40	0.20
III	4 × 4	0.60	0.30
IV	12 × 4	0.90	0.40
V	12 × 4 × 4	—	—
VI	19 × 12	—	—
VII	19 × 12 × 4	—	—

8. *Baking, if required (b_k).* Baking of plated or chemically surface-treated parts is sometimes required. For instance, cadmium or chrome-plated steel parts require a baking operation to remove hydrogen embrittlement.

The operation elements and time values are given in Table 9.5. It should be noted that only the handling time is allowed as labor costs for the baking operation.

9. *Inspection (i).* Parts receive two major inspections, in addition to the in-process visual inspections which are part of normal operations. All incoming parts are inspected to assure that they are free from visible defects and suitable for plating. At this time, parts are removed from the shipping container and placed in totes. Final inspection includes visual and thickness tests, using micrometer or betascope, as appropriate. Final inspections are normally done on an AQL (Acceptable Quality Level) basis, although 100% inspection is

ANALYSIS OF MINUTES PER LOT, BATCH, AND PART VALUES 135

TABLE 9.6. Typical Sampling Plan (AQL = 1.0%)

Lot Size	Sample Size	Allowable Defectives Without Rejecting Lot
1–13	All	None
14–150	13	None
151–500	50	1
501–1200	80	2

Note: AQL stands for acceptable quality level. AQL of 1.0% means that an average rate of defectives of 1 per 100 is acceptable.

The sampling plan is statistically designed to specify sample size and rejection criteria which assure that accepted lots will, on the average, have less than the AQL rate of defectives.

This sampling plan is taken from MIL-STD-105.

sometimes specified. Table 9.6 shows the required sample sizes for the usual AQL. Special tests may also be specified and, if so, must be separately estimated.

9.2.3. Calculation Method and Example

Figure 9.1 is a worksheet for calculating labor times for plating and chemical surface-treatment operations. Information required includes the type of process, number and type of parts, and whether abrasive cleaning and baking, class of masking, and fixturing, is required. Type of part and required additional work per part are described in the tables as referenced on the worksheet.

Sign in/out and miscellaneous time associated with each log is essentially independent of the process, and is shown on the worksheet.

Time per batch is shown in Table 9.1A as well as in Summary Table 9 of Section 3. Number of batches is determined either by direct reference to Table 9.2 or by using that table or Table 9.1B to determine the maximum batch size and then figuring the number of batches required to process the total number of parts.

Time per part is the sum of the handling, abrasive cleaning, baking, masking, and fixturing times. These times are determined by use of Table 9.1A.

Total time to process the lot is the sum of lot time plus number of batches times the time per batch plus number of parts times the time per part.

Figure 9.2 is a worksheet filled in with times and values for Chemical Conversion Coating on Aluminum (Chromate or Iridite).

DATE: _____
BY: _____

PROCESS: _____
APPL. SECTION _____
TYPE OF PART: _____
NO. OF PARTS: _____
NO. PARTS/BATCH: _____
NO. OF BATCHES: _____

1. LOT SIGN IN/OUT AND MISCELLANEOUS TIME	8.9 (SECT. 9.2.2)
TIME PER BATCH _____ NO. OF BATCHES × _____ 2. BATCH TIME, TOTAL	_____

ADDITIONAL TIME PER PART
   HANDLING: _____
   INSPECTION:
     INCOMING _____
     FINAL _____ = _____ PER PART × _____ TOTAL PARTS
   ABRASIVE CLEANING _____
   BAKING _____
   MASKING:
     CORKS _____ = _____ CORKS × 0.10 PER CORK
     TAPE _____ = _____ IN. × 0.03 PER IN.
     WAX _____ = _____ IN.^{2} × 0.09 PER IN.^{2}
   TOTAL PER PART _____
   NO. OF PARTS × _____
 3. ADDITIONAL TIME PER PART
	PARTS INSP.*
	
	 _____

TOTAL TIME, MINUTES 1 + 2 + 3 TOTAL TIME, HOURS	_____ _____

*TYPICAL INSPECTION QTY (AQL=1.0%)
   LOT SIZE   INSP. QTY.
     1−13   All
    14−150   13
   151−500   50
   501−1200   80
	NOTE: ALL REQUIRED DATA
	 FOUND IN TABLE 9.1 OR
	 JOB DESCRIPTION.

FIGURE 9.1. Electroplating estimating worksheet.

DATE: 1/27/85
BY: _____

PROCESS: __Chemical Conversion Coat, Aluminum__
APPL. SECTION __9.3.7__
TYPE OF PART: __III (4"x 4")__
NO. OF PARTS: __100__
NO. PARTS/BATCH: __80__
NO. OF BATCHES: __2__

1. LOT SIGN IN/OUT AND MISCELLANEOUS TIME	__8.9__	(SECT. 9.2.2)

TIME PER BATCH __13__
NO. OF BATCHES × __2__
2. BATCH TIME, TOTAL   __26.0__

ADDITIONAL TIME PER PART
   HANDLING: __.13__
   INSPECTION:
      INCOMING   __.33__   __13__ PARTS INSP.*
      FINAL   __.12__ = __.89__ PER PART × __100__ TOTAL PARTS
   ABRASIVE CLEANING __—__
   BAKING __—__
   MASKING:
      CORKS   __—__ = _____ CORKS × __0.10__ PER CORK
      TAPE   __.24__ = __8__ IN. × __0.03__ PER IN.
      WAX   __—__ = _____ IN.^{2} × __0.09__ PER IN.^{2}
     TOTAL PER PART __.82__
     NO. OF PARTS × __100__
3. ADDITIONAL TIME PER PART   __82.0__

TOTAL TIME, MINUTES 1 + 2 + 3   __116.9__
TOTAL TIME, HOURS   __1.9__

*TYPICAL INSPECTION QTY (AQL=1.0%)   NOTE: ALL REQUIRED DATA
                                                   FOUND IN TABLE 9.1 OR
  LOT SIZE   INSP. QTY.   JOB DESCRIPTION.

LOT SIZE	INSP. QTY.
1–13	All
14–150	13
151–500	50
501–1200	80

FIGURE 9.2. Electroplating estimating worksheet.

9.3. PROCESS DESCRIPTIONS

9.3.1. Anodize Aluminum—Clear

Purpose
  Corrosion resistance
  (Salt spray 240 hours)
  Paint base
Applicable Specifications
  MIL-A-8625, anodic
  coatings for aluminum,
  type II, class 1
  (AMS 2471)
Thickness
  600 milligrams per ft.^{2}.
Time Values Based on:
  36 × 30 × 36 tank size
  or 15 ft.^{2} area of parts
  (10 amp/ft.^{2} at 150 amp)

Baths	Minutes/Batch Bath	Minutes/Batch Work
1. Alkaline cleaner	4.0	2.5
2. Rinse	0.5	0.5
3. Deoxidize	7.0	2.5
4. Rinse	0.5	0.5
5. Anodize	30.0	2.5
6. Rinse	0.5	0.5
7. Seal	15.0	2.5
8. Rinse	0.5	0.5
9. Water displacer	0.5	0.5
10. Degreaser	2.0	2.5
Total	60.5	15.0

9.3.2. Anodize Aluminum with Dye

Purpose
  Corrosion resistance
  (Salt spray 240 hours)
Applicable Specifications
  MIL-A-8625, anodic
  coatings for aluminum,
  type II, class 2
  (AMS 2471)
Thickness
  2500 milligrams per ft.^{2}
Time Values Based on:
  36 × 30 × 36 in. tank
  size or 15 ft.^{2} area of parts
  (10 amp/ft.^{2} at 150 amp)

Baths	Minutes/Batch Bath	Minutes/Batch Work
1. Alkaline cleaner	4.0	2.5
2. Rinse	0.5	0.5
3. Deoxidize	7.0	2.5
4. Rinse	0.5	0.5
5. Anodize	30.0	2.5
6. Rinse	0.5	0.5
7. Dye	15.0	2.5
8. Rinse	0.5	0.5
9. Seal	15.0	2.5
10. Rinse	0.5	0.5
11. Water displacer	0.5	0.5
12. Degreaser	2.0	2.5
Total	76.0	18.0

PROCESS DESCRIPTIONS

9.3.3. Anodize Aluminum, Hard, Clear

Purpose
  Wear resistance
  Corrosion resistance
Applicable Specifications
  MIL-A-8625, anodic coating for aluminum, type III, class 1 (AMS 2468)
Thickness
  0.002 in. or as specified
Time Values Based on:
  30–35 amp/ft.^{2}

Baths	Minutes/Batch	
	Bath	Work
1. Alkaline cleaner	4.0	2.5
2. Rinse	0.5	0.5
3. Deoxidize	7.0	2.5
4. Rinse	0.5	0.5
5. Anodize	45.0	2.5
6. Cold water rinse	0.5	0.5
7. Warm water rinse	0.5	0.5
8. Water displacer	0.5	0.5
9. Degreaser	0.5	0.5
Total	60.5	12.5

9.3.4. Anodize Aluminum, Hard, with Dye

Purpose
  Wear resistance,
  Corrosion resistance
Applicable Specification
  MIL-A-8625, anodic coatings for aluminum type III, class 2 (AMS 2468)
Thickness
  0.002 in. or as specified
Time Values Based on:
  30–35 amp/ft.^{2}

Baths	Minutes/Batch	
	Bath	Work
1. Alkaline cleaner	4.0	2.5
2. Rinse	0.5	0.5
3. Deoxidize	7.0	2.5
4. Rinse	0.5	0.5
5. Anodize	45.0	2.5
6. Cold water rinse	0.5	0.5
7. Warm water rinse	0.5	0.5
8. Dye	15.0	2.5
9. Rinse	0.5	0.5
10. Water displacer	0.5	0.5
11. Degreaser	2.0	2.5
Total	76.0	15.5

9.3.5. Black Oxide Coat Iron and Steel

Purpose
  Decorative finish

Baths	Minutes/Batch	
	Bath	Work
1. Alkaline cleaner	4.0	2.5
2. Rinse	0.5	0.5

140 ELECTROPLATING AND CHEMICAL SURFACE TREATMENT OF METALS

Applicable Specification
MIL-C-13924, black oxide coating for ferrous metals; class 1
(AMS 2485)
Thickness
Not applicable
Time Values Based on:
30 × 18 × 36 in. size

Baths	Minutes/Batch	
	Bath	Work
3. Acid Pickle	15.0	2.5
4. Rinse	0.5	0.5
5. Rinse	0.5	0.5
6. Black oxide	4.0	2.5
7. Rinse	0.5	0.5
8. Wax coating	2.5	2.5
9. Dry	20.0	2.5
Total	47.5	14.5

9.3.6. Cadmium Plate Steel

Purpose
Corrosion resistance
(Salt spray 192 hours)
Applicable Specifications
QQ-P-416, type I, without supplementary phosphate treatment
(AMS 2400)
Thickness
0.0003 to 0.0010 in.
Time Values Based on:
36 × 30 × 36 in. tank size

Baths	Minutes/Batch	
	Bath	Work
1. Electrocleaner	3.0	2.5
2. Rinse	0.5	0.5
3. Acid dip	0.5	1.0
4. Rinse	0.5	0.5
5. Cad. plate	15.0	2.5
6. Rinse	0.5	0.5
7. Post treat	0.5	1.0
8. Rinse	0.5	0.5
9. Water displacer	0.5	0.5
10. Degreaser	2.0	2.5
Total	23.5	12.0

9.3.7. Chemical Conversion Coat Aluminum (Chromate)

Purpose
Paint Base
Corrosion resistance
(Salt spray 168 hours)
Applicable Specifications
MIL-C-5541, chemical films for aluminum and aluminum alloys

Baths	Minutes/Batch	
	Bath	Work
1. Alkaline cleaner	4.0	2.5
2. Rinse	0.5	0.5
3. Deoxidize	7.0	2.5
4. Rinse	0.5	0.5
5. Iridite	2.0	2.5
6. Rinse	0.5	0.5

PROCESS DESCRIPTIONS

(AMS 2473)
Thickness
 Not applicable
Time Values Based on:
 30 × 18 × 36 in. tank size

Baths	Minutes/Batch	
	Bath	Work
7. Water displacer	0.5	0.5
8. Degreaser	2.0	2.5
Total	17.0	12.0

9.3.8. Chrome Plate Copper

Purpose
 Wear resistance
 Decorative finish
Applicable Specifications
 QQ-C-320, class I, bright and satin (AMS 2406)
Thickness
 0.001 to 0.010 in.
Time Values Based on:
 18 × 12 × 18 in. tank size, or 50 in.^{2} area of parts (1 amp/in.^{2} at 50 amp)

Baths	Minutes/Batch	
	Bath	Work
1. Electrocleaner	3.0	2.5
2. Rinse	0.5	0.5
3. Acid dip	0.5	1.0
4. Rinse	0.5	0.5
5. Chrome plate	180.0	2.5
6. Rinse	0.5	0.5
7. Water displacer	0.5	0.5
8. Degreaser	2.0	2.5
Total	187.5	10.5

9.3.9. Chrome Plate Steel

Purpose
 Wear resistance
 Decorative finish
Applicable Specifications
 QQ-C-320, class I, bright and satin (AMS 2406)
Thickness
 0.001 to 0.010 in.
Time Values Based on:
 18 × 12 × 18 in. tank size, or 50 in.^{2} area of parts (1 amp/in.^{2} at 50 amp)

Baths	Minutes/Batch	
	Bath	Work
1. Electrocleaner	3.0	2.5
2. Rinse	0.5	0.5
3. Acid dip	0.5	1.0
4. Rinse	0.5	0.5
5. Chrome reverse etch	1.0	1.5
6. Chrome plate	180.0	2.5
7. Rinse	0.5	0.5
8. Water displacer	0.5	0.5
9. Degreaser	2.0	2.5
Total	188.2	12.5

9.3.10. Copper Plate Copper

Purpose
  Conductivity
  Base for further plating
Applicable Specification
  None
Thickness
  0.001 to 0.002 in.
Time Values Based on:
  36 × 30 × 36 in. tank size

Baths	Minutes/Batch	
	Bath	Work
1. Electrocleaner	3.0	2.5
2. Rinse	0.5	0.5
3. Acid dip	0.5	1.0
4. Rinse	0.5	0.5
5. Copper plate	60.0	2.5
6. Rinse	0.5	0.5
7. Water displacer	0.5	0.5
8. Degreaser	2.0	2.5
Total	67.5	10.5

9.3.11. Gold Plate Aluminum

Purpose
  Conductivity
  Corrosion resistance
Applicable Specifications
  None
Thickness
  50 to 100 micro inches
Time Values Based on:
  18 × 12 × 18 in. tank size

Baths	Minutes/Batch	
	Bath	Work
1. Alkaline cleaner	4.0	2.5
2. Rinse	0.5	0.5
3. Deoxidize	7.0	2.5
4. Rinse	0.5	0.5
5. Acid dip	0.5	1.0
6. Rinse	0.5	0.5
7. Zinc immersion	1.5	2.0
8. Rinse	0.5	0.5
9. Copper strike	0.5	1.0
10. Rinse	0.5	0.5
11. Silver strike	0.5	1.0
12. Rinse	0.5	0.5
13. Gold plate	10.0	2.5
14. Rinse, drag out	0.5	0.5
15. Rinse	0.5	0.5
16. Water displacer	0.5	0.5
17. Degreaser	2.0	2.5
Total	30.5	19.5

9.3.12. Gold Plate Copper

Purpose
  Conductivity, solderability, corrosion resistance
Applicable Specifications
  None
Thickness
  50 to 100 micro inches
Time Values Based on:
  18 × 12 × 18 in. tank size

Baths	Minutes/Batch	
	Bath	Work
1. Electrocleaner	3.0	3.5
2. Rinse	0.5	0.5
3. Cyanide dip	0.5	1.0
4. Rinse	0.5	0.5
5. Silver strike	1.0	1.5
6. Rinse	0.5	0.5
7. Gold plate	10.0	2.5
8. Rinse, drag out	0.5	0.5
9. Rinse	0.5	0.5
10. Water displacer	0.5	0.5
11. Degreaser	2.0	2.5
Total	19.5	13.0

9.3.13. Nickel Plate Copper

Purpose
  Corrosion resistance
  Abrasion resistance
  Underplate for further plating
  Decorative finish
Applicable Specifications
  QQ-N-290, Class I, or II, type I to VII (semi-bright) (AMS 2404)
Thickness
  0.0001 to 0.0005 in.
Time Values Based on:
  18 × 12 × 18 in. tank size

Baths	Minutes/Batch	
	Bath	Work
1. Electrocleaner	3.0	2.5
2. Rinse	0.5	0.5
3. Acid dip	0.5	1.0
4. Rinse	0.5	0.5
5. Nickel plate	10.0	2.5
6. Rinse	0.5	0.5
7. Water displacer	0.5	0.5
8. Degreaser	2.0	2.5
Total	17.5	10.5

9.3.14. Electroless Nickel Plating

Purpose
  Corrosion resistance
  Wear resistance

Baths	Minutes/Batch	
	Bath	Work
1. Degreaser	0.5	0.5
2. Alkaline cleaner	1.0	1.5

Applicable Specifications
  MIL-C-26074, coatings, electroless nickel, Class I, (include heat treatment for other classes) (AMS 2403)
Thickness
  Grade A—0.001 in. minimum
  Grade B—0.0005 in. minimum
  Grade C—.0015 in. minimum
Time Values Based on:
  0.001 in.

	Minutes/Batch	
Baths	Bath	Work
3. Electrocleaner	1.0	1.5
4. Rinse	0.5	0.5
5. Acid dip	1.5	2.0
6. Rinse	0.5	0.5
7. Neutralizing rinse	0.5	0.5
8. Rinse	0.5	0.5
9. Electroless nickel	60.0	2.5
10. Cold water rinse	0.5	0.5
11. Warm water rinse	0.5	0.5
12. Water displacer	0.5	0.5
13. Degreaser	2.0	2.5
Total	70.5	14.0

9.3.15. Passivate Stainless Steel

Purpose
  Corrosion resistance
Applicable Specifications
  MIL-S-5002, surface treatments for metal parts
Thickness
  Not applicable
Time Values Based on:
  18 × 12 × 18 in. tank size

	Minutes/Batch	
Baths	Bath	Work
1. Vapor degrease	1.5	2.0
2. Passivate	60.0	2.5
3. Rinse	0.5	0.5
4. Water displacer	0.5	0.5
5. Degreaser	2.0	2.5
Total	64.5	8.0

9.3.16. Phosphate Treat Steel

Purpose
  Corrosion resistance
  Paint adhesion
  Dry film lubricant adhesion
Applicable Specifications
  MIL-C-490, Grade I

	Minutes/Batch	
Baths	Bath	Work
1. Electrocleaner	4.0	2.5
2. Rinse	0.5	0.5
3. Oxalic acid	1.5	2.0
4. Rinse	0.5	0.5
5. Phosphate treat	4.0	2.5

PROCESS DESCRIPTIONS

Thickness
300 milligrams per ft.^{2}
Time Values Based on:
30 × 18 × 36 in. tank size

Baths	Minutes/Batch	
	Bath	Work
6. Rinse	0.5	0.5
7. Phosphate seal	1.0	1.5
8. Water displacer	0.5	0.5
9. Degreaser	2.0	2.5
Total	14.5	13.0

9.3.17. Rhodium Plate Copper

Purpose
  Corrosion resistance
  Wear resistance
Applicable Specifications
  None
Thickness
  Nickel base 0.0004 in.
  Rhodium—10–20 micro inches
Time Values Based on:
  18 × 12 × 18 in. tank size

Baths	Minutes/Batch	
	Bath	Work
1. Electrocleaner	3.0	2.5
2. Rinse	0.5	0.5
3. Acid dip	0.5	1.0
4. Rinse	0.5	0.5
5. Nickel plate	10.0	2.5
6. Rinse	0.5	0.5
7. Rhodium plate	15.0	2.5
8. Rinse, drag out	0.5	0.5
9. Rinse	0.5	0.5
10. Water displacer	0.5	0.5
11. Degreaser	2.0	2.5
Total	33.5	14.0

9.3.18. Silver Plate Aluminum

Purpose
  Decorative finish
  Conductivity
  Corrosion resistance
  Solderability
Applicable Specifications
  QQ-S-365, type III, (bright) (AMS 2412)
Thickness
  0.0005 to 0.0010 in.

Baths	Minutes/Batch	
	Bath	Work
1. Alkaline cleaner	4.0	2.5
2. Rinse	0.5	0.5
3. Deoxidize	7.0	2.5
4. Rinse	0.5	0.5
5. Acid dip	0.5	1.0
6. Rinse	0.5	0.5
7. Zinc immersion	1.5	2.0
8. Rinse	0.5	0.5

146 ELECTROPLATING AND CHEMICAL SURFACE TREATMENT OF METALS

Time Values Based on:
18 × 12 × 18 in. tank size

Baths	Minutes/Batch	
	Bath	Work
9. Copper strike	0.5	1.0
10. Rinse	0.5	0.5
11. Silver strike	0.5	1.0
12. Rinse	0.5	0.5
13. Silver plate	20.0	2.5
14. Rinse	0.5	0.5
15. Water displacer	0.5	0.5
16. Degreaser	2.0	2.5
Total	40.0	19.0

9.3.19. Silver Plate Copper

Purpose
  Decorative finish
  Conductivity
  Corrosion resistance
  Solderability
Applicable Specifications
  QQ-S-365, type III, (bright) (AMS 2412)
Thickness
  0.0005 to 0.0010 in.
Time Values Based on:
  18 × 12 × 18 in. tank size

Baths	Minutes/Batch	
	Bath	Work
1. Electrocleaner	3.0	2.5
2. Rinse	0.5	0.5
3. Cyanide	0.5	1.0
4. Rinse	0.5	0.5
5. Silver strike	0.5	1.0
6. Silver plate	20.0	2.5
7. Rinse, drag out	0.5	0.5
8. Rinse	0.5	0.5
9. Water displacer	0.5	0.5
10. Degreaser	2.0	2.5
Total	28.5	12.0

9.3.20. Tin Plate Aluminum and Hot Oil Fuse

Purpose
  Solderability
Applicable Specifications
  MIL-T-10727, Type I
Thickness
  0.0005 to 0.0010 in.

Baths	Minutes/Batch	
	Bath	Work
1. Alkaline cleaner	4.0	2.5
2. Rinse	0.5	0.5
3. Deoxidize	7.0	2.5
4. Rinse	0.5	0.5
5. Acid dip	0.5	1.0

PROCESS DESCRIPTIONS

Time Values Based on:
18 × 12 × 18 in. tank size

		Minutes/Batch	
Baths		Bath	Work
6. Rinse		0.5	0.5
7. Zinc immersion		1.5	2.0
8. Rinse		0.5	0.5
9. Copper strike		0.5	1.0
10. Copper plate		5.0	2.5
11. Rinse		0.5	0.5
12. Tin plate		40.0	2.5
13. Rinse		0.5	0.5
14. Hot oil fuse		2.0	2.5
15. Vapor degrease		1.5	2.0
Total		65.0	21.5

9.3.21. Tin Plate Copper

Purpose
Solderability
Applicable Specifications
MIL-T-10727, type I
Thickness
0.0005 to 0.0010 in.
Time Values Based on:
18 × 12 × 18 in. tank size

Baths	Minutes/Batch	
	Bath	Work
1. Electrocleaner	3.0	2.5
2. Rinse	0.5	0.5
3. Acid dip	0.5	1.0
4. Rinse	0.5	0.5
5. Tin plate	40.0	2.5
6. Rinse	0.5	0.5
7. Water displacer	0.5	0.5
8. Degreaser	2.0	2.5
Total	47.5	10.5

9.3.22. Zinc Plating

Purpose
Corrosion resistance
Applicable Specifications
QQ-Z-325, type I, class 2
Thickness
Class 1—0.001 in.

Baths	Minutes/Batch	
	Bath	Work
1. Electrocleaner	1.0	1.5
2. Rinse	0.5	0.5
3. Acid dip	1.0	1.5
4. Rinse	0.5	0.5

148 ELECTROPLATING AND CHEMICAL SURFACE TREATMENT OF METALS

Class 2—0.0005 in.
Class 3—0.0002 in.
Time Values Based on:
  30 × 18 × 26 in. tank size

Baths	Minutes/Batch Bath	Minutes/Batch Work
5. Rinse	0.5	0.5
6. Zinc Plate	50.0	2.5
7. Rinse	0.5	0.5
8. Acid bright dip	0.5	0.5
9. Rinse	0.5	0.5
10. Chromate	0.5	0.5
11. Rinse	0.5	0.5
12. Water displacer	0.5	0.5
13. Degreaser	2.0	2.5
Total	58.5	12.5

SECTION TEN

PLASTICS

10.1.	Types of Plastics	150
10.1.1.	Thermoplastic	151
10.1.2.	Thermosets	153
10.1.3.	New Materials and Technologies	155
10.2.	Fabrication Methods	155
10.2.1.	Injection Molding	158
10.2.2.	Compression Molding	158
10.3.	Costs	160

The use of plastics in the electronics industry has increased rapidly in the past decade in response to two factors. First, new materials have become available that are much more useful in almost every respect. They are stronger, retain their strength better at elevated temperatures, have better dimensional stability, have greater impact resistance, are more moldable, have higher density, greater chemical resistance, lower coefficients of friction, and higher dielectric properties. Second, the requirements on materials, especially structural materials, in electronics have relaxed because of the decreasing size, weight, and power consumption of the basic electronics equipment. Although plastics are still generally more flammable, have lower operating temperature limits, and greater coefficients of thermal expansion than metals, there is a trend toward their use.

Table 10.1 shows the use of plastics in the electronics industry in thousands of tons, for the years 1982 and 1983. It is clear from these values that the use of plastics in electronics is growing, even in the face of fluctuations of the economy.

TABLE 10.1. Use of Plastics in the Electrical and Electronics Industry in 1982 and 1983[a]

	1000 Metric Tons	
Material	1982	1983
ABS	17	30
Cellulosics[b]	1	1
Epoxy (electrical laminates)	11	12
Modified PPO	11	12
Nylon (including wire and cable)	20	23
Phenolic	54	59
Polyacetal	2	2
Polycarbonate	20	26
Polyester, reinforced	53	60
Polyester, Thermoplastic	9	10
Polyethylene, HD[c]	50	58
Polyethylene, LD[c]	126	122
Polypropylene[c]	5	5
Polystyrene	91	102
Polyvinyl chloride[c]	149	157
Styrene acrylonitrile	2	2
Urea	11	14
Other	9	11
Total	641	706

Source: *Modern Plastics*.
[a]Excludes automotive.
[b]Excludes appliance use.
[c]Wire and cable.

The following sections describe the general types of engineering plastics currently in common use in the U.S. electronics industry. In addition, commonly used injection molding manufacturing processes and approximate costs are given. However, this is a very rapidly changing field. Prices of existing materials change almost daily, and new materials are continually becoming available. It would be well to consult a manufacturer or other knowledgeable, up-to-date source on significantly large orders.

10.1. TYPES OF PLASTICS

There are two basic types of plastics—thermoplastics and thermosetting. Thermoplastic materials soften when heated and can be heated and cooled several

TYPES OF PLASTICS 151

times before becoming brittle. Scrap can therefore be ground up and reused, reducing the amount of raw stock required. Thermoplastics are generally more expensive than thermosets, but are easier to mold because they are not as sensitive to temperature and pressure variations.

Thermosetting plastics, or thermosets, undergo a chemical change during the molding process, and cannot be remelted. Since the scrap cannot be recycled, more material must be purchased per pound of delivered product than for thermoplastics. However, the raw material cost is generally lower for materials of comparable properties in the final product.

Table 10.2 shows some of the typical applications of engineering plastics in the electronics industry. This table and the following descriptions are intended only to give a general overview of the field. More detailed information is available from sources such as *Modern Plastics Encyclopedia* (McGraw-Hill) or from the various suppliers.

10.1.1. Thermoplastics

Some of the more common thermoplastic materials used in the electronics industry are:

ABS	Polyacetals (Celcon)
Acrylics	Polycarbonates (Lexan)
Nylons	Polyesters (PBT)
Polyamide-imides (Torlon)	Polyethylenes
	Polyphenylene Sulfides (Ryton)

ABS. ABS is an alloy of acrylonitrile, butadiene, and styrene. The three basic raw materials can be combined in varying proportions to create a family of materials with a wide range of physical properties.

Flame-retardant ABS is used in electronic components such as computer terminals and business machine housings. Structural foam ABS and flame-retardant extruded sheet ABS are also finding wide acceptance for electronics housings.

Acrylics. Acrylics are used where light transmittance is important. In the electronics industry, acrylics are used for lighting covers and lenses.

Nylons. Nylons are a family of polyamides with a wide range of physical properties. The electrical and electronics industries are the second largest users of nylons, the largest being automotive. Nylons are used in plugs, connectors, wiring devices, terminals, cable ties, coil forms, wire jacketing, antenna mounting devices, as well as in mechanical components such as gears and bearings. The UL 94 V-O (high resistance to burning) rating has become increasingly important.

TABLE 10.2. Typical Applications of Plastics in the Electrical/Electronics Industry

Material (Typical Trade Name)	Housings	Connectors, Wiring Devices, Coil Forms	Switches, Circuit Breakers, Knobs, Handles	Gears, Bearings	Lighting Comp. Lenses	PC Boards	Wiring Jacket and Insulation	Encapsulation
ABS	X							
Acrylic					X			
Nylon		X		X			X	
Polyacetal (Celcon)		X		X				
Polyamideimide				Very high quality				
Polycarbonate (Lexan)		High strength			High strength			
Polyester (PBT)		X		X				
Polyethylene							Low loss	
Polyphenylene Sulphide		X	X					
Allyl (DAP)		High reliability	High reliability			High reliability		
Amino/Urea (Melamine)		X	X					
Epoxy		X						X
Fluoroplastic (Teflon)		High temperature		High temperature			High temperature and voltage	
Phenolic		X	X	X		X		
Silicone							X	X

Polyacetal (Celcon). Acetal copolymer is a mid-range engineering plastic with good general properties, including chemical resistance and dimensional stability, especially with respect to temperature. It is good in systems that must be heat aged. Although not as fire resistant or temperature resistant as the polyesters (PBT), it is both easier to manufacture and somewhat less expensive.

Polyamide-imides (Torlon). This material is a high-performance plastic that can be injection molded with close tolerances into complex precision parts. It has good dimensional stability and otherwise superior mechanical properties that persist at temperatures up to about 500°F. In addition, it is comparatively safe with respect to fire. The various forms of the material have found extensive application in the aerospace industry in electronic and other types of equipment by replacing metals for significant weight reduction.

Polycarbonates (Lexan). Polycarbonates exhibit exceptionally high strength and good electrical properties. Flame-retardant grades are used in printers, connectors, circuit boards, wiring blocks, and lighting components.

Polyesters (PBT). Polybutylene terphthalate (PBT) exhibits good electrical properties which are nearly independent of temperature and humidity. In addition, its inherent crystalline structure makes it resistant to most chemicals. The combination of high dimensional stability, thermal stability, and chemical resistance find applications for PBT in the usual range of high-quality injection molded electrical and electronic parts, as well as in integrated circuit carriers and high-voltage components.

Polyethylene. Polyethylene is the largest-volume thermoplastic product. Its primary use in the electrical/electronics industry is in cable jacketing and other wire applications.

Polyphenylene Sulphide. This material has a comparatively simple chemical structure, which leads to a high degree of crystallinity, hence high physical stability. Primary applications for the injection molding compounds are electrical and electronic uses, such as connectors, coil forms, bobbins, high-voltage components, and structural components.

10.1.2. Thermosets

Among the more common thermosets are:

Allyls	Fluoroplastics
Aminos and Ureas	Phenolics
Epoxies	Silicones

Allyls. Allyl resins retain their good electrical qualities under conditions of high temperature and high humidity. This property, together with good physical properties, finds broad application for the allyls in the military and other high-reliability electronics market. The most common forms of the allyls are diallyl phthalate and diallyl isophthalate, most of which are sold under military specification for use in critical electrical/electronic applications requiring high reliability under long-term, adverse environmental conditions. It is used for connectors, insulators, potentiometers, circuit boards, switches, and circuit breakers.

Aminos and Ureas. Various fillers are added to the amino resins to form a family of useful materials, among them melamine. Primary applications for alpha-cellulose-filled urea aminos are wiring devices, including circuit breakers, housings, knobs, and handles. Wood- and glass-filled melamines find industrial and military electronics applications, and mineral-filled melamines find exclusive use in military specifications.

Epoxies. As a family, the epoxies exhibit unusually good electrical, thermal, and chemical resistances. In the electrical/electronics industry, epoxy formulations with low and high viscosity, filled and unfilled, fast and slow curing, at high and low temperatures, are in use. They are used to encapsulate transistors, integrated circuits, switches, coils, and insulators. New casting processes are being developed that are expected to extend their utility to bushings, switchgear, and connectors.

Fluoroplastics (Teflon). Fluoroplastics, of which Teflon is the best known representative, offer unique characteristics. They are very unreactive chemically, being generally unaffected by chemicals, except for a few exotic fluorinated solvents. The fluoroplastic polymers have outstanding electrical properties, including very high resistance, good dielectric, and good power factors. The materials are not particularly strong, but have useful mechanical properties from cryogenic temperatures to about 260°C. PTFE (sold as Teflon, Halon, and Fluon) has a coefficient of friction lower than that of almost any other material. It will not support combustion, and on exposure to flame decomposes leaving little or no residue.

The material has an unusually high melt viscosity, and so cannot be processed by usual injection molding or extrusion methods. It is formed by methods similar to powder metallurgy, involving pressures of 2000 to 10,000 psi and sintering temperatures of 360 to 380°C. Key applications in the electrical field are high-temperature and high-voltage cable insulation and molded electrical components.

Phenolics. Phenolics are a venerable family of plastics long used in electrical applications. They are characterized by high heat resistance, chemical resistance, good dielectric properties, surface hardness, dimensional stability, easy moldability, and relatively low cost. Although formerly relatively brittle, improved chemistry and quality control have resulted in much higher impact resistances.

Silicones. Silicone fluids are best known for their extreme temperature properties, and are used as transformer coolants and heat transfer aids. Silicon resins were originally developed for thermally stable electrical insulation. Uses now include encapsulation and vehicles for high-temperature points, including those used on wood stoves. Like the resins and fluids, silicone elastomers have outstanding temperature capability, as well as high electrical resistance. As a result, they are in common use for power cable and high-voltage wires. Room temperature vulcanizing (RTV) silicone rubbers are in wide use for potting, encapsulating, and embedding electrical and electronic devices, circuits, and systems.

10.1.3. New Materials and Technologies

The substitution of plastics for metals generally saves weight and costs, but sometimes introduces problems due to the lack of conductivity of the plastics. In particular, replacing a metal housing or chassis with a plastic one may satisfy the mechanical requirements of the system, but provides no shielding, whether for local interference within a given unit, RFI of broader scale, or the very high RF levels associated with EMI/EMP. A new sub-industry has emerged, providing ways of shielding equipment in plastic housing through use of conductive coatings.

Conductive plastics have been developed and are just beginning to enter the market. Some materials are made conductive by the addition of conductive fillers, generally flake or powdered metals, and some researchers have actually developed conductive polymers. Applications of the conductive plastics are drawing attention. For instance, one manufacturer has constructed and patented commercial-size batteries with conductive polymer electrodes. Such batteries are projected to be three times smaller and up to ten times lighter than conventional batteries.

Plastics with very high temperature capabilities are also in the development stages. Tests by Bell Laboratories of integrated circuit chip carriers made of a glass-fiber-filled polyphenylene sulfide and others made of a fused silica and glass fiber filled silicone epoxy were found to be extremely reliable in accelerated environmental tests.

Production technologies are also advancing rapidly. CAD/CAM systems are being used to automatically compensate for plastic shrinkage in the design of die sets. Eight week turnaround from design to finished molds is not uncommon in die making.

10.2. FABRICATION METHODS

There are many methods used to fabricate plastic parts, as indicated by the range of processes listed in Table 10.3. Only the most common will be considered here: injection molding and compression molding. The basic equipment

TABLE 10.3. Common Polymers and Applicable Manufacturing Methods

Type of Plastics	Compression Molding	Transfer Molding	Injection Molding	Extrusion	Blow Molding	Thermoforming (Vacuum Forming)	Calendering	Rotational Molding	Dip Molding	Powder Coating	Sheet For Forming	Casting	Laminating	Expanding (Foaming)	Encapsulation	Sintering	Glass Fiber Reinforcing
Thermosets																	
Allyls (DAP & DAIP)	X	X										X			X		X
Aminos (Urea And Melamine Formaldehyde)	X	X	X[a]										X	X[g]			
Epoxides	X											X	X	X	X		X
Phenolics	X	X	X	X								X	X	X	X		X
Polyesters	X										X	X	X	X			X
Polyimides	X			X									X			X	
Polyurethanes	X	X										X	X	X	X		X
Silicones	X	X	X									X	X		X		
Thermoplastics																	
Acrylics			X	X	X	X					X	X	X	X[e]			
Acrylonitrile Butadiene Styrene (ABS)			X	X	X	X		X[c]			X			X[c]			X
Cellulose Acetate	X		X	X		X	X			X	X						
Cellulose Acetate Butyrate (CAB)	X		X	X		X	X	X		X	X						
Cellulose Propionate	X		X	X		X				X	X						

156

Polymer	1	2	3	4	5	6	7
Chlorinated Polyether		X	X	X			
Chlorinated Polyethylene		X	X	X			
Ethylene Vinyl Acetate (EVA)		X	X	X			
Fluorocarbons: PTFE	X		X^b				
PCTFE		X^b	X^b				
FEP		X	X	X	X		X
Ionomers		X	X	X			
Nylons		X	X^a	X	X^d		X
Polyacetal		X	X	X		X	X
Polycarbonate		X	X	X	X^c	X	X
Polyethylenes		X	X	X		X	X
Polyethylene Terephthalate		X	X	X			
Polyphenylene Oxide (PPO)	X	X	X		X	X	
Polyphenylene Sulphide		X	X	X			
Polypropylene		X	X	X	X	X	
Polystyrenes		X	X	X	X	X	
Polysulphones		X	X	X	X	X	
Polyvinyl Chloride (PVC)	X	X	X	X	X	X^e	X
Styrene Acrilonitrile (SAN)	X	X	X	X	X	X	X
TPX (Methylpentene Polymer)		X	X	X			
Polyurethane (Thermoplastic)		X	X		X		
Phenoxy		X	X	X			

Notes: [a]Some grades.
[b]With difficulty.
[c]Not known to have been exploited commercially at this time.
[d]Nylons 11 and 12.
[e]Produced in Japan.
[f]Produced in USA.
[g]Urea formaldehyde.

is the same for the usual ranges of both thermoplastic and thermosetting materials. There are some differences in the thermal requirements; for instance, thermoplastics require electric heaters to maintain proper temperatures in the fill tube and molds whereas the thermosets give off so much heat as they cure that water jacket cooling is required.

10.2.1. Injection Molding

The majority of plastic electronic components are produced by some form of injection molding. As shown in Figure 10.1 a ram forces a slug of stock material into a heating chamber and then into the mold cavities. A feed screw mechanism recharges the heating cylinder, and mixes the material during heating. The material left in the space between the heating chamber and the mold is called sprue, and the scraps between the various product cavities in the mold are called runners.

Injection molding is particularly suited to the making of electronics components because the usual range of parts sizes are easily handled, the parts require a minimum of additional finishing, and production rates are high, from 40 to 200 parts per hour per mold cavity. However, the press support equipment tends to require considerable maintenance. Table 10.4 compares the injection molding processes applied to the normal range of thermosets and thermoplastics.

10.2.2. Compression Molding

Compression molding is used to form complex parts in a single press stroke and can also be used to make embedded components. Instead of being fed continuously, the stock is provided in the form of pre-weighed or pre-compacted slugs called preforms. As shown in Figure 10.2, a hydraulic ram drives the preform slug into the die cavity. Although the mold cycle itself is fast, the complete process includes time to prepare and handle the preform slugs.

Figure 10.3 illustrates how the process can be used to produce one or several parts per press stroke.

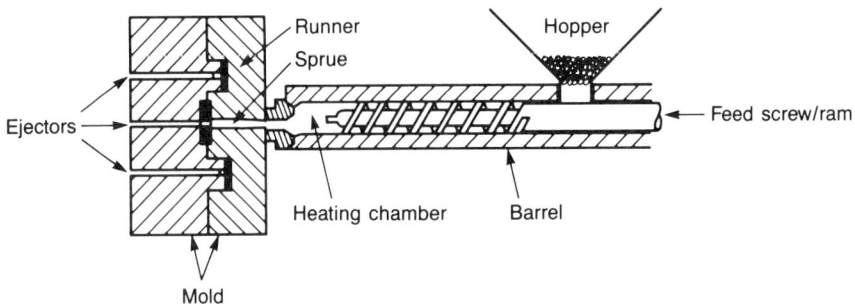

FIGURE 10.1. Typical injection mold die.

TABLE 10.4. Feed Processes for Injection Molding of Thermoset and Thermoplastic Material

	Cycle	Feed Process		
Material Group	Time (Seconds)	Loose Powder	Screw	Plunger
Thermoset	60–80	Yes	Yes	Yes
Thermoplastic	20–25	No	Yes	Yes

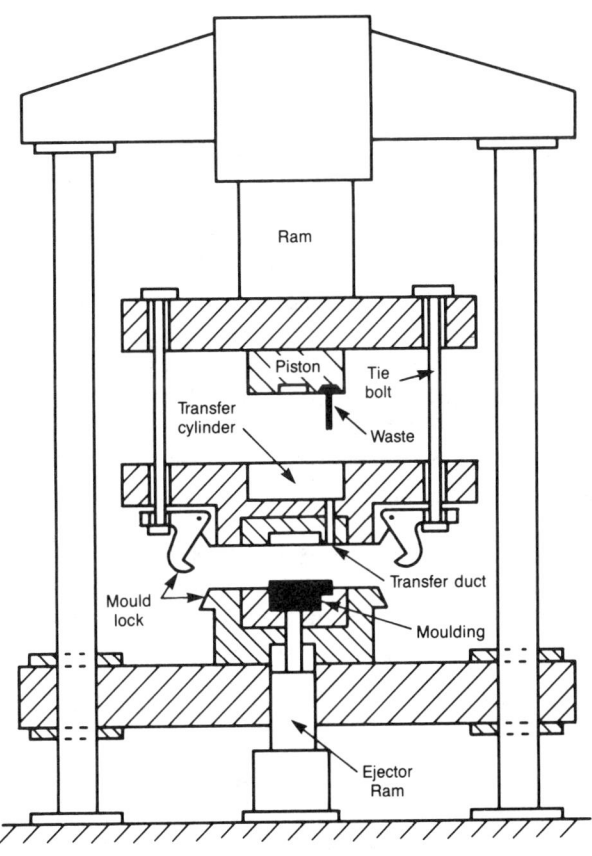

FIGURE 10.2. Typical downstroking press for compression molding of electronic parts.

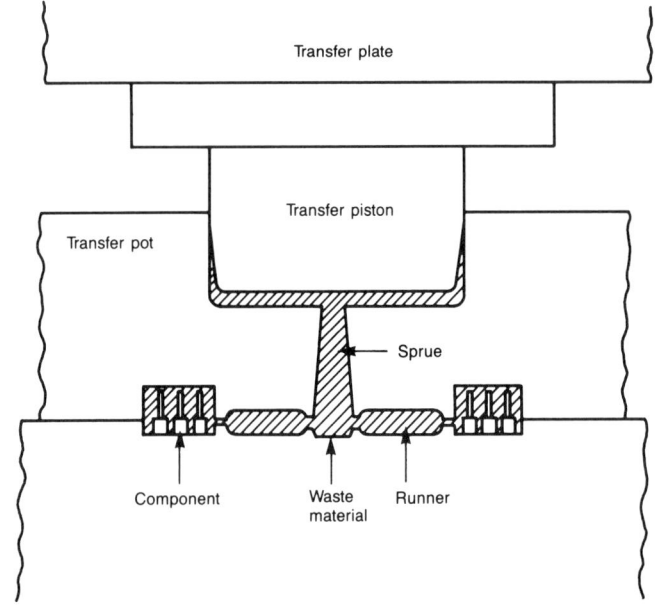

FIGURE 10.3. Transfer die set for manufacturing multiple electronic parts.

10.3. COSTS

The cost of making plastic parts depends primarily on material, labor, and tooling. All three are difficult for the non-expert to evaluate. Because costs are so variable among the different materials and processes, it is extremely difficult to establish an accurate estimating formula for plastic parts, especially for short runs of engineering plastics. It is highly recommended that comparative estimating be used, and used carefully, to ensure that similarity is obtained. The help of reputable suppliers should be obtained where a procurement of significant size is involved.

The cost estimating method given here is only intended for use with plastic components, which make up a relatively small part of the cost of an overall system. In particular, the information is not adequate for selecting a particular plastic material or manufacturing method. It is assumed that the plastic part has been engineered and fully specified.

For instance, the material costs given are for representative materials from each of the major families of engineering plastics. Prices vary widely within each group according to the particular chemical formulation and filler materials used to obtain the necessary properties. Furthermore, material selection is based not only upon raw material costs, but also on many other factors, such as yield rates and scrap utilization, press cycle times, required pre-treatments

COSTS 161

(such as drying or the making of preform slugs), heating and cooling requirements during the molding cycle, and the abrasion of tooling, especially by glass or mineral filler materials.

A worksheet, such as that illustrated by Figure 10.4, often helps in estimating cost. An explanation of each line entry follows:

1. PART DESCRIPTION: _____
   NUMBER REQUIRED: _____ WT PER PIECE _____ #
   MATERIAL   : _____ COST _____ #/#
   FABRICATION METHOD: _____
   SCRAP RATIO : _____ IN-PROCESS % SCRAP _____ % QA SCRAP

2. MATERIAL

   _____ $/# × _____ #/PC × $\dfrac{1}{1-\_\_\% \text{ IN-PROCESS SCRAP}}$ = _____ $/PC
   × ___ PCS
   × $\dfrac{1}{1-\_\_\%\text{QA SCRAP}}$ = $ _____

3. LABOR
   PERFORM:
   _____ PCS ÷ _____ PCS/PREFORM SLUG = _____ SLUGS
                    × 0.1 MIN/SLUG =   _____ MIN
   PRESS/MOLD:
   _____ PCS ÷ _____ PCS/CYCLE × _____ MIN/CYCLE = _____ MIN
   OTHER: (FINISH, ASSEMBLY, INSPECT)
                                    TOTAL LABOR = _____ MIN
                              × $\dfrac{1}{1-\_\_\%\text{ QA SCRAP}}$ = _____ MIN

4. TOOLING:
   _____ $/DIE ÷ _____ PCS/DIE = _____ #/PC
                    × _____ PCS (THIS RUN) = $ _____

5. TOTALS:
              TOTAL COST = $ _____ = $ _____ /PC
              TOTAL LABOR = _____ MIN = _____ MIN/PS

FIGURE 10.4. Plastic parts cost estimating worksheet.

Line 1. The top portion of the cost estimating worksheet can be filled in almost completely with information from an engineering specification of the part. If expected scrap ratios are not provided, the following may be used: in-process scrap ratios (material lost in sprues and runners) are typically 5% (3–8%). In addition, typically 2% (1–3%) of the molded parts will be rejected at QC inspection.

Line 2. Material costs for typical engineering plastics are shown graphically in Figure 10.5. These prices represent truckload quantities of pellet form material,

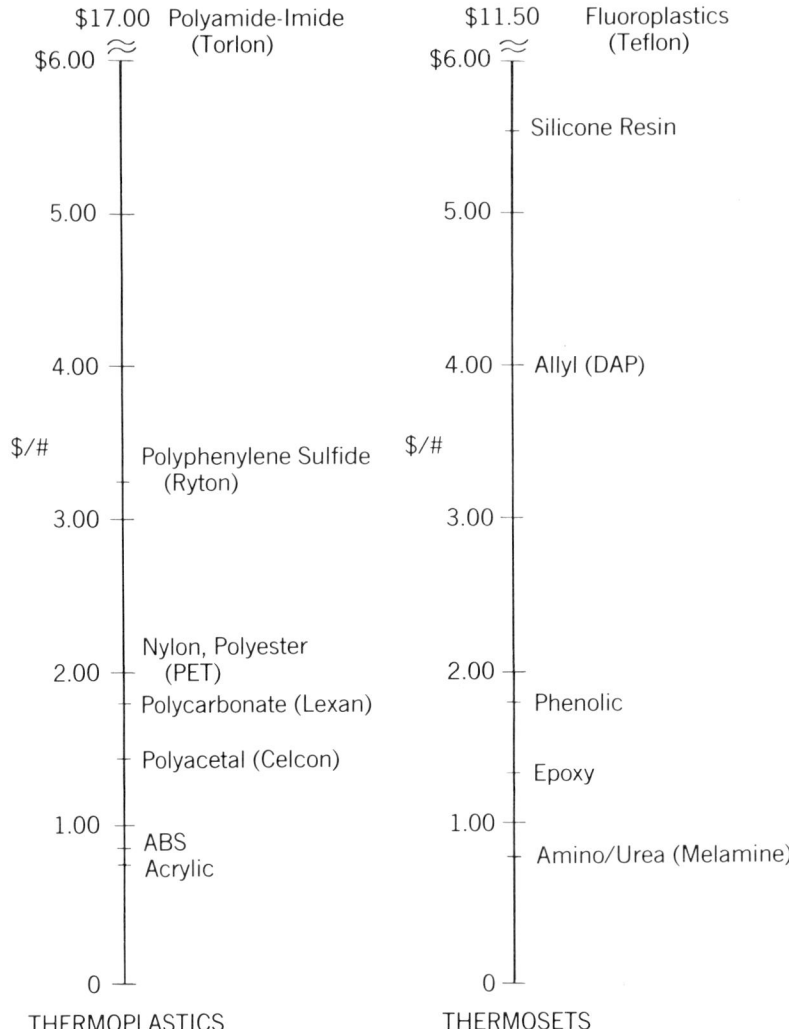

FIGURE 10.5. Approximate prices of typical engineering plastics (truckload quantities of pellet stock or liquid) (October 1982).

except in the case of epoxy where the normal form is liquid. Again, prices vary considerably within each family of material. Quotes on the specific material required are highly recommended.

The usual run of engineering plastics for injection molding costs between about $0.78 and $4.00 per pound in truckload (40,000 pound) quantities, with most materials under $2.00. The fluoroplastics (Teflon) and polyamide-imides (Torlon) cost considerably more, in the $10 to $20 range.

Line 3. Labor is related to the press cycle time and the capacity of the die set (pieces per cycle), although the support labor in preparation, material handling, finishing, inspection, and so on can be considerable, especially for short production runs. In general, thermoplastics have press cycle times of about 0.4 minute (20–25 seconds) while thermosets take longer, on the order of 1.2 minutes (60–80 seconds) per cycle. Press design is an area of continuing development, and cycle times are steadily decreasing.

For complex, low-volume parts, normally the dies are built with one cavity, hence one part is produced per press cycle. For higher volumes, several small parts are produced at once. Up to 64 cavities per mold are possible, but this would only be justified for very large-volume production of small parts.

Where preform slugs are required, the preforms are sized to provide enough material for one press cycle, regardless of the number of parts involved. About 0.1 minute per slug is required on a preform press.

Line 4. Tooling costs are substantial. A die set to produce the simplest parts will cost at least $10,000. Die sets for simple, small parts cost between $10,000 and $20,000; for large complex parts the cost can easily run to $100,000–$150,000. The major factors influencing die costs are the required accuracy and complexity of shape of the part. Mechanical features such as part ejectors, hollow core supports, and clamps to hold the die closed during injection are relatively small contributors to die cost. The advent of CAD/CAM shows imminent promise of reducing design and fabrication costs, but these have not be quantified reliably as yet.

In addition to the material, labor, and tooling that are always part of the cost of plastic parts, there may be other fabrication steps. The molded parts may require finishing, several different parts may be assembled, and special inspection may be required. Time and material costs for these steps must be added, if required.

SECTION ELEVEN

PRINTED CIRCUIT BOARD FABRICATION

11.1.	Double-Sided Etched Circuit Board Fabrication	165
11.1.1.	Process Description	165
11.1.2.	Double-Sided Etched Circuit Board Process Steps	166
11.1.3.	Double-Sided Etched Circuit Board Fabrication Costs	169
11.1.4.	Calculation Method	170
11.2.	Multi-Layer Etched Circuit Board Fabrication	171
11.2.1.	Process Description	171
11.2.2.	Summary of Multi-Layer Etched Circuit Board Process Steps	174
11.2.3.	Multi-Layer Etch Circuit Board Costs	175
11.2.4.	Calculation Method	175
11.3.	Electrical Testing	175

Essentially all etched circuit boards for modern high-quality electronics are either double sided or multi-layered. Fabrication of these boards requires the use of a wide variety of techniques, many of which are automated to some extent. The following sections describe processes in use by leading manufacturers.

The basic unit of etched circuit board fabrication is the panel. A panel runs in sizes between 15 by 15 in. and 12 by 18 in. The panel may yield a single large circuit board or it may be cut up into several smaller boards. Within normal ranges, the cost to process a panel is predicted by area alone. Excessive numbers of holes or very tight circuit path width and spacing requirements would raise costs.

DOUBLE-SIDED ETCHED CIRCUIT BOARD FABRICATION

Sections 11.1 and 11.2 provide information on double-sided and multi-layer boards, respectively. Each of these areas includes a description of the fabrication process, a list of process steps and the associated labor, a discussion of cost, and a method for calculating the cost of a circuit board. Section 14.3 covers the cost of electrical testing of printed circuit boards for shorted or open circuit paths.

11.1. DOUBLE-SIDED ETCHED CIRCUIT BOARD FABRICATION

11.1.1. Process Description

Several different processes are available for making etched circuit boards, and each process has several variations. This section will briefly describe the processes in common use by leading companies at the time of revision of this book. Other processes may be used by a particular manufacturer but, if competitive, such other process will have approximately the same cost.

The subtractive process is the most commonly used method of manufacturing etched circuit boards. The raw material for all high-reliability boards at this time is an epoxy-fiberglass substrate with copper cladding. During the manufacturing process, the unwanted copper is etched away, leaving the desired circuits paths and contact pads. The desired copper is protected by a resist material, which may be applied in several ways. For volume production, the resist is generally either an ink-like material applied by silk screening or a light-sensitive dry film material laminated to the board by heat and pressure. The light-sensitive resist is then exposed photographically and developed to leave the circuit pattern on the board.

The most common etched circuit board has circuit paths on both sides, with copper plated-through holes to provide conducting paths between the sides. The most common substrate is 0.062 in. thick FR-4 epoxy fiberglass, with 0.0014-in. copper cladding on both sides (this is referred to as one ounce per square foot cladding, while the clad board is referred to as "0.062, and one over one"). To produce the plated-through holes, the boards are first drilled and a copper strike applied with an electroless process. The copper strike is only 90 to 100 micro inches thick, but it provides a conductive coating for later electroplating. A dry film resist is laminated to the board, exposed through a diazo mask, and developed to leave a resist mask which covers the non-circuit areas.

The exposed circuit areas and holes have about 0.0014 in. of new copper electroplated to them. Non-plated-through-holes are either covered by the dry film resist ("tented") or not drilled until after electroplating. A layer of solder is plated over the new copper. The solder both serves as etch resist and tins the exposed copper for later soldering of components to the finished board.

The dry film is removed, and the board placed in an etching bath, which

166 PRINTED CIRCUIT BOARD FABRICATION

removes the exposed copper. The etchant also undercuts the solder-protected areas approximately 0.001 in.

Gold is plated to the contact fingers only. The boards travel through an automatic gold-plating line on edge, so tape masking of the area just inboard of the gold contact areas is sufficient. Application of the tape may be either manual or semi-automatic. The automatic gold-plating line first etches back the solder to expose the base copper, then plates a layer of nickel, and finally plates 25 to 50 micro inches of gold.

The tape is stripped, the board is cleaned, and then it is run through a reflow process which melts the solder briefly. This improves the appearance of the solder and consolidates any loose edges or flakes caused by undercutting during etching.

Silk screening is used to apply a solder mask coating and the lettering and component outlines or other artwork that is to appear on the board. The solder mask is applied to all areas of the board that are not component soldering pads, except for very large open areas. It prevents solder from creating bridges during the wave soldering of components to the finished board.

Final machining includes contour milling to establish final perimeter dimensions and to make any interior cut outs that may be required. Connector tabs are chamfered and slotted to the required mechanical specifications.

Table 11.1 is a summary of the steps in double-sided board manufacture.

Inspections are required often during manufacture. Typical AQL is 0.065% (i.e., random sampling designed to allow no more than 0.065% defective boards, on the average). Inspection adds significantly to the cost of high reliability etched circuit boards as can be seen by the distribution of labor in the following sections. Of the labor involved in the production process, 20.8% is devoted to inspection.

Other work is required in addition to the fabrication steps described here. In any manufacturing operation, a certain amount of work goes into preparations for production. This is discussed in Section 11.1.3. Also, test of finished boards for shorted or open circuit paths is often performed. The work associated with this test is described in Section 11.3.

11.1.2. Double-Sided Etched Circuit Board Process Steps

The following is a description of the production process steps identified in Table 11.1. Percentages indicate each step's relative share of the total process labor.

1. (1.1%) Cut panels. Base material comes in 36 × 42 or 48-in. sheets, which are sheared into the approximately 15 × 15 in. panels that are the basic units of fabrication.

2. (1.1%) Prepare panels. Drill and dowel pin together up to three panels for stacked drilling.

DOUBLE-SIDED ETCHED CIRCUIT BOARD PROCESS STEPS

TABLE 11.1. Double-Sided Etched Circuit Board Fabrication

Pre-Production
1. Release for manufacture
2. Prepare N/C tapes

Production
1. Cut panels
2. Drill and dowel
3. N/C drill
4. Deburr
5. Inspect
6. Copper depo
7. Inspect
8. Clean
9. Apply, expose, develop resist
10. Inspect and touch-up
11. Copper plating
12. Solder plating
13. Strip resist
14. Secondary drill
15. Inspect drilling
16. Inspect and touch-up
17. Etch
18. Remove touch-up paint
19. Inspect etch
20. Shear thief area
21. Tape mask
22. Deplate solder
23. Plate nickel
24. Plate gold
25. Inspect gold
26. Demask and clean
27. Reflow solder
28. Inspect
29. Apply solder mask
30. Print lettering
31. N/C rout perimeter
32. Deburr
33. Drill large holes
34. Chamfer
35. Slot
36. N/C rout internal cutouts
37. Final clean
38. Final inspect
39. Audit
40. Stamp and ship

Post-Production Test (Optional)
1. Set-up and program ATE
2. Test board

---

3. (3.6%) Machine drill. Stacked panels are N/C drilled to provide holes for component mounting and thru side electrical interconnections.

4. (2.2%) Deburr. Remove burrs from drilling. Boards are passed through a self-contained machine that applies abrasive deburr wheels, followed by water wash to remove debris.

5. (0.9%) Inspect. Inspect selected boards for proper location of holes (with mask) and size of holes (with go/no-go pin gauge). AQL = 0.065%.

6. (2.2%) Copper strike. Electroless deposition of 90–100 micro inches of copper. This provides a base for later electro deposition of copper through holes.

7. (1.1%) Inspect. Visual inspection for non-plated holes using light table. AQL = 0.065%.

8. (4.5%)	Clean. Machine cleaning of copper surface in preparation for application of dry film resist.
9. (12.5%)	Apply dry film resist. Dry film resist is laminated to the board with heat and pressure. Both sides are applied in one pass through a laminator. The dry film is then exposed to ultraviolet light through reusable diazo mask. A developer washes away the exposed dry film, leaving a negative dry film mask laminated on the panel.
10. (3.4%)	First inspection and touch up. All panels are inspected for open or shorted paths, and touched up where necessary.
11. (4.5%)	Copper plating. Electroplate copper over exposed areas including through holes. Typically add 0.0014 in. (1 oz. per in.^{2}) on surface and 0.0012 in. in holes. Typically 60 to 90 minutes plating time.
12. (4.5%)	Solder plating. Electroplate solder over exposed areas, typically 0.0003 to 0.0004 in. Provides etch resist and tinning of copper. Typically 10 to 12 minutes plating time.
13. (2.2%)	Strip dry film. Run panels through chemical bath to soften the dry film, with an ultrasonic cleaner to remove it. Stripper is automatic, requires only loading and unloading of carousel rack.
14. (0.9%)	Secondary drill. If necessary, drill holes that are not to be plated through. If possible, it is cheaper to "tent" holes drilled in step 3, or apply dry film over them. Tenting not applicable to some situations, such as large holes. May be done at this point or between steps 19 and 20.
15. (0.2%)	Inspection of secondary drilling. Similar to step 5.
16. (4.5%)	Second inspection and touch up. Similar to step 10, with additional inspection for solder coverage.
17. (2.2%)	Etch. Remove unprotected copper (all cladding except circuit paths). Primary protection is solder plating.
18. (1.1%)	Remove touch up paint. [Paint used to touch up solder plating etch resist (step 16) must be removed.]
19. (1.1%)	Inspect etch. Inspect for shorts and excessive undercutting. Undercutting greater than 20% of line width is cause for rejection.
20. (3.4%)	Shear thief area. Remove additional plating area on perimeter of panel.
21. (4.5%)	Tape. Mask for gold plating operation. Semi-automatic taping.
(4.5%)	*Note:* Steps 22 to 24 are performed on an automatic plating machine.
22.	Deplate. Remove solder plating to provide secure base for gold plating process.

DOUBLE-SIDED ETCHED CIRCUIT BOARD PROCESS STEPS 169

23. Nickel plate. Electroplate 0.0002 in. of nickel base for gold plating process.

24. Gold plate. Electroplate 25 to 50 micro inches of gold. Gold plating is applied to finger contact areas to assure good connections.

25.
(1.1%) Inspect gold. Inspect adherence using tape test and thickness using betascope.

26.
(4.5%) Clean, flux, and reflow. Reflow solder plating to improve appearance and to coat undercut edges. Reflow also consolidates loose edges and flakes of solder plating, which cuts down on shorts. Heat for reflow is applied by infrared lamps. Liquid flux is water soluable.

27.
(1.1%) Inspect. Inspect selected panels. AQL = 0.065%

28. Apply solder mask. Silk screen application of epoxy paint based solder mask over all areas of circuits except those to be soldered. Prevents bridging of traces during wave soldering operation, and provides some moisture resistance. Heat cure or ultraviolet cure.

29.
(6.8%) Printing. Silk screen printing of letter and component outlines.
Note: Steps 28 and 29 performed by same personnel.

30. Rout. NC routing of perimeter of boards to final dimensions.

31. Deburr. Deburr edges of boards with sand paper (manually).

32. Drill large holes, as required.

33. Chamfer. For connector tabs, to improve insertion.

34. Slot. For connector keys, narrow slots that cannot be milled.

35.
(8.0%) NC mill internal cut outs.
Note: Steps 30 through 35 performed by same personnel.

36.
(2.2%) Final clean. Remove fingerprints, fiberglass dust, and so forth.

37.
(6.8%) Final inspect.

38.
(1.1%) Audit. Make necessary records.

39.
(2.2%) Stamp and ship.

11.1.3. Double-Sided Etched Circuit Board Costs

(a) Production. The process described in Section 11.1.1 requires 0.3 workminutes per in.^{2} of finished circuit board, in production quantities of 50 to 1000. The step-by-step percentages shown in Section 11.1.2 are not necessary for normal cost estimating but may be used to estimate the effects of new techniques or unusual board specifications. For instance, if gold-plated connectors are not required, the labor required is reduced by the amount shown for steps 21 through 25.

(b) Pre-Production. Work is also required prior to manufacture. Before release for manufacture, the customer order is examined for compatibility with in-house specifications, artwork is checked, and production control paperwork is initiated. For new jobs, 2.0 work-hours are required to perform these tasks and release the order for manufacturing. Less time is required for repeat orders. About 0.4 work-hours suffice to retrieve existing documentation and prepare new production control paperwork.

In addition, N/C drill and router tapes are required for new orders. To set up, mark drill sequences, digitize, insert auxiliary commands such as tool changes, optimize the tool path, produce tape, and verify the tape requires 4.0 work-hours per 15 by 15 in. panel of normal circuit density. Smaller circuits may be estimated by about the ratio of perimeters, rather than areas. Smaller boards require more time than a simple area ratio would suggest, as some of the work is essentially independent of board size. In addition, small boards are manufactured in multiple boards per panel. Relative dimensions can be offset by the computer from the original digitizing, but some consideration of layout spacing and orientation is required. Thus, a 7½ by 7½ in. circuit board would be manufactured in four copies per 15 by 15 in. panel. Compared to a single 15 by 15 in. circuit, each 7½ by 7½ in. board would require one-fourth the production cost (ratio of areas), but one-half the set-up cost (ratio of perimeters).

(c) Post-Production Test. Electrical testing of boards to detect shorted or open circuit paths is becoming common. Without such testing, defects that are not detected during the several visual inspections will not show up until the board has been loaded with components and placed on circuit-level test. At this point, defects are much harder to isolate and much more expensive to repair. It is often more economical to test each board electrically prior to further assembly. This is done at the customer's option and expense.

The costs of such testing are included in Figure 11.1, and described in Section 11.3.

11.1.4. Calculation Method

Figure 11.1 is a sample cost estimating worksheet for double-sized boards. The time required to fabricate a lot of boards is the sum of pre-production time, production time, and electrical test time, as described in Sections 11.1.2, 11.1.3, and 11.3. If process steps used are different from those in Section 11.1.1 and 11.1.2, appropriate adjustments can be made as outlined in Section 11.1.3.

MULTI-LAYER ETCHED CIRCUIT BOARD FABRICATION 171

```
                                    DATE: _____
                                    BY:   _____
┌─────────────────────────────────────────────────────────────────────┐
│ PRE-PRODUCTION TIME                                                 │
│    RELEASE FOR MANUFACTURE = _____ MIN. (NEW ORDERS = 120 MIN.    │
│                                            RE-ORDERS  =  24 MIN.)   │
│    PREPARE NC TAPES (NEW ORDERS ONLY)                               │
│       PERIMETER OF BOARD _____ INCHES                         │
│       TIME FACTOR ×          4.0     MIN/INCH                       │
│       PREPARATION TIME =  _____ MIN.                          │
│ 1. PRE-PRODUCTION TIME                        = _____ MIN.      │
├─────────────────────────────────────────────────────────────────────┤
│ PRODUCTION TIME                                                     │
│    STANDARD TIME       0.30     MIN/SQ. IN.                         │
│    ADJUSTMENTS       _____                                      │
│    UNIT TIME           _____ MIN/SQ. IN.                        │
│    AREA OF BOARD × _____ SQ. IN.                            │
│    TIME PER BOARD = _____ MIN/BOARD                          │
│    NO. OF BOARDS × _____                                      │
│ 2. PRODUCTION TIME                            = _____ MIN.      │
├─────────────────────────────────────────────────────────────────────┤
│ ELECTRICAL TEST FOR SHORTS/OPEN (IF SPECIFIED)                      │
│    FABRICATE FIXTURE AND PROGRAM ATE                                │
│    NO. TEST POINTS    _____                                     │
│    PER TEST POINT ×    3.75    MIN/POINT                            │
│       PREP TIME                     = _____                      │
│    TIME PER BOARD (HANDLING +                                       │
│       TESTER CYCLE TIME)              0.5   MIN/BOARD               │
│    NO. OF BOARDS                  × _____                        │
│    TEST TIME                      = _____ MIN.                   │
│ 3.    TOTAL FOR ELECTRICAL TEST               = _____ MIN.      │
├─────────────────────────────────────────────────────────────────────┤
│    TOTAL TIME 1 + 2 + 3                       = _____ MIN.      │
│                                               = _____ HRS.      │
└─────────────────────────────────────────────────────────────────────┘
```

FIGURE 11.1. Double-sided etched circuit boards—cost estimating worksheet.

11.2. MULTI-LAYER ETCHED CIRCUIT BOARD FABRICATION

11.2.1. Process Description

Multi-layer boards are made by laminating several single- or double-sided boards together. These are interleaved with layers of resin impregnated material ("prepreg") that provides the bonding. The resulting sandwich is laminated under heat and pressure to form the multi-layer board.

Multi-layer boards may offer significant space and weight savings and in many cases permit replacement or simplification of complex wiring or connection systems. Multi-layers are particularly effective in the interconnection of

172 PRINTED CIRCUIT BOARD FABRICATION

integrated circuits. Internal and external shielding or ground layers can be incorporated directly into the circuit board to provide electrical isolation, heat sinking, and controlled impedance conductors or strip lines.

Multi-layer boards are generally no thicker than single-layer boards. The total thickness is held to 0.062 in. by controlling the thicknesses of the individual circuit and prepreg layers. Twelve- to 16-layer boards have been manufactured; a reasonable limit for normal production runs is six layers.

Figure 11.2 shows the alternation of circuit layers and prepreg layers. Typically the inner layers are copper clad on both sides of their FRP base material, while the outer layers are clad on one side only. The inner layers are etched prior to lamination.

Etching of the inner-layer circuit paths is somewhat different from etching of the outer layers. Because plating through holes is not a concern at the time of inner-layer etching, a simpler process suffices. Dry film resist is applied, exposed, and developed to mask the circuit paths, and the unprotected copper is etched away. For outer layers, the non-circuit areas are masked, followed by the copper strike, copper plate, solder plate, and then etched as described above. After the inner circuit layers are etched, the circuit layers are interleaved with prepreg and the board is laminated.

As shown in Figure 11.3, connections between layers are made by drilling through the laminated board and plating through the holes, establishing contact with the exposed edges of the circuit layers. Registration, or alignment, of the various layers is critical, and poses the greatest manufacturing problem, which increases with the number of layers. Special inspections, including use of

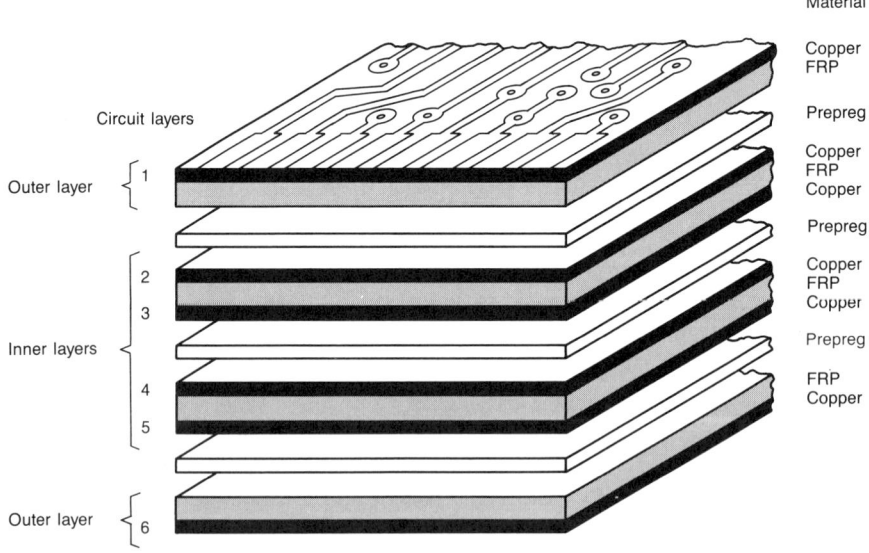

FIGURE 11.2. Multi-layer etched circuit board cross section (exploded view).

MULTI-LAYER ETCHED CIRCUIT BOARD FABRICATION

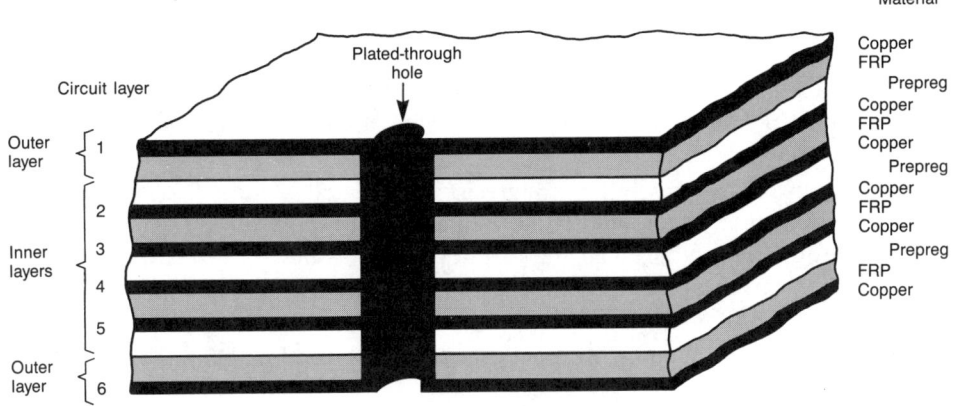

FIGURE 11.3. Multi-layer etched circuit board cross section through hole.

x-rays, are performed prior to drilling to verify proper registration of the inner layers.

After drilling, the laminated board is processed as described in Section 11.1 to etch the outer layers, plate through the holes, and so on.

Table 11.2 summarizes the steps in the fabrication of multi-layer boards.

TABLE 11.2. Multi-Layer Etched Circuit Board Fabrication

Pre-Production
1. Release for manufacture
2. Prepare N/C tapes

Pre-Lamination and Lamination
1. Cut panels
2. Apply dry film resist
3. Inspect and touch up
4. Etch
5. Remove touch-up paint
6. Strip dry film resist
7. Inspect etch
8. Drill registration holes
9. Cut prepreg
10. Drill prepreg
11. Cut exterior layers
12. Drill exterior layers
13. Assemble books
14. Laminate
15. Inspect for registration

Post-Lamination

Processing from this point is identical to that for double-sided boards, Table 11.1, steps 3–40.

Post-Production Test (If Specified)
1. Set-up and program ATE
2. Test boards

11.2.2 Summary of Multi-Layer Etched Circuit Board Process Steps

1. Cut panels. Shear copper-clad fiberglass base material to desired size. Thickness of layers chosen to provide 0.062 in. overall thickness of the finished board. Single-sided boards are used in interior and exterior layers.
2. Apply dry film resist. Dry film resist is laminated to one side of the board with heat and pressure. The dry film is then exposed to ultraviolet light through a reusable diazo mask. A developer washes away the exposed dry film, leaving a positive dry film mask on the panel. (Dry film resist covers the areas that will become circuit paths.)
3. Inspect and touch up. Inspect for open or shorted paths in the dry film resist and touch up where necessary.
4. Etch. Remove unprotected copper, leaving circuit paths.
5. Remove touch-up paint.
6. Strip dry film resist. Run panels through chemical bath to soften the dry film. An ultrasonic cleaner lifts the dry film. The stripper is automatic, requiring only manual loading and unloading of the carousel feeders.
7. Inspect etch. Inspect for shorts, opens, and excessive undercutting. Undercutting greater than 20% of line width is cause for rejection.
8. Drill registration holes. These permit pinning layers together prior to lamination.
9. Cut prepreg to size and drill. Prepare layers of prepreg material. Thicknesses are chosen to contribute to 0.062 in. overall thickness.
10. Cut exterior layers to size. Exterior layers are prepared like interior layers (steps 1 and 8) except they are not etched until after the multi-layer board has been laminated.
11. Assemble books. Prepare boards for lamination by laying up proper sequence of exterior layer, prepreg, interior layer, prepreg, and so on using pins in registration holes for alignment. Books are assembled between sheets of aluminum for protection.
12. Laminate. The assembled boards are laminated by a controlled pressure and heat profile. The cycle typically takes three to four hours.
13. Inspection for registration. Portions of the exterior boards are peeled back to reveal inner-layer registration marks (note that the outer layers are unetched). X-rays of this area are used to check registration. Typically one of 16 boards is x-rayed.

From this point, the boards follow the double-sided board process described earlier. Due to registration problems, boards are not stacked for drilling. Otherwise, the production process from here is essentially identical to that described for double-sided boards.

11.2.3. Multi-Layer Etch Circuit Board Costs

As in Section 11.1.3, the values given below apply generally to work done in a facility optimized for production quantities of 50–1000 boards. Larger or smaller facilities will normally have smaller or larger costs, respectively.

(a) Pre-Lamination and Lamination. Inner circuit layers require about 0.072 work-minute per in.^{2}* for steps 1 through 8 of Table 11.2. Prepreg (steps 9 and 10) and exterior layers (steps 11 and 12) require only cutting and drilling, which require 0.005 work-minute per in.^{2}.

Assembly of the various layers into "books" for lamination requires about 0.13 work-minute per sheet. Lamination is a fixed-duration cycle of heat and pressure. The press operator's time, when distributed over the area of boards available in the press, is about 0.001 work-minutes per in.^{2}.

After lamination, 1.06 work-minutes per board are required to verify the registration of the inner layers.

(b) Post-Lamination. The remainder of the fabrication process is identical to that for double-sided boards, as described by steps 3 through 40 of Table 11.1.

Steps 1 and 2 are not required, and so the time per in.^{2} is reduced to 0.29 work-minutes.

(c) Pre-Production and Post-Production. Pre-production costs are similar to those for double-sided boards, but apply also to the individual layers, as shown in Figure 11.4.

Post-production test for shorts and opens is described in Section 11.3. The process for multi-layer boards is identical to that for double-sided boards, however, more test points are typically required.

11.2.4. Calculation Method

Figure 11.4 is a sample cost estimating worksheet for multi-layer boards. The time required to fabricate a lot of boards is the sum of the pre-production time, pre-lamination and lamination time, post-lamination time, and electrical test time as described in Sections 11.2.2, 11.2.3, and 11.3. If process steps used are different from those assumed here, appropriate adjustments can be made as outlined in Section 11.1.3.

11.3. ELECTRICAL TESTING

Over 20% of the direct labor in printed circuit board fabrication is devoted to in-process inspection and test, and the resulting boards are normally of very

---

*Area figured for one side only, for example, a 10 by 10 in. inner layer would require 100 in.^{2} × 0.072 = 7.2 work-minutes.

PRE-PRODUCTION				
RELEASE FOR MFG			_____ MIN.	(NEW=90 MIN., REPEAT=24 MIN.)
+ 30 MIN./CIRCUIT LAYER		=	_____ MIN.	(NEW ORDERS ONLY)
PREPARE N/C TAPES (NEW ORDERS ONLY)				
4.0 MIN/PERIMETER INCH × _____ INCHES		=	_____ MIN.	
1. TOTAL PRE-PRODUCTION			=	_____ MIN.

PRE-LAMINATION & LAMINATION					
PREPARE LAYERS					
MATERIAL	AREA, IN^2 (ONE SIDE)	# SHEETS	MIN/IN^2/ SHEET	MIN.	
INNER CIRCUIT	_____ ×	_____ ×	.072	= _____	
PRE-PREG	_____ ×	_____ ×	.005	= _____	
OUTER CIRCUIT	_____ ×	_____ ×	.005	= _____	
ASSEMBLE BOOKS	# SHEETS _____ ×		.13	= _____	
LAMINATE	AREA ONE SIDE _____ ×		.001	= _____	
INSPECT FOR REGISTRATION			=	1.06	
			× _____ BOARDS		
2. TOTAL PRE-LAMINATION & LAMINATION			=	_____ MIN.	

POST-LAMINATION				
STANDARD (SEE SECTION 11.1, STEPS 3-40)				
AREA ONE SIDE _____ ×	.29	= _____		
ADJUSTMENTS: AREA ONE SIDE _____ × _____		= _____		
SUBTOTAL		= _____		
	× _____ BOARDS			
3. TOTAL POST-LAMINATION			=	_____ MIN.

ELECTRICAL TEST FOR SHORTS/OPENS (IF SPECIFIED)			
FABRICATE FIXTURE & PROGRAM ATE			
# TEST POINTS _____ ×	3.75	= _____	
TEST BOARDS # BOARDS _____ ×	.50	= _____	
4. TOTAL FOR ELECTRICAL TEST		=	_____ MIN.

TOTAL TIME 1 + 2 + 3 + 4	= _____ MIN.
	= _____ HRS.

FIGURE 11.4. Multi-layer etched circuit boards cost estimating worksheet.

high quality. Even so, because of the very high value that is added to convert a finished board into a working circuit and the very high cost of fault isolation of finished circuits, it is common practice to conduct a post-production electrical test of printed circuit boards to detect shorted and open circuit paths.

Electrical testing is performed by programmable automatic test equipment (ATE), which is connected to the board under test by a "bed-of-nails" multiple-point contact test fixture. Preparing the ATE to test a particular board design includes both fabricating the fixture (inserting spring-loaded contact pins in the proper locations and wiring them to a suitable cable connector) and programming the ATE test sequence. Together these require 3.75 minutes per test point. Testing each board requires 0.5 minute, which includes handling the board, operating the test fixture, and running the test program.

SECTION TWELVE

TERMINAL BOARD FABRICATION

12.1. Calculation Method 177

The fabrication of terminal boards requires the use of operations already covered in detail in previous sections of the book. The most common of these operations are combined here into one worksheet, which can be used to estimate various types of terminal boards.

The standards are based on 0.125-in. thick phenolic laminate sheet stock.

12.1. CALCULATION METHOD

Figure 12.1 is a cost estimating worksheet for terminal board manufacture. Figure 12.2 shows a sample calculation, for a 4 by 4 in. board with 25 terminals.

TERMINAL BOARD FABRICATION

REF. SECTION	OPERATION	SET-UP MINUTES	MIN./ OPERATION	NO. OPERATION	TOTAL MIN.
7.1	Cut blank to size, 2 edges per inch	8.08	.02		
8.6	Deburr 4 edges, per inch	6.23	.02		
11.6	Drill holes				
	Set-up, per hole size	23.82	—		
	Drill, per hole		0.10		
8.6	Deburr holes, per hole, each side	6.23	0.06		
14.1	Silk Screen	63.30	0.73		
15.2	Stake Terminal	8.05	0.04		
15.1	Handle (5 operations/board)		0.06	5	0.30
	Total set-up	115.71			
	Total oper. per board × _____ boards/lot = _____				_____
	Total time/lot = _____				

FIGURE 12.1. Terminal board cost estimating worksheet.

REF. SECTION	OPERATION	SET-UP MINUTES	MIN./ OPERATION	NO. OPERATION	TOTAL MIN.
7.1	Cut blank to size, 2 edges per inch	8.08	.02	8	.16
8.6	Deburr 4 edges, per inch	6.23	.02	16	.32
11.6	Drill holes				
	Set-up, per hole size	23.82	—		
	Drill, per hole		0.10	25	2.50
8.6	Deburr holes, per hole, each side	6.23	0.06	25	1.50
14.1	Silk Screen	63.30	0.73	1	.73
15.2	Stake Terminal	8.05	0.04	25	1.00
15.1	Handle (5 operations/board)		0.06	5	0.30
	Total set-up	115.71			6.51
	Total oper. per board × 50 boards/lot = 325.5				
	Total time/lot = 441.2 = 7.4 Hrs				

FIGURE 12.2. Terminal board cost estimating worksheet—sample calculation.

SECTION THIRTEEN

PAINTING OPERATIONS

13.1.	Surface Preparation	180
13.2.	Mask and Demask Part for Painting	180
13.3.	Set-Up Time per Paint Type	181
13.4.	Primer	181
13.5.	Surfacer	182
13.6.	Lacquer	182
13.7.	Enamel and Modified Vinyl	183
13.8.	Plastic Protective Film, Strippable	183
13.9.	Fungicide (Spray Application)	184
13.10.	Miscellaneous Detail Values	184
13.11.	Calculation Method	188

This section covers spray painting using an individual spray booth type of facility. The Summary Table of Section 3 and the analysis of this section list the typical operations required to paint metal chassis, panels, and cabinets. The Summary Table also gives set-up time and drying time. The drying time is for oven drying at temperatures of 200–300°F, according to the specific paint requirements.

Method:

The spray booth method requires manual parts handling as opposed to the continuous conveyor method, where the operator merely sprays the parts as it passes before him. The time values include the following elements.

1. Handle parts to and from the turntable in the spray booth. This includes picking up parts from material handling boxes or floor pallets as required by their size.

180 PAINTING OPERATIONS

2. Spray part on turntable.
3. Handle parts to and from a drying oven or heat lamp bank conveyor.

Major Time Variables:

1. Part size—Varies both handling time and spray time to cover a given area.
2. Part configuration—Accessibility, shape, contours, and hole patterns will also vary the paint time required for specific parts. The time values are for "average" parts.
3. Type finish—Prime coats require one pass. A high-gloss finish coat usually requires two passes.
4. Type paint—Thin primers give the fastest coverage. The higher the viscosity, the slower the coverage. Another factor is the number of spray passes and air dries required by some wrinkle and hammertone finishes prior to the final baking operation.

Definition of Time Values:

The painting time values are for an average cube-shape part of the dimensions given. Each part is assumed to have at least four sides to be painted. There are a maximum of six outside surfaces to be painted on a cube. The time values are sufficient for this area. It is also conceivable to have four or five inside surfaces to be painted in the same cube. The times will *not* cover both inside and outside surfaces.

13.1. SURFACE PREPARATION

Section 3, Summary Table 13, lists a typical surface preparation for each of the common base metals. The summary values are found in Section 3 under electroplating and metal finishing, Summary Table 9. The detailed analysis for these values is given in Section 9.

13.2. MASK AND DEMASK PART FOR PAINTING

Adhesive-type masking tape is used for this element. Work tasks include: rough measure tape, tear off from roll, and apply to part. It also includes removing the masking tape from the part after the paint operation. These values are averages for the purpose of quick, easy application to estimates. It is advisable, where possible, to use the detail values of Section 13.10 to itemize the number of tapes, plugs, shields, and so on.

13.3. SET-UP TIME PER PAINT TYPE

The equipment necessary for this operation includes a paint booth, turntable, and a spray gun. The following elements must be accomplished in setting up a new paint type or paint color:

Secure paint.
Open and mix paint.
Thin paint if necessary, and transfer to tank.
Secure air and paint lines.
Clear air lines and attach spray gun.
Attach nozzle.
Adjust and try out.
After job, clean above items with solvent.
Put away.

    Total set-up time   22.1 minutes

13.4. PRIMER

Spray paint applications consist of parts handling plus the actual spray time to cover the given part. Each of the values given in Section 3, Summary Table 13, includes the following handling times. The spray time by part type and part size may be segregated by subtracting these handling times from the Summary Table values.

Run time Analysis	Size			
	Minutes/Unit/Coat			
Parts handling to paint booth turntable:	To 3 in.	To 8 in.	To 20 in.	To 30 in.
1. Pick up and lay on turntable.	0.10	0.13	0.16	0.21
2. Pick up and lay aside spray gun.	not rqd.	0.10	0.10	0.10
3. Lay aside part.	0.10	0.13	0.16	0.21
Sub Total	0.20	0.36	0.42	0.52
Parts handling to drying rack or oven:				
1. Pick up and position part to rack.	0.10	0.13	0.16	0.21
2. Lay aside part.	0.10	0.13	0.16	0.21
Sub Total	0.20	0.26	0.32	0.42

	Size			
	Minutes/Unit/Coat			
	To 3 in.	To 8 in.	To 20 in.	To 30 in.
Total Handling per Part	0.40	0.62	0.74	0.92
Spray time:				
Wash Primer	0.10	0.15	0.30	0.35
Zinc Chromate	0.40	0.60	1.00	1.25
Total Time per Part				
Wash Primer	0.50	0.77	1.04	1.27
Zinc Chromate	0.80	1.22	1.74	2.17

13.5. SURFACER

A surfacer is used where a high-gloss finish is desired. It acts as a filler to cover minor tooling and welding marks on cabinet work. It may be applied by brush or by spray. A smooth base for the finish coat is achieved by hand rubbing or power buffing the surfacer coat.

Run Time Analysis	Size			
	Minutes/Unit/Coat			
	To 3 in.	To 8 in.	To 20 in.	To 30 in.
Handle part (ref. Section 13.4)	0.40	0.62	0.74	0.92
Spray paint	0.60	0.90	1.00	1.25
Power buff	1.40	2.00	2.40	3.85
Total	2.40	3.52	4.14	6.02

13.6. LACQUER

Lacquer is used as either a flat or gloss finish coat. The cost of lacquer finish is normally less than enamel. On the other hand, a lacquer finish is not as durable as a baked enamel.

Lacquer takes less drying time and/or facilities, and its cost is about 25% less than enamel. If lacquer is air dried, much less storage area is required for the drying parts. If oven-dried, the oven is tied up for a much shorter time than with enamel, thus more parts can be dried per day. The lower facilities requirement reflects itself in a lower operating cost for lacquered items.

	Size			
Run Time Analysis	Minutes/Unit/Coat			
	To 3 in.	To 8 in.	To 20 in.	To 30 in.
Flat-parts handling	0.40	0.62	0.74	0.92
Spray time	0.40	0.60	1.00	1.25
Total	0.80	1.22	1.74	2.17
Gloss-parts handling	0.40	0.62	0.74	0.92
Spray time	0.70	1.20	1.80	2.30
Total	1.10	1.82	2.54	3.22

13.7. ENAMEL AND MODIFIED VINYL

Most commercial enamels are baked to give a hard, durable finish. The gloss, wrinkle, or hammertone surface texture is achieved by the paint make-up and the baking cycle. Wrinkle and hammertone enamels usually receive a double coat. The first pass is air dried for about ten minutes and the second pass for about 20 minutes prior to baking.

	Size			
Run Time Analysis	Minutes/Unit/Coat			
	To 3 in.	To 8 in.	To 20 in.	To 30 in.
Flat Parts handling	0.40	0.62	0.74	0.92
Spray time	0.40	0.60	1.00	1.25
Total	0.80	1.22	1.74	2.17
Gloss, wrinkle, hammertone:				
Parts Handling	0.40	0.62	0.74	0.92
Spray time	0.70	1.20	1.85	2.30
Total	1.10	1.82	2.54	3.22

Modified vinyl paints are replacing enamels in some applications. Application times for modified vinyls are the same as those for enamels.

13.8. PLASTIC PROTECTIVE FILM, STRIPPABLE

This type of film is used to protect painted surfaces during final assembly operations. It is also used to protect a unit while in shipment to the customer.

The film may be applied by brush or spray. It dries to a soft, translucent skin over the equipment, and is easily peeled off when desired. The time values below are for a spray type application and do not include peeling the hardened film off the unit.

	Size			
Run Time Analysis	Minutes/Unit/Coat			
	To 3 in.	To 8 in.	To 20 in.	To 30 in.
Handling	0.40	0.62	0.74	0.92
Spray time	0.30	0.50	0.85	1.00
Total	0.70	1.12	1.59	1.92

13.9. FUNGICIDE (SPRAY APPLICATION)

Most fungicide applications are made by dipping components prior to assembly or spot brushing completed assemblies. Occasionally, completed assemblies will be sprayed.

	Size			
Run Time Analysis	Minutes/Unit/Coat			
	To 3 in.	To 8 in.	To 20 in.	To 30 in.
Handling	0.40	0.62	0.74	0.92
Spray time	0.50	0.80	1.30	1.60
Total	0.90	1.42	2.04	2.52

13.10. MISCELLANEOUS DETAIL VALUES

These values are often required to estimate partial operations or specific detail parts. All of the elements are not necessarily used on one part.

		Size			
		Minutes/Operation			
1.	Handling Time:	To 3 in.	To 8 in.	To 20 in.	To 30 in.
	The small part sizes include handling the part to and from a box on the bench. The larger size parts include handling to skid or tote boxes positioned on the floor.				
	Per part	0.20	0.26	0.32	0.42

MISCELLANEOUS DETAIL VALUES

	Minutes/Operation
2. File or Burr Edge: A file or emery cloth is usually used for this operation. The time values include burring both the top and bottom edges of a flat piece of metal. In other words, to burr both sides of a 10-in. edge, allow 0.20 minutes.	
Per in.	0.02
3. Sand or Grind by Power: A portable electric tool is used for this operation. It may be the vibrator type or the rotary type. This operation is usually performed on large cabinets and consoles where weld and forming marks must be removed prior to painting.	
Per ft.^{2}	1.09
4. Sand by Hand: Hand sanding is sometimes required because of inaccessibility by power tools.	
Per ft.^{2}	2.18
5. Blow off Surface: After sanding, a compressed air hose is used to blow all dust particles from the unit.	
Per ft.^{2}	0.05

PAINTING OPERATIONS

	Minutes/Operation
6. Wash Surface with Solvent: Hand washing is necessary where parts are too large to be dipped in the vapor degreaser. Per ft.^{2}	0.05
7. Spray Paint: This is an average value for primer spray painting. It does not include handling the spray gun or the part being painted. Where two passes are required for a high-gloss finish coat, double the following values. Per ft.^{2}	0.06
8. Brush Paint: Areas which cannot be reached by a spray gun must be touched up by the conventional brush method. The time value includes dipping the brush in a paint can, removing excess, and applying to the desired area. Per ft.^{2} Flat	0.24
Per ft.^{2} w/Frequent holes, edges and lips	1.00

MISCELLANEOUS DETAIL VALUES

	Minutes/Operation
9. Apply and Remove Masking Tape (Per 10 in.):	
Apply 10 in.	0.42 minutes
(Where required) trim 10 in.	0.44 minutes
Remove and clean with solvent 10 in.	0.25 minutes
Total per 10 in.	
(no trimming)	0.67 minutes
(with trimming)	1.11 minutes
Total per in.	
(no trimming)	0.07 minutes
(with trimming)	0.11 minutes

10. Assemble and Remove Masking Plugs and Stencils:

Plugs, stencils, and shields are the most productive type of masking devices. They are usually made up special for the part so as to minimize the assembly and disassembly time required for masking. The time value includes both assemble and disassemble the masking device from the part.

Per plug, stencil, or shield	0.10 minutes

13.11. CALCULATION METHOD

A format for estimating cost of painting operations is provided in Figure 13.1.

PAINTING COST ESTIMATING WORKSHEET	DATE: _____
FOR: _____	BY: _____

1. JOB ASSIGNMENT ========= min.
SURFACE PREPARATION
Operation:
File/Burr Edge   0.02 min/in.   × _____ in. = _____
Power Sand   1.09 min/ft.^{2} × _____ ft.^{2} = _____
Hand Sand   2.18 min/ft.^{2} × _____ ft.^{2} = _____
Blow Off Surface   0.05 min/ft.^{2} × _____ ft.^{2} = _____
Wash With Solvent   0.05 min/ft.^{2} × _____ ft.^{2} = _____
Brush Touch-up   .24–1.00 min/ft.^{2} × _____ ft.^{2} = _____
Apply & Remove Tape, Clean   0.07 min/in.   × _____ in. = _____
Trim Tape   0.04 min/in.   × _____ in. = _____
Apply & Remove Plugs, Stencils, Blanks   0.10 min/each × _____   = _____
Other _____   _____ × _____ = _____
Total Per Piece   _____ min/pc
×   _____ pcs.
=   _____ min.
2. SURFACE PREPARATION
3. SET UP SPRAY EQUIPMENT
Per Coating Type = 22.10 min × _____ types =   _____ min.
SPRAY, Including Handling
Coating Type (List Each Coat)   Min/Pc/Coating
_____   _____
_____   _____
_____   _____
_____   _____
_____
TOTAL   ×   _____ min/pc/coating
_____ pcs = _____ min.
4. SPRAY TIME
5. OTHER (Describe)
_____ min.
TOTAL TIME 1 + 2 + 3 + 4 + 5   _____ min.
= _____ hrs.

FIGURE 13.1. Printing cost estimating worksheet.

SECTION FOURTEEN

SILK SCREEN PRINTING AND ENGRAVING

14.1.	Silk Screen Printing Process Description	189
14.2.	Photographic Operations to Prepare Silk Screen Stencils	191
14.3.	Silk Screen Printing, Including Decals	193
14.4.	Engraving	194
14.5.	Calculation Method	196

Each of the operations covered in this section is a distinct art or line of business in itself. In many shops these operations are incidental to the main production effort and are maintained only as service functions. On the other hand, there are many small concerns that specialize in fields such as industrial art work and photography, silk screen printing, and engraving.

A very brief coverage pertaining more to the type of service function within a larger company is given here. The specific method, facilities, and so on are given with each operation analysis.

14.1. SILK SCREEN PRINTING PROCESS DESCRIPTION

Silk screen printing is a stencil process. A fine silk or synthetic fabric "screen" is used to carry a plastic film stencil. Ink squeegeed onto the screen passes to the work through cutouts in the film stencil. Fine detail can be resolved using fine-mesh screens and modern films.

Silk screen film hardens on exposure to light. A photographic film positive of the artwork is used to expose the stencil film. Stencil film behind the artwork lines is not hardened and is dissolved during the developing process. As a result, ink will coat the finished work in the same pattern as the original artwork.

Making the photographic film positive from the original artwork requires two photographic steps: first a negative, then the positive. It is possible to screen print half-tone artwork or photographs as well as simple lettering or solid-line artwork; however, a few additional minutes are required to prepare a half-tone negative compared with the simpler process for line artwork.

In summary, the following steps are required to prepare a silk screen for printing (see Figure 14.1).

Photographic negative
Photographic positive
Stencil film negative

The stencil film negative is attached to the screen during the developing process. When dry, the stencil is ready for printing.

Summary Table 14 in Section 3 contains two times, one for preparation of a silk screen stencil from line artwork, and one for half-tone artwork. More detailed times can be determined from the following sections but are not generally necessary.

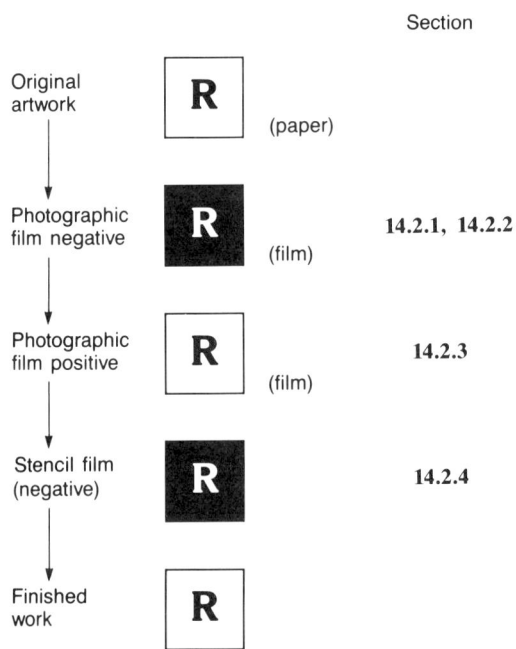

FIGURE 14.1. Silk screen stencil preparation and printing.

14.2. PHOTOGRAPHIC OPERATIONS TO PREPARE SILK SCREEN STENCILS

Photographic processes are used to prepare silk screen stencils from both line and continuous-tone copy. Where multi-color silk screening is done, a separate screen must be made for each color.

A film positive is required to make the silk screen; an inter-negative step is necessary.

The basic equipment used in the photo lab are:

Dark room with developer, fix and rinse baths
Industrial copy camera
Contact printer
Enlarger
Viewing table
Drying cabinet

14.2.1. Negative from Line Copy

This operation covers the preparation of a negative from line copy. The elemental breakdown is given in Table 14.1.

        Total expose and develop   13.4 minutes

14.2.2. Half-Tone Negatives

The elemental breakdown in Table 14.1 is for making a half-tone negative from continuous-tone copy. In other words, this operation converts the shades of gray and black in the original photograph to the dot pattern of the half-tone.

        Total time per job   21.7 minutes

14.2.3. Positive

To prepare a film positive from either type of negative, Table 14.1 shows:

        Total expose and develop   11.9 minutes

14.2.4. Fabricate Silk Screen Stencil

Table 14.2 gives the steps and times necessary to fabricate a silk screen stencil, given a film positive of the artwork, as described above.

        Total time per job   23.5 minutes

TABLE 14.1. Photographic Operations

		Section Number		
		14.2.1	14.2.2	14.2.3
	Machine	Work-Time (Minutes)		
Element	Time (Minutes)	Line Negative	Halftone Negative	Positive from Negative
Expose Film:				
1. Assembly copy to camera copy board	—	1.00	1.00	—
2. Turn light switch and adjust	—	0.30	0.30	0.30
3. Set lens opening	—	0.50	0.50	—
4. Set timer	—	0.30	0.30	0.30
5. Adjust lens board to center image	—	0.50	0.50	—
6. Cut film and position to vacuum holder	—	0.70	0.70	0.70
7. Position halftone contact screen over film	—	—	1.00	—
8. Position negative over film	—	—	—	0.50
9. Position glass over negative	—	—	—	0.50
10. Expose	0.6–5.0	0.60	2.50	0.60
11. Disassemble film from camera	—	0.30	0.30	0.30
12. Disassemble copy from board	—	0.50	0.50	—
Subtotal		4.70	7.60	3.20
Develop Film:				
1. Place film in developer bath	2.5	2.50	2.50	2.50
2. Rinse	.2	.20	.20	.20
3. Place film in fix bath	3.0	3.00	3.00	3.00
4. Place film in wash bath	10.0	2.50	2.50	2.50
5. Dry in film dryer	25.0	0.50	0.50	0.50
Subtotal		8.70	8.70	8.70
Total Expose and Develop:		13.40	16.30	11.90
Re-Exposure Allowance:				
1. Halftones-add a 33% allowance for highlight adjustment re-exposures		—	5.40	—
Total Minutes per Job		13.40	21.70	11.90

TABLE 14.2. Fabricate Silk Screen Stencil

		Minutes/Stencil
1.	Sensitize:	
	Cut film to size	1.00
	Immerse in sensitizing solution	
	Rub surface to remove air bubbles	1.00
2.	Transfer film to transparent vinyl support, emulsion side down	0.50
	Squeegee excess sensitizer and air bubbles from film	0.50
	Wipe dry with cloth	0.50
3.	Expose:	
	Place vinyl support with film on emulsion side of positive	0.50
	Place support and positive in vacuum frame	0.50
	Cut to length and assemble opaque tape around border of positive	1.00
	Expose film	5.00
4.	Wash out:	
	Immerse film and support in 110° water	0.50
	Soak .5 minute and peel backing away	1.00
	Agitate and dissolve unexposed part of gelatin	1.00
5.	Attach developed film to screen:	
	Clean silk screen with cleaner or solvent and brush under running water	4.00
	Drain and place developed film and support on table	1.00
	Place silk screen over film	0.50
	Place paper towel over screen to absorb moisture remove	1.00
	Weigh frame down and allow to air dry	2.00
	Peel vinyl support from screen stencil	2.00
	Total	23.50 Minutes

14.3. SILK SCREEN PRINTING, INCLUDING DECALS

14.3.1. Set-Up Silk Screen for Production

		Minutes/Color
1.	Receive work order, drawings and blank boards	2.00
2.	Set up to clean boards	1.00
3.	Set up silk screen:	
	Draw silk screen from storage rack	0.30
	Clamp to two bench hinges	1.00

SILK SCREEN PRINTING AND ENGRAVING

	Minutes/Color
Mark alignment pin holes on ink board	0.30
Drill and pin ink board	2.00
Position stock sample to pins	0.10
Line up ink board to screen pattern	0.50
Staple base down	0.30
Screen first piece	
Mix ink	1.00
Squeegee	1.00
Check registration	3.00
4. Set up touch-up operation:	
Obtain ink, brush, negative, positive, or sample	2.00
Total	14.50 minutes

14.3.2. Make Prints

		Run Time/Color (Minutes)
1.	Handle stock to board, align	0.13
2.	Squeegee ink through screen	0.40
3.	Handle stock to drying rack	0.10
4.	Remove from drying rack	0.10
	Total	0.73 minutes/print

14.3.3. Decals

Decals are made by the silk screen printing process described above, with two additional steps. To make a decal, a layer of clear lacquer is first applied to special, glue-impregnated paper. This lacquer serves to carry the artwork when the decal is applied. The artwork color(s) are applied over the base lacquer, and then sealed with a top coat of clear lacquer. Cost estimates of making decals therefore include as many stencil preparations (Section 14.2) as there are colors in the artwork, plus as many printing cycles (Section 14.3.1 and 14.3.2) as colors in the artwork, plus two cycles for the clear lacquer layers.

14.4. ENGRAVING

Engraving is a common method used for placing identification characters on both control panels and name plates. The letters are cut into the metal with a revolving or vibrating cutting tool, and on control panels, are usually filled with a contrasting color or lacquer. The engraving process is more expensive than silk screening but also more durable.

14.4.1. Set-Up Analysis

Equipment—Gorton Pantograph, Model 3Z, or equivalent.

		Minutes/Job
1. Set up master copy.		
(a) Unlock stylus and (4) type keys (two lines)		0.50
(b) Assemble type—0.3 minutes per letter		
(c) Lock (4) type keys		0.80
(d) Disassemble prior holding fixture		1.00
(e) Assemble and adjust new holding fixture		1.50
2. Set Pantograph Ratio		
(a) Compute ratio from blueprint		2.00
(b) Unlock (2) bolts on arms		0.40
(c) Position arms to ratio desired		1.00
(d) Lock (2) bolts		0.40
3. Adjust (2) micrometer adjustments for straight line reproduction		1.20
4. Adjust (1) micrometer adjustment for depth of cut		0.70
5. Assemble and disassemble cutting tool		0.70
Total		10.20
+ per Letter		0.30

14.4.2. Run Time Analysis

	Minutes/Unit
Handle part to and from fixture. This element includes time to pick up part from tote, position and lock in a holding fixture, disassemble from fixture, deburr rough edges with emery and lay aside to tote.	1.10

14.4.3. Engrave and Lacquer Letters

Refer to Table 14.3.

TABLE 14.3. Engrave and Lacquer

Element		Letter Size—Minutes/Letter				
	From	0 in.	0.126	0.251	0.376	0.501
	To	0.125 in.	0.250	0.375	0.500	0.625
Engrave letter						
Position stylus to master		0.03	0.03	0.03	0.03	0.03
Lower depth lever		0.01	0.01	0.01	0.01	0.01
M/T—cut letter		0.09	0.13	0.16	0.20	0.24
Raise depth lever		0.01	0.01	0.01	0.01	0.01
Total engrave		0.14	0.18	0.21	0.25	0.29
Fill with lacquer						
Squeegee lacquer into letter		0.02				
Wipe with solvent		0.01				
Total lacquer		0.03	0.03	0.03	0.03	0.03
Total engrave and lacquer		0.17	0.21	0.24	0.28	0.32

14.5. CALCULATION METHOD

14.5.1. Silk Screen Printing

Summary Table 14 in Section 3, summarizes the standard for estimating the cost of silk screen printing. An average size screen is about 8 × 10 in., but the normal range is wide, and within reason size does not greatly affect the time required to screen print.

Figure 14.2 is a sample cost estimating form. With the standard times

```
                                              DATE: _____
                                              BY: _____
         SILK SCREEN PRINTING COST ESTIMATING WORKSHEET
```

1. PREPARE STENCIL
    LINE COPY: NO. COLORS _____ × 48.8 = _____ min.
    TONE COPY: NO. COLORS _____ × 57.1 = _____ min.	

2. SET-UP NO. COLORS* _____ × 14.5 = _____ min.	

3. PRINT
    NO. PCS _____ × NO. COLORS* _____ × 0.73 = _____ min.
    TOTAL (1 + 2 + 3)   = _____ min.
                                                 = _____ hrs.	

Note: When making decals, add 2 for clear lacquer layers.

FIGURE 14.2. Silk screen printing cost estimating worksheet.

CALCULATION METHOD

DATE: _____
BY: _____

ENGRAVING COST ESTIMATING WORKSHEET

1. SET-UP 10.2 min.	
NO. LETTERS _____ × 0.3 = _____ min.	
SUBTOTAL = _____ min.	
2. ENGRAVE AND LACQUER	
NO LETTERS _____ × _____ * = _____ min/pc	
HANDLING = 1.1 min/pc	
PER PC.	
NO. PCS. × _____	
SUBTOTAL = _____ min.	
TOTAL (1 + 2) = _____ min.	
= _____ hrs.	

*FROM THIS TABLE:

LETTER SIZE	MIN/LETTER
TO ⅛"	.17
TO ¼"	.21
TO ⅜"	.24
TO ½"	.28
TO ⅝"	.32

FIGURE 14.3. Engraving cost estimating worksheet.

pre-printed, the total time is simply the sum of the times for preparing the stencils, setting up, and printing.

14.5.2. Engraving

As above, Summary Table 14 in Section 3 summarizes the standard for engraving, and Figure 14.3 is a sample cost estimating form. The total time for engraving is the sum of the set-up, engraving, and lacquering times, as shown.

SECTION FIFTEEN

RIVETING AND MECHANICAL ASSEMBLY

15.1.	Parts Handling	198
15.2.	Rivet, Stake, Eyelet	201
15.3.	Screw and Nut Operations	203
15.4.	Miscellaneous Assembly Operations	204
15.5.	Tool Handling Values	206
15.6.	Tape and Tags	206
15.7.	Cement and Glyptol	207
15.8.	Walking	207
15.9.	Dynamic Balance Operations	207

The parts handling, riveting, and screw-and-nut operations are given by detailed time and motion analysis elements. The parts handling table, for example, provides pick up and lay aside values which are tailored to a riveting operation. The "transport empty" element is not included for the smaller size parts because it is done simultaneously with the lay aside of the prior part. In this way the tables can be used to develop standards for various types of operations. Summary Table 15 in Section 3, provides totaled "average" values for most of the operations in this section.

15.1. PARTS HANDLING

The following is an explanation of part types.

PARTS HANDLING

Type Part	Maximum Volume (inches)	Description
I	½ × ½ × ½	Hardware parts Screws, nuts, springs, clips, rivets, eyelets, bushings, washers, and so on.
II	2 × 2 × 2	Small piece parts Brackets, resistors, capacitors, integrated circuits, tubes, switches, small terminal boards, and so on.
III	4 × 4 × 4	Brackets and boards Terminal and printed circuit boards, brackets, transformers, relays, and so on.
IV	8 × 8 × 8	Subassemblies Small chassis, panels, large transformers, speakers, electromechanical assemblies, and so on.
V	12 × 12 × 12	Small chassis Also large heavy transformers, panels, bases, and so on.
VI	18 × 18 × 18	Medium chassis and panels Large 19-in. rack panels, rack mounting chassis, and so on.
VII	24 × 24 × 24	Large Chassis Chassis, boxes, cases, small cabinets, awkward assemblies, and so on.

		Type Part and Approximate Size						
Element	Move (in.)	I ½ in.	II 2 in.	III 4 in.	IV 8 in.	V 12 in.	VI 18 in.	VII 24 in.
Pick Up Part (P/U)								
Transport empty	10	0.007	0.007	0.007	0.007	0.007	0.007	0.007
	20	0.009	0.009	0.009	0.009	0.009	0.009	0.009
	30	0.012	0.012	0.012	0.012	0.012	0.012	0.012
	40	0.014	0.014	0.014	0.014	0.014	0.014	0.014
				0.043	0.046	0.063	0.069	0.075
Add (2) steps and return								
Grasp		0.009	0.005	0.004	0.005	0.006	0.006	0.007

(continued)

RIVETING AND MECHANICAL ASSEMBLY

Element	Move (in.)	Type Part and Approximate Size						
		I ½ in.	II 2 in.	III 4 in.	IV 8 in.	V 12 in.	VI 18 in.	VII 24 in.
Pre-position		0.008	0.006	0.006	0.009	0.013	0.017	0.023
Transport loaded	10	0.007	0.007	0.007	0.007	0.009	0.010	0.013
	20	0.009	0.009	0.009	0.009	0.013	0.015	0.018
	30	0.012	0.012	0.012	0.012	0.015	0.018	0.023
	40	0.014	0.014	0.014	0.014	0.017	0.022	0.028
Release to bench		0.002	0.002	0.002	0.002	0.002	0.002	0.002
Position to one stud or hole		0.006	0.008	0.010	0.013	0.015	0.018	0.024
Position to two studs or holes			0.016	0.020	0.025	0.030	0.037	0.048
Lock (1) fixture clamp			0.015	0.017	0.020	0.023	0.029	0.035
Lock (2) fixture clamp			0.030	0.048	0.041	0.046	0.058	0.069
Average P/U		0.030	0.025	0.029	0.062	0.074	0.164	0.194
Lay Aside Part (L/A)								
Unlock (1) fixture clamp			0.015	0.017	0.020	0.023	0.029	0.035
Unlock (2) fixture clamp			0.030	0.048	0.041	0.046	0.058	0.069
Grasp		0.009	0.005	0.004	0.005	0.006	0.006	0.007
Transport loaded	10	0.007	0.007	0.007	0.007	0.009	0.010	0.013
	20	0.009	0.009	0.009	0.009	0.013	0.015	0.018
	30	0.012	0.012	0.012	0.012	0.015	0.018	0.023
	40	0.014	0.014	0.014	0.014	0.017	0.022	0.028
Add (2) steps and return				0.025	0.026	0.043	0.051	0.062
Release toss		0	0	0	0	0	0	0
Stack in tote		0.006	0.005	0.006	0.007	0.010	0.013	0.010
Transport empty	10	0.007	0.007	0.007	0.006	0.007	0.007	0.007
	20	0.009	0.009	0.009	0.009	0.009	0.009	0.009
	30	0.012	0.012	0.012	0.012	0.012	0.012	0.012
	40	0.014	0.014	0.014	0.014	0.014	0.014	0.014
Average L/A		0.016	0.012	0.018	0.023	0.033	0.040	0.113
Example: Average P/U and L/A for riveting operations		(0.046)	(0.037)	(0.047)	(0.085)	(0.107)	(0.204)	(0.307)

15.2. RIVET, STAKE, EYELET

Element	Move (in.)	I ½ in.	II 2 in.	III 4 in.	IV 8 in.	V 12 in.	VI 18 in.	VII 24 in.
P/U and L/A base part		0.046	0.037	0.047	0.085	0.106	0.204	0.307
P/U and position terminal, eyelet, clip, socket, and so on To the above base:								
Type part	I	0.030	0.030	(0.030)	0.038	0.041	0.051	0.066
	II		0.032	0.036	0.041	0.048	0.058	0.072
	III			0.051	0.055	0.056	0.066	0.076
	IV				0.075	0.086	0.099	0.109
	V					0.123	0.137	0.151
	VI						0.154	0.190
Additional alignment		0.006	0.006	0.012	0.012	0.017	0.029	0.029
Trip machine cycle and move base to next hole: Rivet or eyelet machine (include hopper feed rivet or eyelet)		0.032	0.037	(0.039)	0.041	0.046	0.058	0.074
Hand arbor press		0.067	0.074	0.078	0.084	0.089	0.100	0.116
Air press		0.083	0.090	0.093	0.100	0.105	0.018	0.120
Wrap and unwrap Painted parts:								
Unwrap		0.02	0.04	0.06	0.08	0.12	0.17	0.23
Wrap		0.04	0.06	0.08	0.12	0.17	0.23	0.35
Average terminal Staking value (Element 2 + 4)					(0.069)			

15.3. SCREW AND NUT OPERATIONS
Elemental Table for Screws

Element	Type of Screw					
	Machine	Machine	Set Screw	Thumb Screw	Self-Tapping	Wood
Assemble To:	Nut or Tapped Hole	Elastic Stop Nut	Tapped Hole	Tapped Hole	Undersized Hole	Undersized Hole
Pick up, transport 20 in. finger start 2 threads: *Diameter length*						
⅛ in. to ¼ in.	0.070	0.070	(0.070)	0.081	0.070	0.070
to 1 in.	(0.062)	(0.062)	0.062	0.081	(0.062)	0.062
¼ in. to ¼ in.	0.075	0.075	0.075	0.081	0.075	0.075
to 1 in.	0.067	0.067	0.067	(0.081)	0.067	(0.067)
½ in. to ¼ in.	0.074	0.074	0.074	0.081	0.074	0.074
to ½ in.	0.074	0.074	0.074	0.081	0.074	0.074
Locate tool to slot:						
Screw driver	(0.030)	(0.030)	(0.030)		(0.030)	(0.030)
Air gun	0.022	0.022	0.022		0.022	0.022
Turn down flush: With fingers						
1 thread	0.008			0.015		
5 threads	0.040			(0.075)		
Tighten	0.018			(0.018)		
Screw driver						
1 thread	0.015	0.017	0.015		0.030	0.018
5 threads	(0.075)	(0.086)	(0.075)		(0.150)	0.092
10 threads	0.150	0.173	0.150		0.299	(0.184)
Air gun						
1 thread	0.0018	0.0018	0.0018	0.0018	0.0018	0.0018
10 threads	0.018	0.018	0.018	0.018	0.018	0.018
20 threads	0.037	0.037	0.037	0.037	0.037	0.037
Tool handling—pick up and lay aside:						
Screw driver	0.035					
Air gun—suspended	0.029					
On bench	0.041					
Total average Values (example)	(0.167)	(0.178)	(0.175)	(0.714)	(0.242)	(0.281)

SCREW AND NUT OPERATIONS

Elemental Table for Nuts

		Type of Nut		
Element	Assemble To:	Hex	Elastic Stop	Wing
		Screw or Stud	Screw or Stud	Screw or Stud
Pick up, transport 20 in. finger or spintite start (2) threads:				
to ⅛ in. diameter		0.070	0.070	
to ¼ in. diameter		(0.067)	(0.067)	0.067
to ½ in. diameter		0.074	0.074	(0.074)
Locate tool to nut:				
Spintite		(0.030)	(0.030)	
Air gun		0.022	0.022	
Open end wrench		0.033	0.033	
Turn down flush:				
With fingers				
1 thread		0.008		0.015
5 threads		0.040		(0.075)
Tighten		0.018		(0.018)
Spintite				
1 thread		0.015	0.017	
5 threads		(0.075)	(0.086)	
10 threads		0.150	0.173	
Air gun				
1 thread		0.0018	0.0018	
10 threads		0.018	0.018	
20 threads		0.037	0.037	
Open end or socket wrench per thread				
Ratchet type		0.033	0.033	
Fixed type		0.068	0.068	
Tool handling—pick up and lay aside:				
Spintite		(0.035)		

(continued)

Elemental Table for Nuts

		Type of Nut		
Element	Assemble To:	Hex Screw or Stud	Elastic Stop Screw or Stud	Wing Screw or Stud
Air gun				
Suspended		0.029		
On bench		0.041		
Open end wrench		0.035		
Pliers		0.039		
Total average values (example)		(0.207)	(0.183)	(0.167)

15.4. MISCELLANEOUS ASSEMBLY OPERATIONS

The values included in this section do not include tool handling. These values are based on 10–20 in. move distances.

	Minutes/Operation	
Bushing		
Pick up and assemble to screw or stud		0.035
"C" Washer		
Pick up and assemble with spreading pliers		0.115
Cotter pin		
Assemble and bend tabs		0.138
Decal		
Soak, apply, wipe dry		0.575
Disassemble Hardware		
Screw		0.138
Nut		0.115
Screw, nut, lock washer		0.207
Fuse		
Disassemble cap	0.07	
Assemble fuse	0.05	
Re-assemble cap	0.10	0.219
Glyptol Hardware		
Pick up dropper, fill, return to pot		0.039
Apply per spot		0.030

MISCELLANEOUS ASSEMBLY OPERATIONS

	Minutes/Operation
Grommets, Rubber	
To ½ in. diameter	
Soft	0.069
Medium	0.081
Hard	0.115
To 1 in. diameter	
Soft	0.115
Medium	0.173
Hard	0.230
Knobs	
Push on type—assemble	0.056
Set screw type—assembly and line up	
(1) Screw—already started	0.115
—start and lock	0.185
(2) Screws—already started	0.181
—start and lock	0.321
Label, small	
Assemble with glue	0.288
Pressure sensitive	
Strip back and assemble	2.30
Pilot lamp	
Bayonet type	0.039
Screw type	0.064
Stamp, rubber	
Small-repetitive usage	0.046
Intermittant for employee ID	0.115
Tap holes—retap to remove paint	
Locate tap in hole 0.020	
Twist per thread 0.017	
Total for (6) threads in and out	0.224
Terminal lug	
Pick up and assemble to screw or stud	0.037
Line up before tighten	0.012
Tube-assemble to socket	0.069
Twist ears	
Electrolytic capacitor ear with twisting tool	0.025
Unwrap	
Small switch, and so forth, boxed	0.115
Meter-boxed	
Remove, sort, lay aside hardware	0.460
Washer	
Pick up and assemble to screw	0.037

15.5. TOOL HANDLING VALUES

	Pick Up	Lay Aside	Total
Screw driver	0.023	0.012	0.035
Spintite	0.023	0.012	0.035
Pliers	0.025	0.012	0.037
Air gun			
Suspended	0.021	0.008	0.029
On bench	0.025	0.016	0.041
Open-end or socket wrench	0.023	0.012	0.035
Solder iron and solder	0.029	0.023	0.052
Wire strippers	0.023	0.012	0.035

15.6. TAPE AND TAGS

	Minutes/Operation
Electrical Tape—assemble to cable	
Tool Handling	0.035
P/U and cut off tape	0.058
Cover 1st inch along cable	0.138
(3 turns)	0.230
Add per additional inch of cable covered	0.092
Masking Tape	
Tear off and apply to panel—per in.	0.015
Remove from panel, per inch	0.018
Pressure Sensitive Identification Tapes (¼ × 1 in.)	
Assemble to wire	
P/U from backing card, wrap	
(3 turns)	0.155
Disassemble (3) turns	0.132
Identification Tag	
Write part number, serial number, and so on, on tag—per character	0.017
Tie tag to unit	0.288
Loop tag to unit	0.115

15.7. CEMENT AND GLYPTOL

	Minutes/In.^{2}	
	Hard Surface	Cloth Surface
Brush viscous liquid to surface		
No restrictions	0.006	0.009
Some restriction	0.009	0.015
Very restricted (cut outs, etc)	0.014	0.022
Handle brush	0.058	0.058
Position cloth to cemented area		
No restrictions	0.004	
Some restrictions	0.006	
Very restricted	0.009	
Apply cement to cloth tape		0.052
Position and dress tape to unit		0.058

15.8. WALKING

	Minutes
Single step—one leg only moves	0.017
Both legs move	0.023
Two full 30-in paces.	
General	0.030
Restricted	0.035
Over two paces for each 30 in. of travel (Pace)	
General	0.014 + 0.009 per pace
Restricted	0.014 + 0.012 per pace
Up steps	0.017
Down steps	0.012

15.9. DYNAMIC BALANCE OPERATIONS

The purpose of this operation is to dynamically balance a rotor, armature, and so on dynamically, to reduce vibration to meet the minimum requirements of MIL-STD-167 on mechanical vibration. The part to be balanced is supported in a Schenck or American Hoffman type dynamic balance. Either a direct end drive or a belt is used to rotate the component while sensors built in the machine identify the amount of imbalance. Read-out dials tell the operator how much weight should be added or removed and the location where the adjustment should take place.

Weight can be added by threading fasteners into previously drilled and tapped holes, or by placing weights (often washers) under existing fasteners or studs. Weight can be removed by grinding off stock or by drilling holes. Some rotors are manufactured with extra stock that may be ground off to assist in balancing. Moment corrections on large parts are made while the part is still in the balance machine. Small parts are removed to a work area where the addition or subtraction of weight takes place.

This section gives reasonable times for non-automatic low-volume operations. For high-volume production runs involving live automation, the balance time would be integrated with other operations in the process.

Military Work

Set-Up Analysis

Install component in balance machine. Install belt or direct drive. Includes receiving job assignment, reviewing production documents and completing necessary reports at end of job.

Run Analysis

Start drive motor, raise RPM, record unbalance, brake motor. Add temporary weight if necessary and repeat test cycle.

Remove component from balance machine and make permanent weight correction. Weigh temporary weights if necessary. Add weights or drill or grind away weight as required.

Reinstall component in balance machine and finish balance. Start drive motor, record unbalance, brake motor. Add temporary weights if necessary.

Remove component from balance machine and make final permanent weight corrections. Weigh temporary weights if necessary.

Reinstall component in balance machine and run up to speed to verify final balance condition.

Normally a rough and a finish cycle and a final check are required on a military item.

Total time

	Minutes	
	Component Weight	
	Through 40 Pounds	Over 40 To 600 Pounds
Set-Up	25	30
Run 1	40	70
Run 2	30	60
Run 3	20	40
Run 4	10	15
Run 5	10	20
Total	135 Minutes	235 Minutes

DYNAMIC BALANCE OPERATIONS

For commercial work, the set-up time and documentation are less time consuming. Balance requirements are often lower. Production runs are often longer.

Commercial Work	Minutes	
	Component Weight	
	Through 40 Pounds	Over 40 To 600 Pounds
Set-Up Analysis		
Install component in balance machine. Install belt or direct drive. Includes receiving job assignment, reviewing production documents and completing necessary reports at end of job.	15	20
Run Analysis		
Start drive motor, raise RPM, record unbalance, brake motor. Add temporary weight if necessary and repeat test cycle.	20	35
Remove component from balance machine and make permanent weight correction. Weigh temporary weights if necessary. Add weights or drill or grind away weight as required.	20	35
Reinstall component in balance machine and run up to speed to verify final balance condition.	10	20
Normally a single cycle plus a final check are all that is required.		
Total time	65 Minutes	110 Minutes

Summary.

New Parts up to 40 Pounds	New Parts from 41 to 600 Pounds	New Parts over 600 pounds and Irregular Moments About Axis of Rotation
	Military (Average time to balance)	
135 Minutes	235 Minutes	480 minutes
	Commercial (Average time to balance)	
65 Minutes	110 minutes	240 minutes

SECTION SIXTEEN

COIL WINDING OPERATIONS

16.1.	Coil Winding Terms	210
16.2.	Example	211
16.3.	Set-Up Analysis	212
16.4.	Run Time Analysis	213

Coils are generally considered in the same class of many electronic components—a commodity product. They are purchased competitively as standard stock items in lot quantities. In those instances where coil fabrication is part of the manufacturing operations, or where special designs are required, this section may be used for estimating cost. The process represents small order quantities using traditional manufacturing methods of manual and semi-automatic operations.

The standards consist of one detail check-off list of elemental time values. The elements are listed in the approximate order of usage and include those used for both flat and universal coils. Flat coils may be wound individually on a bobbin or in multiples of 5 to 10 on a paper tube.

16.1. COIL WINDING TERMS

1. Flat Wound

   All turns run in the same direction and are parallel to each other.

2. Close Wound

   Same as above but with the additional requirements that each turn is placed close to and against the prior turn.

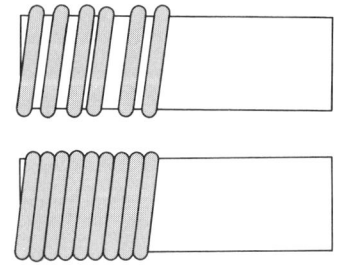

3. **Space Wound**
   Same as flat wound but with a controlled space between each turn.

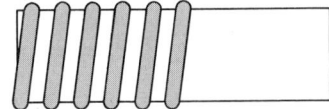

4. **Universal Wound**
   This coil is identified by the criss-cross pattern formed by each succeeding layer. A special cam on universal winding machines forms the pattern.

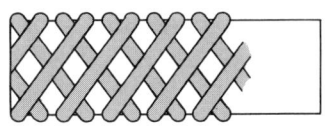

16.2. EXAMPLE

The following example shows the application of detail values to a typical small coil.

Windings	Turns	Type	Gauge	Compound
Primary	27	Flat close	36AWG	Lacquer
Primary	27	Flat close	36AWG	Lacquer
Secondary	2	Flat close	36AWG	Lacquer

The time for winding consists of the following elements extracted from the detail table, Section 16.4.

Reference Element	Element	Minutes/ Element	Number Elements	Total Minutes
	Prepare to Wind:			
2	Assemble form to wing nut arbor	0.03	1	0.03
	Assemble nut	0.05	1	0.05
6	Anchor start lead—wrap to terminal	0.03	3	0.09
9	Index wire guide	0.015	3	0.045
10	Position wire around guide button	0.03	3	0.09
11	Set counter	0.045	3	0.135
12	Hand wind three turns	0.062	3	0.186
	After Wind:			
20	Release brake, turn arbor to position	0.025	3	0.075

(continued)

Reference Element	Element	Minutes/ Element	Number Elements	Total Minutes
22	Push back guide	0.015	3	0.045
23	Cut leads to approximate length	0.005	3	0.015
	Handle scissors	0.03	3	0.09
26	Wrap finish lead to terminal	0.03	3	0.09
	Trim all leads	0.025	6	0.15
	Handle scissors	0.03	3	0.09
28	Solder leads	0.04	6	0.24
	Handle solder iron	0.045	1	0.045
32	Disassemble form from arbor	0.02	1	0.02
	Disassemble nut	0.04	1	0.04
37	Brush coil with lacquer per ½ in.	0.035	2	0.07
	Handle brush	0.03	1	0.03
	Total manual time for (3) windings			1.626
14	Total machine time	0.001	56	0.056
				1.682

16.3. SET-UP ANALYSIS

	Minutes/Job	
	Hand Wind	Machine Wind
Material		
Trip to tool crib or storeroom for wire, tubes, bobbins, forms, arbor adapter.	5.00	5.00
Machine		
Calculate gear ratio for required turns per inch.	0	2.50
Calculate cam size for coil width, pi, or universal cross-over pattern.	0	2.50
Cam-disassemble, reassemble.	0	0.20
Nut-disassemble, reassemble.	0	0.30
Allen screw-unlock, relock four at 0.20	0	0.80
Readjust at 50% occurrence	0	0.70
Gear-disassemble, assemble three at 0.20	0	0.60

RUN TIME ANALYSIS

	Minutes/Job	
	Hand Wind	Machine Wind
Allen screw-disassemble, reassemble three at 0.30	0	0.90
Line up and adjust.	0	1.00
Spindle-assemble arbor base.	0.30	0.30
Winding finger Screw-unlock, relock three	0	1.20
Tail stock Nut-unlock, relock	0	0.30
Counter-adjust to stop.	0.40	0.40
Division control-adjust.	0	0.20
Tension and spool holder	0	3.00
Subtotal minutes/job	5.70	19.90
Multiplier	.022	.022
Set-up hours/job	0.1 hr.	0.4 hr.
Add per additional coil for multiple winding		
Adjust winding finger.		1.20
Tension and spool holder		3.00
Subtotal minutes/job		4.20
Multiplier		.022
Set-up hours per additional coil on arbor		0.1

16.4. RUN TIME ANALYSIS

Reference Element	Element	Minutes/ Element
	Prepare to Wind:	
1	Advance arbor (multiple pi only)	0.005
2	Assemble tube or bobbin to arbor Random position	
	Chuck arbor	0.04
	Wing nut arbor	0.03
	Add to assemble nut	0.05
	Expansion arbor	0.04
	Spring arbor	0.03

(continued)

Reference Element	Element	Minutes/Element
	Align tube to a specific lead position	
	Chuck arbor	0.045
	Wing nut arbor	0.04
	Add to assembly nut	0.05
	Expansion arbor	0.05
	Spring arbor	0.04
3	Assemble spacer to arbor	0.02
4	Close tail stock	0.03
5	Dress and secure leads of previous winding (this element is necessary for coils having windings of two or more wire sizes)	
	Tape to arbor: dispense and position tape	0.04
	Apply per lead	0.01
	Wrap lead to terminal	0.035
6	Anchor start lead to wind. Thread to hole through coil form and dress	
	Coil on arbor	
	18–22 gauge	0.13
	24–36 gauge	0.11
	38–40 gauge	0.12
	Wrap to terminal	
	18–22 gauge	0.07
	24–28 gauge	0.05
	30–40 gauge	0.03
	Wrap to tube and tape (Wrap lead two turns, dispense and position tape)	0.10
	Wrap to arbor and tape (Wrap lead two turns, dispense and position tape)	0.08
7	Double back start lead to reinforce	0.035
8	Twist precut and stripped reinforcement lead to coil lead	0.05
9	Index wire guide	0.015
10	Position wire around guide buttons	0.03
11	Set counter	
	Veeder root type-clear to zero	0.025
	Self-braking type-set up to 99 turns	0.045
	Add per additional digit	0.015
12	Hand wind first few turns—1 turn	0.044
	2 turn	0.053
	3 turn	0.062
	4 turn	0.072

Reference Element	Element	Minutes/ Element
13	Cement start of winding	
	Handle applicator	0.025
	Apply per lead or per pi	0.015
	Wind:	
14	Machine Wind	
	Using machine lead spacer:	
	Start, stop, and brake machine	0.045
	Space Wind	
	14–16 gauge, 300 rpm	0.003
	18–20 gauge, 500 rpm	0.002
	22–24 gauge, 1000 rpm	0.001
	26–up gauge, 1500 rpm	0.0007
	Close or universal	
	14–18 gauge, 200 rpm	0.005
	18–20 gauge, 300 rpm	0.003
	22–24 gauge, 500 rpm	0.002
	26–up gauge, 1000 rpm	0.001
	Groove wind on threaded core	
	16–18 gauge, 200 rpm	0.005
	20–22 gauge, 300 rpm	0.003
	24–up gauge, 500 rpm	0.002
15	Hand Wind	
	Hand feed wire to required turns/in.:	
	Start, stop, and brake machine	0.02
	Close wind, or groove wind	
	14–16 gauge, 60 rpm	0.017
	18–20 gauge, 80 rpm	0.013
	22–up gauge, 100 rpm	0.01
	Taps and Insulation Strips:	
16	Release brake, turn arbor to position	0.025
17	Make tap	
	Twist tape and knot with crochet needle	0.13
	Twist tape and wrap one turn to terminal	0.15
	Loop tap-Anchored with tape	0.19
	Anchored with tape and cement	0.21
	Wrap to terminal	0.18
	Add for second tape around tap	0.07

(continued)

Reference Element	Element	Minutes/ Element
18	Cambric type insulation strip	
	Insert under lead	0.07
	Spread and position preformed tap	0.11
	Form and position flat tap	0.12
	Hand wind one turn. Dress cambric and tap together	0.06
19	Kraft paper type insulation strip	
	Position to coil	0.04
	Brush with cement—per ½ in.	0.035
	Handle brush	0.03
	After Wind:	
20	Release brake, turn arbor to position	0.025
21	Anchor finish lead to coil with tape dispense and assemble	0.06
22	Push back guide	0.015
23	Cut leads: to approximate length	0.005
	to exact length	0.02
	handle scissors	0.03
24	Thread finish lead:	
	One hole through coil form and dress:	
	18–22 gauge	0.13
	24–36 gauge	0.11
	38–40 gauge	0.12
25	Twist precut and stripped insulated lead to coil lead	0.05
26	Wrap finish lead to terminal	
	18–22 gauge	0.07
	24–28 gauge	0.05
	30–40 gauge	0.03
	Trim leads wrapped to terminal	0.01
	Handle cutters or tweezers	0.025
27	Unwrap start leads	
	From arbor or tube	0.015
	From terminal	0.03
28	Solder start, finish, or splice leads per joint	0.04
	Handle solder iron	0.045
29	Disassemble lead of previous winding	
	Taped to arbor	0.02
	Taped to tube	0.03
	Unwrap from terminal	0.03

Reference Element	Element	Minutes/ Element
30	Open tail stock	0.03
31	Disassemble arbor from chuck	0.04
32	Disassemble coil from arbor	
	Chuck arbor	0.03
	Wing nut arbor	0.02
	Add disassemble nut	0.04
	Expansion arbor	0.035
	Spring arbor	0.02
	Spacer-remove Simo with coil	—
33	Disassemble lead from slit in spacer	0.012
	Work Done During Winding or as Separate Operations	
34	Prep. insulated reinforcement lead	
	Artos cut, twist, tin to 15 in.	0.09
35	Precut Kraft paper insulation	
	Average	0.02
36	Dip coil with compound, wax, and so on	
	Lay aside	0.04
37	Brush coil with compound, cement, wax, and so on, lay aside	
	Per ½ in. width	0.035
	Handle brush	0.03
38	Color code coil per spot	0.03
	Handle brush per color	0.03
39	Saw paper section windings	
	1 in. diameter	0.15
	2 in. diameter	0.30
	3 in. diameter	0.60
40	Pull tap after sections are sawed	
	Sort layers, pull loop with crochet needle, cut loop	0.15
	Add tape per element 21.	
41	Tape cutting (included in above taping values)	
	Cut on cutting board	0.005
	Pull from roll dispenser and tear off	0.02
	Pull cloth tape from roll and cut off with scissors	0.025

SECTION SEVENTEEN

WIRE PREPARATION

17.1.	Preparation Insulated Wire, Machine	219
17.2.	Preparation Insulated Wire, Hand	219
17.3.	Stake Taper Pin to Wire	221
17.4.	Cut to Length Bus Wire and Sleeving	222
17.5.	Cut and Bend Resistors, Capacitors, and Other Components	222
17.6.	Cut to Length Coaxial and Shielded Cable on Bench Cutter	223
17.7.	Preparation Shielded Cable with Ground Lead	224
17.8.	Preparation Coaxial Cable	225
17.9.	Stamp and Assemble Identification Tape to Cables	227
17.10.	Assemble Plug or Connector to Harness Assembly	227
17.11.	Buzz Wire to Identify Continuity	229
17.12.	Pull Tubing Over Cable	230
17.13.	Twist Wires	230
17.14.	Spot Tie Harness	230
17.15.	Harness Fabrication	231
17.16.	Fabricate Harness Nail Board	234

The tables of this section summarize elemental time values into times per operation. The wire preparation operations are then combined with crimping and soldering operations, in Section 19, to obtain the average wiring values used on the Summary Table 17 & 18 in Section 3.

17.1. PREPARATION INSULATED WIRE, MACHINE

Set-up:	Set-Up (Minutes)
Artos Wire Cutter and Stripper	
Change wire size and length.	2.52/change
Average: one change per 10 wires on list.	0.25/wire
Bench Solder Pot	
For twist and tin operation. Does not include melt down	Negligible
Wire Stamping Machine (Kingsley Type)	
Set heat control, foil feed, type characters, spacing adjustment	2.52/change
Average: one change per 10 wires on list	0.25/wire
Total (average)	0.50/wire

Run Time:
Detail time values are given in Table 17.1. The times are based on the following methods.

Artos Cut and Strip. The Artos machine will automatically unreel, cut to length, and strip either solid or stranded hook-up wire. The time values pay only for the manual time required to pick up the wire and tie into a bundle with an identification tag.

Twist and Tin Strands. Untie bundle of wires, pick up 2 to 6 wires, according to gauge and length, twist strands, dip in flux, dip in solder pot, wipe, invert, repeat for second end, rebundle and tie at completion of lot.

Stamp Wire. Automatic type—Stamps wire as it is fed to the Artos machine. No manual time required.
 Semi-Automatic—Pick up and position wire to machine; stamp 15 in. wire approximately every 2 in.; stamp 60 in. wire approximately every 4 in.; remove wire and bundle.

17.2. PREPARATION OF INSULATED WIRE, HAND

	Wire Length	
Set up Analysis:		
Add 2.66 minutes per change in wire specification.	15 in.	60 in.
	Minutes per Piece	
1. Pull wire from reel and cut it	0.121	0.175
2. Identify one wire end with tape	0.207	0.207

TABLE 17.1. Machine Prep Insulated Wire

Operation	Number of ends	Wire Length (In.) From 2½ To 3	3 15	15 20	20 30	30 45	45 60
Gauge 14-20							
Machine cut and strip	1	0.023	0.023	0.046	0.046	0.069	0.092
	2	0.023	0.023	0.046	0.046	0.069	0.092
Twist strands	1	0.040	0.029	0.037	0.037	0.037	0.043
	2	0.069	0.049	0.064	0.064	0.064	0.071
Tin strands	1	0.030	0.020	0.024	0.035	0.035	0.040
	2	0.044	0.028	0.035	0.064	0.064	0.070
Stamp wire—Simultaneously with machine cut	1	——— Concurrent with cut and strip ———					
	2	0	0	0	0	0	0
Stamp wire—separate operation	1	0.104	0.150	0.184	0.219	0.265	0.311
	2	0.104	0.150	0.184	0.219	0.265	0.311
Total—cut and stamp Simultaneously	1	0.093	0.071	0.107	0.117	0.140	0.175
	2	0.136	0.100	0.145	0.175	0.198	0.233
Total—stamp separate operation	1	0.197	0.221	0.291	0.336	0.374	0.485
	2	0.239	0.250	0.389	0.393	0.462	0.544
Gauge 22-26							
Machine cut and strip	1	0.023	0.023	0.046	0.046	0.069	0.092
	2	0.023	0.023	0.046	0.046	0.069	0.092
Twist strands	1	0.040	0.029	0.037	0.037	0.037	0.043
	2	0.069	0.049	0.064	0.071	0.071	0.078
Tin strands	1	0.035	0.020	0.030	0.035	0.035	0.040
	2	0.048	0.028	0.044	0.064	0.064	0.070
Stamp wire—simultaneously with machine cut	1	——— Concurrent with cut and strip ———					
	2	0	0	0	0	0	0
Stamp wire—separate operation	1	0.104	0.150	0.184	0.219	0.265	0.311
	2	0.104	0.150	0.184	0.219	0.265	0.311
Total—cut and stamp simultaneously	1	0.098	0.071	0.116	0.117	0.140	0.174
	2	0.140	0.100	0.154	0.181	0.205	0.240
Total—stamp separate operation	1	0.201	0.221	0.297	0.336	0.405	0.485
	2	0.244	0.250	0.338	0.400	0.469	0.551

	Wire Length	
	15 in.	60 in.
	Minutes per Piece	
3. Identify second wire end with tape	0.207	0.207
4. Strip two wire ends with hand held wire stripper	0.173	0.188
5. Tin two ends with solder iron (per wire)	0.292	0.292
6. Tin two ends by dipping in solder pot	0.148	0.148
7. Pick up wires and bundle securely (No lacing) (50 wires/bundle)	0.380	0.440
Total	1.528	1.657

17.3. STAKE TAPER PIN TO WIRE

Set-Up—Machine:

This is for AMP-type machine taper pins supplied in a continuous chain or a reel. In one staking action the machine automatically cuts the pin free from the chain, stakes it to the wire, and positions the next pin to the anvil. A new set-up is required only if the taper pin size is changed. (Normally, no change in size is necessary.)

      Set-up time 24 minutes

Run Time—Machine:

	First	Second
	Minutes/End	
Separate wire from group	0.006	—
Grasp	0.003	0.003
Position to taper pin on anvil	0.025	0.025
Press foot pedal and machine cycle	0.012	0.012
Remove wire from anvil and aside	0.006	0.006
Repeat for second end		
Total—each end	0.052	0.046

Run Time—Hand:

	First	Second
	Minutes/End	
Pick up, position pin to pliers	0.052	0.052
Pick up, position wire to pin	0.046	0.046

WIRE PREPARATION

	First	Second
	Minutes/End	
Squeeze	0.017	0.017
Release pliers, aside wire	0.012	0.012
Repeat for second end		
Total—each end	0.127	0.127

17.4. CUT TO LENGTH BUS WIRE AND SLEEVING

Set-Up—Machine:

    Artos type bus cutter (for bus wire)
    Bus cutter or band saw (for sleeving)
      Average per 10 on list   6.90 minutes
      Average per wire   0.690 minutes

Run Time—Machine   0.006 minutes/unit
Run Time—Hand

    Cut to length with pliers   0.023
    Tool Handling   0.035

17.5. CUT AND BEND RESISTORS, CAPACITORS, AND OTHER COMPONENTS

Set-Up—Machine:
  Pig Tailor machine

Cut and bend ends simultaneoulsy. Set blade positions.
Average per component   0.75 minutes

Single-lead trimming machine
  Cuts only one lead at a time.
  Turn hole plate to desired lead length
    Set-up Time   0.25 minutes

Run Time—Machine

	Minutes/Unit	
	Resistors Capacitors	Transistors
Pig Tailor		
Pick up component and assemble to machine	0.029	

CUT TO LENGTH COAXIAL AND SHIELDED CABLE

	Minutes/Unit	
	Resistors Capacitors	Transistors
Press foot pedal and cycle time	0.012	
Remove from machine	0.017	
Total	0.058	
Single-Lead Trimmer		
P/U component	0.029	0.029
Insert 1st lead to cutter and trim	0.017	0.017
Insert 2nd lead to cutter and trim	0.017	0.017
Insert 3rd lead to cutter and trim		0.017
Total	0.063	0.080
Run Time—Hand		
Cut to length with pliers		
two ends at 0.025 in.	0.058	
three ends at 0.025 in.		0.086
Tool handling	0.035	0.035
Sub Total	0.093	0.121
Bend leads per PC board mounting holes	0.046	0.069
Total	0.139	0.190

17.6. CUT TO LENGTH COAXIAL AND SHIELDED CABLE ON BENCH CUTTER

Run Time—Hand	
Pull wire from reel, pass under cutter blade to stop	0.062
Step on treadle to cut	0.014
Aside wire to bench	0.009
Total	0.085 minutes
Add per additional inch:	
1 to 48 in.	0.005 minutes/inch
49 in. and up (2-person operation)	0.009 minutes/inch
Example—for 15 in. of wire	
Allowance for first piece	0.085
Plus 0.005 minutes per in. or 15×0.005	0.075
Time to cut	0.160 minutes

224   **WIRE PREPARATION**

Example—for 60 in. piece
    Allowance for first piece   0.085
    Plus 0.009 minutes per in. or 60 × 0.009   0.540
    Time to cut   0.625 minutes

17.7. PREPARATION SHIELDED CABLE WITH GROUND LEAD

The following is data for plastic-covered shielded cable.

Run Time—Hand:

	Minutes/End	
	One Conductor	Additional Conductor
Strip plastic cover 2 in. with hand stripping tool *0.035 + 0.173	0.208	
Push back braid	0.046	
Pick hole with pick 0.035 + 0.069	0.104	
Pull conductor through hole	0.345	0.345
Strip conductor 0.035 + 0.063	0.098	0.063
Cut to length pigtail 0.035 + 0.035	0.070	
Ground Lead:		
Cut to length	0.035	
Strip ends 2 at 0.063	0.126	
Tin ends 2 at 0.046	0.092	
Prorate tool handling 0.104/10	0.010	
Splice to braid	0.288	
Solder on bench iron	0.092	
Dress back along cable	0.035	
Pick up sleeve with pliers from solvent 0.035 + 0.023	0.058	
Assemble sleeve over braid	0.150	
Lay aside	0.035	
Total (1) Conductor, (1) End	1.792	0.408
Total (2) Conductor, (1) End	2.200	
Total (3) Conductor, (1) End	2.608	

*First value 0.035 or 0.046 is tool handling.

17.8. PREPARATION COAXIAL CABLE

Construction. Center conductor, center insulation, braid shield, outer insulation.

Size. Time values are for the miniature type ⅛" diameter to the standard sizes 3/16 to 5/16" diameter.

17.8.1. Strip and Tin Coaxial Cable

	Minutes/End	
	One Conductor	Additional Conductor
Measure and strip cover 2 in. with hand stripping tool	0.208	0.063
Strip conductor 0.035 + 0.063	0.098	0.046
Tin conductor-pot dip	0.046	
	0.352	0.109

17.8.2. Assemble Ground Lead To Coaxial Cable

		Minutes/End (One Conductor)
Comb 1 in. of braid out to parallel strands		0.403
Trim excess and form pigtail		0.173
Cut to length		0.035
Strip end with hand tool	2 at 0.063	0.126
Tin end-pot dip	2 at 0.046	0.092
Prorate tool handling	0.104/10	0.010
Splice to braid		0.288
Solder on bench iron		0.092
Dress back along cable		0.035
Pick up sleeve with pliers from solvent	0.035 + 0.023	0.058
Assemble sleeve over braid (tight)		0.150
Lay aside		0.035
		1.497 minutes

17.8.3. Assemble RF Connector with Nut, Grommet, and Washer, to Coaxial Cable

	Minute/End
Cut plastic bag to unwrap connector and contents (tool handling 0.035 cut 0.035)	0.070
Disassemble Connector:	
Position connector to fixture	0.115
Tool handling open end wrench	0.035
Disassemble nut	0.253
Disassemble connector, from fixture	0.075
Disassemble with pick tool handling	0.035
Washer	0.046
Grommet	0.046
Ferrule	0.046
Sub Total	0.721
Assemble Cable to Connector:	
Pick up and position cable	0.035
Thread	
Nut	0.115
Washer	0.115
Grommet (tight)	0.207
Dress braid forward and together	0.045
Thread ferrule (tight)	0.322
Fan braid over ferrule	0.190
Pick up terminal with pliers (0.030 + 0.050)	0.092
Fill with solder	0.161
Solder conductor terminal	0.115
Cool and line up terminal	0.184
Dress braid over ferrule	0.045
Sub Total	1.626
Reassemble Connector:	
Assemble connector over cable and terminal	0.069
Line up terminal. Press tight	0.230
Press grommet into connector (tight)	0.633
Press washer into connector	0.115
Press nut into connector and seat	0.045
Run down nut finger tight	0.161
Position to fixture	0.069

ASSEMBLE PLUG OR CONNECTOR TO HARNESS ASSEMBLY

	Minute/End
Lock second arm	0.045
Tighten nut with wrench (0.030 + 0.280)	0.357
Disassemble from fixture	0.075
Lay aside cable	0.023
Sub Total	1.822
Grand Total	4.169 minutes

17.9. STAMP AND ASSEMBLE IDENTIFICATION TAPE TO CABLES

	Reference Section	Set-Up Time (Minutes)	Minutes Per Tape
Stamp Tape: Same machine as wire stamping. Set up.	17.1	2.52	0.104
Cut to length tape (0.030 + 0.050)			0.092
Assemble to connector	—	—	0.138
Total		2.52	0.334

17.10. ASSEMBLE PLUG OR CONNECTOR TO HARNESS ASSEMBLY

Stamp and assemble identification tape.

17.10.1. Round Connector, 90° Shell

For an AN type (Amphenol, Cannnon, etc.) with a base with hollow terminals and shell bends wires at 90° to base.

	Minutes		
Disassemble and position to cable:	Small 5−10 Pin	Medium 11−25 Pin	Large 26−40 Pin
P/U. connector, unwrap.	0.115	0.115	0.115
Disassemble screw. 2 at 0.10	—	0.230	0.230
L/A ring and (2) halves. 3 at 0.05	0.173	0.173	0.173
Assemble connector base to fixture	0.058	0.058	0.058

228 WIRE PREPARATION

	Minutes		
	Small 5–10 Pin	Medium 11–25 Pin	Large 26–40 Pin
Clamp:			
Disassemble screw. 2 at 0.10	—	0.230	0.230
L/A (2) halves 2 at 0.105	—	0.115	0.115
Grommet assemble to cable clamp	0.230	0.345	0.345
Thread to cable:			
Clamp with grommet and washer (tight)	0.345	0.576	0.104
Connector mounting ring	0.058	0.058	0.058
Sub Total	0.979	1.900	1.428
Button up after wiring:			
Form and dress wires	0.115	0.230	0.460
Tie wires together	0.345	0.345	0.690
Assemble 90° shell halves	0.230	0.460	0.920
Seat and run down mounting ring	0.230	0.230	0.230
Seat clamp and grommet, run down	0.345	0.690	1.270
Assemble clamp halves 2 at 0.10	0.230	0.230	0.230
Assemble screw 2 at 0.15	0.173	0.345	0.345
Identification tape, stamp and assemble	0.345	0.345	0.345
Disassemble from fixture—L/A cable	0.115	0.115	0.230
Sub Total	2.128	2.990	4.720
Total per cable assemble	3.107	4.890	6.148
Add for each previously dressed conductor installed into connector	0.34 minutes/conductor installed		

17.10.2. Rectangular Connector 90° Cover

Cinch Jones type. Base with flat hole type terminals. 90° cover secured to base with drive pins.

	Minutes	
	Small 5–10 Pin	Medium 10–25 Pin
Disassemble and position to cable:		
Pick up, unwrap.	0.115	0.115
Disassemble screw at 0.10	0.115	0.230

	Minutes	
	Small 5−10 Pin	Medium 10−25 Pin
Aside clamps at 0.05	0.115	0.115
Disassemble pin with hammer and pliers at 0.20	0.230	0.460
Disassemble cover	0.115	0.115
Assemble connector to fixture	0.158	0.058
Assemble grommet to cover	0.230	0.345
Thread cover with grommet to cable-wire exit at 90° to base	0.345	0.460
Sub Total	1.423	1.898
Button up after wiring:		
Form and dress wires	0.115	0.230
Tie wires together	0.345	0.345
Seat cover and grommet to base	0.345	0.690
Assemble pins at 0.20	0.230	0.460
Assemble clamps at 0.05	0.115	0.115
Assemble screw at 0.15	0.345	0.345
Identification tape—stamp and assemble	0.345	0.345
Disassemble from fixture and lay aside cable	0.115	0.115
Sub Total	1.955	2.645
Total per Cable Assembly	3.378	4.543

17.11. BUZZ WIRE TO IDENTIFY CONTINUITY

Reach for leads and clip one to wire or terminal
Touch unidentified ends with second lead until buzzer sounds

Set up per assembly   0.10 minutes
Per wire in assembly   0.01 minutes

17.12. PULL TUBING OVER CABLE

	Constant	Add per Ft.
Values for medium tight fit of tubing over cable wires—1 to 2 ft. cable:		
Cut length of tubing	0.115	—
Thread wires through tubing	—	0.748
Total	0.115	0.748
3 to 5 ft. cable:		
Cut length of tubing	0.115	0.058
Tie end wires together	0.460	—
Thread wires through tubing	—	0.748
Total	0.575	0.806
6 to 15 ft. cable:		
Cut length of tubing	0.115	0.058
Anchor cable connector	0.230	—
Tie cord around end wires	0.575	—
Thread cord through tubing (blow through or thread on wire)	—	0.115
Anchor threaded cord	0.345	—
Grease cable wires	0.115	0.058
Pull tubing over wires	—	0.748
Total	1.380	0.979
If heat shrink tubing is used add for shrinking		
Set up	0.117	
Shrink time		1.782

17.13. TWIST WIRES

	Minutes
Per twist	0.02
Per in.—small twisted pairs at 1/in.	0.02
Per ft.—long cables 2–10 ft. at 4 Twists/ft.	0.09

17.14. SPOT TIE HARNESS

Tie with cord, pick up cord, tie clove hitch and overhand knot, cut excess cord with scissors.

Per Tie	0.383
Per Foot	1.149

Tie with nylon wire ties using hand gun.

                       Per tie   0.23
                       Per foot   0.69

17.15. HARNESS FABRICATION

A detailed table of the elements and time values required to form and lace a harness on a nail board is given in Tables 17.2 and 17.3. Table 17.2 shows a sample application for wires up to 15 in. long while Table 17.3 shows the same for wires up to 60 in. long. Some explanatory notes to the table elements follow:

1. Pick up pre-cut wire from rack.
   The total number of in. of all wires in the cable must be multiplied by 0.002 minutes/in. This allows for the extra motions required to pick up and preposition longer wires, for example, 100 wires 10 in. long—2.00 minutes per cable added to the basic pick-up time.
2. Fasten wire to board.
   Spring: press wire end into coils to hold.
   Clip: open alligator clip, insert wire, release clip.
   Hole: thread wire to hole in board and press down to secure.
3. Twist pairs.
   Twisted pairs average one twist per inch, less first and last inch.
4. Lay wire along nail pattern.
   Per inch of wire in cable—0.003 minutes, for example, 100 wires 10 in. long = 3.00 minutes per cable to dress and wire along the cable pattern.
5. Bend wire around nail.
   Each right angle bend for either a wire break out or sharp bend in the harness warrants a "bend" allowance.
6. Tie cord to cable and cut with scissors.
   Prior to lacing, the cord is anchored to the cable with a double knot as follows:
   Pull cord from spool and cut with scissors.
   Form loop and thread under cable.
   Twist loop, thread one end of cord through loop and pull tight.
   Thread single cord under cable.
   Tie first knot.
   Tie second knot.
7. Lace cable per stitch.
   Thread cord under wire.
   Form loop and twist for lock stitch.
   Thread cord through loop.
   Pull tight to cable.

TABLE 17.2. Form and Lace Harness on Nail Board. Sample Application for Wires up to 15 in. Long.

Element		Minutes/ Element	Number of Element[a]	Total Minutes[a]
1.	Pick up wire from bin:			
	Up to 20 × 20 in. harness	0.025		
	Over 20 × 20 in. harness	0.029	50	1.45
	Add per inch of wire in cable (ave.)	0.002	650	1.30
2.	Fasten wire to board per end: spring	0.015	100	1.50
	clip	0.022		
	hole	0.035		
3.	Twist pairs per twist	0.017		
4.	Lay wire along nail pattern:			
	Per inch of wire in harness	0.003	650	1.95
5.	Bend wire around nail: to 18 gauge	0.012	125	1.50
	(average 2½ bend/wire) 16 gauge	0.023		
	14 gauge	0.035		
6.	Tie cord to cable and cut with scissors	0.288	6	1.73
7.	Lace harness per stitch (36″ at ½″ per lace)			
	Single cord— to ¾″ diam. cable	0.081	72	5.83
	over ¾″ diam. cable	0.092		
	Double cord— to ¾″ diam. cable	0.092		
	over ¾″ diam. cable	0.104		
	Add obstruction allowance:			
	Per nail on board	0.006	120	0.72
	Per congested cable junction	0.058	2	0.12
8.	Brush cement to knot	0.050		
	Handle brush	0.055		
9.	Release wire from board per end: spring	0.006	100	0.60
	clip	0.017		
	hole	0.006		
10.	Pull cable from nails per pull	0.009	5	0.05
11.	Lay aside harness: to 20 × 20 in.	0.058	1	0.06
	to 40 × 40 in.	0.115		
	Total minutes per harness			17.114
	Total minutes per wire at 50 wires			0.342

[a]Number of elements and total minutes are shown as an example only.

8. Brush cement to knot.
This value can be used for painting both cement or color spots to wires, harnesses, and components.
Brush handling is to and from a can with a small hole to fit the brush handle.

9. Release wire from board per end.
Spring-pull wire end from between spring coils.

TABLE 17.3. Form and Lace Harness on Nail Board. Sample Application for Wires up to 60 in. Long

Element		Minutes/ Element	Number of Element[a]	Total Minutes[a]
1.	Pick up wire from bin:			
	Up to 20 × 20 in. harness	0.025		
	Over 20 × 20 in. harness	0.029	50	1.45
	Add per inch of wire in cable (ave.)	0.002	2000	4.00
2.	Fasten wire to board per end: spring	0.015	100	1.50
	clip	0.022		—
	hole	0.035	2000	70.00
3.	Twist pairs per twist	0.017		
4.	Lay wire along nail pattern:			
	Per inch of wire in harness	0.003		
5.	Bend wire around nail: to 18 gauge	0.012	125	1.50
	(average 2½ bend/wire) 16 gauge	0.023		
	14 gauge	0.035		
6.	Tie cord to cable and cut with scissors	0.288	18	5.18
7.	Lace harness per stitch (36″ at ½″ per lace)			
	Single cord— to ¾ in. diameter cable	0.081	145	11.75
	over ¾ in. diameter. cable	0.092		
	Double cord—to ¾ in. diameter cable	0.092		
	over ¾ in. diameter cable	0.104		
	Add obstruction allowance:			
	Per nail on board	0.006	360	2.16
	Per congested cable junction	0.058	6	0.35
8.	Brush cement to knot	0.050		
	Handle brush	0.055		
9.	Release wire from board per end: spring	0.006	100	0.60
	clip	0.017		
	hole	0.006		
10.	Pull cable from nails per pull	0.009	15	0.14
11.	Lay aside harness: to 20 × 20 in.	0.058		
	to 40 × 40 in.	0.115	1	0.12
	Total minutes per harness			35.75
	Total minutes per wire at 50 wires			0.715

[a]Number of elements and total minutes are shown as an example only.

    Clip-press alligator clip and remove wire.
    Hole-pull wire from hole in board.
10. Pull cable from nails per pull.
    Pull each leg of cable free from nails using both hands simultaneously.
11. Lay aside cable.
    To tote box at side of operator.

TABLE 17.4. Fabricate Harness Nail Board

	Number Wires in Harness	Set-Up Constant (Minutes)	Minutes/ Wire	Calculated Total Minutes
Average				
Harness lays in one plane	10–49	172	4.86	
	50–199	242	4.14	
	200–399	276	3.48	
Complex				
Disconnect clusters; harness lays in two planes	50–199	311	6.21	
	200–399	345	4.86	

17.16. FABRICATE HARNESS NAIL BOARD

(See Table 17.4.)

Wire list or equal is supplied. Time estimate includes:

1. Plan harness layout, routings, wire lengths.
2. Make harness drawing with wire destinations.
3. Prepare wood jib board with nails, holes, springs, clips, and so forth.

Machines directed by minicomputers are available for fabricating harnesses. The complete machine cost is about $25,000 each. The computers may be programmed to prompt the assembly person with instructions such as "connect wire to terminal 7B and run to terminal 9C." When the two connections are made, the computer automatically performs a test to be sure that the connections have been made correctly before displaying the next instruction. Because of the capital investment and the lengthy set-up time, this system is not normally used for small production runs.

SECTION EIGHTEEN

WIRING AND COMPONENT INSERTION

18.1.	Insulated Wire: Prepare and Install (2) Ends	236
18.1.1.	Crimp or Butt Solder Termination	236
18.1.2.	Taper Pin Termination (Solderless)	238
18.1.3.	Pneumatic Wrap Termination (Solderless)	239
18.2.	Bus Wire-Cut, Crimp, Solder (2) Ends	240
18.3.	Resistor, Capacitor, Transistor: Prepare and Install	240
18.4.	Sleeving—Prepare and Assemble	241
18.5.	Shielded Cable: Prepare and Install (2) Ends	241
18.6.	Coaxial Cable: Prepare and Install (2) Ends	242
18.7.	Crimp Wire to Terminal: Detail Values	243
18.8.	Crimping Obstruction Allowances	245
18.9.	Wire Dressing Values	246
18.10.	Select Wire from Group	246
18.11.	Tool Handling Values	246
18.12.	Develop Wire List from Schematic	247
18.13.	Set-Up Components or Wire at Work Station	247

The first half of this section develops the quick application estimating values which are used on the Summary Table 18 in Section 3. The second half consists of elemental breakdowns for wire crimping, dressing, and miscellaneous wiring and component operation. The values in panentheses on the detail tables are those which have been used in the quick application summaries. Both machine and hand methods are defined for use with large or small production quantities.

18.1. INSULATED WIRE: PREPARE AND INSTALL (2) ENDS

The total time of all operations required to hook up a wire is summarized in the following tables. Typical operations and elements from the sections on wire preparation, soldering, and detail crimping have been combined to give quick application estimating values. The times are based on 18–24 gauge wire and solid terminals where conventional crimping and soldering are called for. The times may be adjusted to a specific wire size, terminal size, stripping method, and so on to utilize the detail tables listed in the reference column. Definitions of some of the table headings are given below.

Point-to-Point Wiring. Connect first end of wire to unit, connect second end, dress and bend wire to desired contour of terminal board, chassis, and so forth.

Crimp and Solder. Connect wire to a solid or hole-type terminal by wrapping with long nose pliers and soldering with a hand solder iron.

Butt Solder. Connect wire to a hollow connector terminal by first prefilling the terminal with solder, then reheating and inserting the wire. Insulating sleeving is usually assembled over the wire and terminal at the joint.

Lay in U-Channel. This is the plastic-type channeling which is used in lieu of a laced harness, usually for six-ft. cabinet wiring. The wires lie loose in the channel and are dressed out through the side notches to their destination. Additional time must be allowed for the mechanical assembly of the channel to the rack and for assembling the lid on the channel after the wires are in place.

Lace Wire into Harness. The time values are averages for lacing a harness on a nail board.

18.1.1 Crimp or Butt Solder Termination

Reference Section	Operation (Run time values) in Minutes)	Machine Preparation Set-Up Minutes	Length 15 in.	60 in.	Hand Preparation Set-Up Minutes	Length 15 in.	60 in.
	Point to point—crimp and solder						
17.1 & 2	Cut, strip, twist, tin, ID stamp	5.04	.250	.544	2.66	1.528	1.65
18.7	Crimp 1st end		.145	.145		.145	.14
18.8	Ave. obstruction allow.		.012	.012		.012	.01

INSULATED WIRE: PREPARE AND INSTALL (2) ENDS

Reference Section	Operation (Run time values) in Minutes)	Machine Preparation			Hand Preparation		
		Set-Up Minutes	Length 15 in.	Length 60 in.	Set-Up Minutes	Length 15 in.	Length 60 in.
18.7	Crimp 2nd end		.122	.122		.122	.122
18.8	Ave. obstruction allow.		.012	.012		.012	.012
18.11	Tool handling .038/5		.007	.007		.007	.007
18.9	Dress wire at .002/inch		.030	.120		.030	.120
19.1	Solder joint 2 at .067		.134	.134		.134	.134
18.11	Tool handling .052/5		.010	.010		.010	.010
	Total per wire	5.04	.722	1.106		2.000	2.219
	Point to point—butt solder						
17.1 & 2	Cut, strip, twist, tin, ID stamp	5.04	.258	.544	2.66	1.528	1.657
19.1	Fill hollow term. with solder 2 at .081		.162	.162		.162	.162
18.11	Tool handling .052/5		.010	.010		.010	.010
18.4	Sleeving—prep. and assemble 2		.240	.240		.344	.344
19.1	Reheat and insert 2 at .092		.184	.184		.184	.184
18.11	Tool handling 0.52/5		.010	.010		.010	.010
18.9	Dress wire at .002/inch		.030	.120		.030	.120
	Total per wire	5.04	.894	1.270	2.66	2.268	2.487
	Lay wire in U channel						
	Prep., crimp, solder (above)		.722	1.106		2.000	2.219
18.9	Dress wire to channel at .003/inch		.045	.180		.045	.180
	Total per wire		.767	1.286		2.045	2.399

Reference Section	Operation (Run time values) in Minutes)	Machine Preparation			Hand Preparation		
		Set-Up Minutes	Length 15 in.	60 in.	Set-Up Minutes	Length 15 in.	60 in.
	Lace wire into harness						
	Prep., crimp, solder (above)	5.04	.722	1.106		2.000	2.219
17.15	Lace harness (pro rata)		.342	.719		.342	.719
	Total per wire	5.04	1.064	1.825		2.342	2.938

18.1.2. Taper Pin Termination (Solderless)

Reference Section	Operation (Run Time Values) In Minutes)	Machine Preparation			Hand Preparation	
		Set-Up Minutes	Length 15 in.	60 in.	Length 15 in.	60 in.
	Point to point					
17.1 and 2	Cut, strip, twist, tin, ID stamp	7.59	0.250	0.551	0.637	0.758
17.3	Stake (2) taper pins to wire		0.101	0.101	0.288	0.288
18.7	Assemble to female term—1st end		0.079	0.079	0.079	0.079
	2nd end		0.071	0.071	0.071	0.071
18.11	Tool handling 0.032/5		0.007	0.007	0.007	0.007
18.9	Dress wire at 0.002/in.		0.035	0.140	0.035	0.140
	Total per wire	7.59	0.543	0.949	1.117	1.343
	Lay wire in U channel					
	Prep, stake, assemble (above)	7.59	0.543	0.948	1.117	1.341
18.9	Dress wire to channel at .003/in.		0.052	0.207	0.052	0.207
	Total per wire	7.59	0.595	1.155	1.169	1.548

INSULATED WIRE: PREPARE AND INSTALL (2) ENDS

Reference Section	Operation (Run Time Values) In Minutes	Machine Preparation			Hand Preparation	
		Set-Up Minutes	Length		Length	
			15 in.	60 in.	15 in.	60 in.
	Lace wire into harness					
	Prep., stake, assemble (above)	7.59	0.543	0.948	1.117	1.341
17.15	Lace harness (pro rata)		0.343	0.719	0.343	0.719
	Total per wire	7.59	0.886	1.667	1.460	2.060

18.1.3. Pneumatic Wrap Termination (Solderless)

Reference Section	Operation (Run Time Values) In Minutes	Machine Preparation			Hand Preparation	
		Set-Up Minutes	Length		Length	
			15 in.	60 in.	15 in.	60 in.
17.1 & 2	Point to point Cut, strip, ID stamp (solid wire)	7.59	0.173	0.403	0.495	0.615
18.7	Pneumatic wrap to term—1st end		0.067	0.067	0.067	0.067
	2nd end		0.048	0.048	0.048	0.048
18.11	Tool handling 0.036/5		0.008	0.008	0.008	0.008
18.9	Dress wire at 0.002/in.		0.035	0.140	0.035	0.140
	Total per wire	7.59	0.331	0.666	0.653	0.878
	Lay wire in U channel					
	Prep and wrap to terminal (above)	7.59	0.331	0.664	0.653	0.876
18.9	Dress wire to channel at 0.003/in.		0.052	0.207	0.052	0.207
	Total per wire	7.59	0.383	0.871	0.705	1.083
	Lace wire into harness					
	Prep and wrap to terminal (above)	7.59	0.331	0.664	0.653	0.876
17.15	Lace harness (pro rata)		0.343	0.719	0.343	0.719
	Total per wire	7.59	0.674	1.383	0.996	1.595

240 WIRING AND COMPONENT INSERTION

18.2. BUS WIRE-CUT, CRIMP, SOLDER (2) ENDS

Reference Section		Set-Up Time	Minutes
17.4	Cut to length	6.90 minutes	0.032
18.7	Crimp 1st end		0.136
18.8	Average obstruction allowance		0.012
18.7	Crimp 2nd end		0.112
18.8	Average obstruction allowance		0.012
18.11	Tool handling 0.038/5		0.007
19.1	Solder joint 2 at 0.054		0.108
18.11	Tool handling 0.054/5		<u>0.011</u>
	Total per bus wire		0.430

18.3. RESISTOR, CAPACITOR, TRANSISTOR: PREPARE AND INSTALL

Reference Section	Operation (Run time value in minutes)	Set-Up (Minutes)	Machine Preparation		Hand Preparation	
			(2) Ld. Comp.	(3) Ld. Comp.	(2) Ld. Comp.	(3) Ld. Comp.
	Crimp and solder to terminal:					
17.5	Cut to length	0.75	0.058	0.063	0.093	0.138
18.7	Crimp 1st end		0.136	0.136	0.136	0.136
18.8	Average obstruction		0.012	0.012	0.012	0.012
	Crimp 2nd end		0.112	0.112	0.112	0.112
	Average obstruction		0.012	0.012	0.012	0.012
	Crimp 3rd end		—	0.112	—	0.112
	Average obstruction		—	0.012	—	0.012
18.11	Tool handling 0.038/5		0.007	0.007	0.007	0.007
18.9	Dress component (ave.)		0.038	0.057	0.038	0.057
19.1	Solder joint at 0.054		0.108	0.162	0.108	0.162
18.11	Tool handle 0.054/5		<u>0.011</u>	<u>0.011</u>	<u>0.011</u>	<u>0.011</u>
	Total per component	0.75	0.494	0.696	0.529	0.771

SHIELDED CABLE: PREPARE AND INSTALL (2) ENDS

Reference Section	Operation (Run time value in minutes)	Set-Up (Minutes)	Machine Preparation		Hand Preparation	
			(2) Ld. Comp.	(3) Ld. Comp.	(2) Ld. Comp.	(3) Ld. Comp.
	Crimp and dip solder to PC board:					
17.5	Cut to length	0.75	0.058	0.080	0.138	0.190
	and bend	0.75	0.058	0.080	0.138	0.190
18.7	Crimp 1st end to board		0.123	0.123	0.123	0.123
	Crimp 2nd end to board		0.080	0.080	0.080	0.080
	Crimp 3rd end to board		—	0.070	—	0.070
18.11	Tool handle 0.038/5		0.007	0.007	0.007	0.007
19.3	Dip solder (prorated)		0.044	0.044	0.044	0.044
	Total per component	0.75	0.370	0.484	0.530	0.704

18.4. SLEEVING—PREPARE AND ASSEMBLE

Reference Section		Minutes	
		Machine	Hand
17.4	Cut to length sleeving	0.006	0.058
	Thread to component lead or wire	0.057	0.057
	Total for component lead	0.063	0.115
	Add to push down over connector terminal after butt soldering and inspection	0.057	0.057
	Total for connector wire	0.120	0.172
	Double for (2) ends	0.240	0.344

18.5. SHIELDED CABLE: PREPARE AND INSTALL (2) ENDS

Reference Section	Operation	Hand Preparation/Minutes	
		15 in.	60 in.
	Single conductor with ground lead:		
17.6	Cut to length	0.300	0.517
17.7	Prepare end with ground lead 2 at 1.792	3.584	3.584
18.7	Crimp center conductor 2 at 0.150	0.300	0.300

242 WIRING AND COMPONENT INSERTION

Reference Section	Operation	Hand Preparation Minutes	
		15 in.	60 in.
18.8	Obstruction allow. 2 at 0.012	0.024	0.024
18.7	Crimp ground lead 2 at 0.150	0.300	0.300
18.8	Obstruction allowance 2 at 0.012	0.024	0.024
18.11	Tool handling 0.038/3	0.013	0.013
18.9	Dress cable per in. at 0.002	0.020	0.096
19.1	Solder joint 4 at 0.067	0.268	0.268
18.11	Tool handling 0.054/3	0.018	0.018
	Total per cable	4.851	5.144
	Double conductor with ground lead:		
17.6	Cut to length	0.300	0.517
17.7	Prepare end with ground lead 2 at 2.200	4.400	4.400
18.7	Crimp center conductor 4 at 0.150	0.600	0.600
18.8	Obstruction allow. 4 at 0.012	0.048	0.048
18.7	Crimp ground lead 2 at 0.150	0.300	0.300
18.8	Obstruction allow. 2 at 0.012	0.024	0.024
18.11	Tool handling 0.038/2	0.019	0.019
18.9	Dress cable per in. at 0.002	0.030	0.120
19.1	Solder joint 6 at 0.067	0.402	0.402
18.11	Tool handling 0.054/2	0.027	0.027
	Total per cable	6.150	6.457

18.6. COAXIAL CABLE: PREPARE AND INSTALL (2) ENDS

Reference Section	Operation	Set-Up Minutes	Hand Preparation (Minutes)	
			15 in.	60 in.
	Ground lead termination:			
17.6	Cut to length		0.130	0.517
17.8.1	Strip and tin center conductor	2 at 0.352	0.704	0.704
17.8.2	Assemble ground lead	2 at 1.500	3.014	3.014
18.7	Crimp center conductor	2 at 0.150	0.300	0.300

CRIMP WIRE TO TERMINAL: DETAIL VALUES

Reference Section	Operation		Set-Up Minutes	Hand Preparation (Minutes)	
				15 in.	60 in.
18.8	Obstruction allowance	2 at 0.012		0.024	0.024
18.7	Crimp ground lead	2 at 0.150		0.300	0.300
18.8	Obstruction allowance	2 at 0.012		0.024	0.024
18.11	Tool handling	0.038/3		0.019	0.019
18.9	Dress cable per inch	at 0.002		0.020	0.096
19.1	Solder joint w/hand Iron	4 at 0.067		0.268	0.268
18.11	Tool handling 0.054/3			0.018	0.018
	Total per cable		0	4.821	5.284
	Connector termination:				
17.6	Cut to length			0.130	0.517
17.8.1	Strip and tin center conductor	2 at 0.352		0.704	0.704
17.8.3	Assemble RF connection, grommet and so on	2 at 4.169		8.338	8.338
17.9	Stamp and Assemble ID tape to cable		2.52	0.334	0.668
	Assemble cable and connectors to mating unit	2 at 0.069		0.138	0.138
18.9	Dress cable per inch	at 0.002		0.020	0.096
	Total per cable		2.52	9.664	10.461
	Connector 1st end: ground lead 2nd end:				
	Connector end (½ above)			4.832	5.231
	Ground lead end (½ above)		0	2.411	2.642
	Total per cable		2.52	7.243	7.873

18.7. CRIMP WIRE TO TERMINAL: DETAIL VALUES

The table values of Table 18.1 constitute one element of the wiring summaries given so far in this section.

The values apply to all wire crimping regardless of whether the wire originates from a transformer, resistor, or another wire, and regardless of where the wire is being hooked up. The time for a specific connection is synthesized by choosing the appropriate crimping value from the table and adding obstruction, dress, selection, and tool handling allowances as required by the connection.

An average pick up distance of 15 in. is used for the first-end values, and 8 in.

TABLE 18.1. Crimp Wire to Terminal: Detail Values

TYPE WIRE	WIRE GAGE	WIRE END	SOLID SHOULDER TERM.			SOLID STRAIGHT TERM.		
			NO. WIRES ON TERM.			NO. WIRES ON TERM.		
			1st	2nd	3rd	1st	2nd	3rd
BARE WIRE (BUS, RESISTOR, ETC.)	18-24	1st	.129	.136	.143	.140	.147	.154
		2nd	.105	.112	.118	.116	.123	.131
	16	1st	.136	.143	.150	.147	.154	.161
		2nd	.113	.110	.127	.124	.131	.138
INSULATED (ALGN INSULATION 1/32-1/16" FROM TERMINAL).	18-24	1st	.138	.145	.152	.150	.156	.163
		2nd	.115	.122	.129	.127	.133	.140
	16	1st	.146	.153	.160	.158	.164	.171
		2nd	.123	.130	.137	.135	.141	.148

TYPE WIRE	WIRE GAGE	WIRE END	HOLE TYPE TERM.			PC BOARD MTG. HOLE		
			HOLE CLEARANCE			HOLE CLEARANCE		
			Loose	Med.	Tight	Loose	Med.	Tight
BARE WIRE (BUS, RESISTOR, ETC.)	18-24	1st	.127	.133	.138	.116	.123	.128
		2nd	.107	.114	.118	.074	.080	.085
	16	1st	.133	.140	.145	.123	.130	.135
		2nd	.115	.122	.127	.080	.087	.092
INSULATED (ALGN INSULATION 1/32-1/16" FROM TERMINAL).	18-24	1st	.136	.143	.147	.125	.132	.137
		2nd	.117	.124	.129	.083	.090	.094
	16	1st	.144	.151	.155	.133	.140	.145
		2nd	.125	.132	.137	.091	.098	.102

TYPE WIRE	WIRE END	TAPER PIN ASSEMBLE TO FEMALE TERM.	PNEUMATIC WRAP
ALL WIRES AND GAGES	1st	.079	.067
	2nd	.071	.048

for grasping the second end. The wire dressing values allow for the additional moves on long wires. A description of the table heading is given below.

Solid Shoulder, or Straight Terminal. Pick up cut, stripped, and tinned wire or component with one hand while holding long nose pliers in the other hand. Position wire to terminal, grasp with pliers and wrap ¾ turn, regrasp, wrap ¾ turn, and squeeze end tight against terminal. Reach for next wire.

Hole Type Terminal. Pick up, stripped and tinned wire or component with one hand while holding long nose pliers in other hand. Thread wire to hole in terminal. Grasp with pliers and wrap ¾ turn, regrasp, wrap ¾ turn, and squeeze end tight against terminal.

Printed Circuit Board Mounting Hole. Pick up component with leads cut and bent to fit mounting holes. Position first end through hole. Position second end through hole. Turn board over with one hand while holding pliers in other hand. Bend both leads against circuit pad with pliers. Turn board right side up again. (For automatic insertion, substitute riveting cycle times from Section 15. Use class II part size and hopper feed cycle in Table 15.2.)

Taper Pin (Solderless Connection). AMP Type (Aircraft Marine Products). Pick up wire with taper pin previously staked to it, while holding insertion tool in other hand. Regrasp wire near pin, preposition flat of pin to slot in tool, and assemble pin to tool. Using tool, press taper pin into female terminal until tool clicks to release. Remove tool while picking up next wire.

Pneumatic Wrap Solderless Connection. (Keller "Wire-Wrap" type tool.) Pick up previously cut and stripped solid conductor wire with one hand while holding air gun in other hand. Insert wire to hole in tool, anchor in slot, and dress wire back along nose piece. Position nose piece over terminal and press trigger to wrap. Remove tool while picking up next wire.

18.8. CRIMPING OBSTRUCTION ALLOWANCES

These allowances are to be added to the basic crimping values of Section 18.7. They are to be used when other terminals, wires, and components impede the normal wrapping and crimping action of long nose pliers.

Closely spaced components and terminals on a board and many congested wires in a chassis are two examples of where the allowances should be used. They are almost never required when assembling taper pin wires with the insertion tool. The clearance dimensions on the following table are for the *least*-obstructed path of motion.

Clearance (Terminal to Obstructing Object)	Allowance (Minutes/Terminal)
To ⅛ in.	0.038
To ¼ in.	0.023
To ⅜ in.	0.012
Over ⅜ in.	0.000

18.9. WIRE DRESSING VALUES

	Minutes
Per move	0.012
Per round bend with finger of hand	0.012
Per square bend with pliers (hand pliers)	0.019
Per in. average for long wires in racks and chassis	0.002
Lay in plastic U channel (per inch)	0.003
Add—Bend wire out through termination slot (per bend)	0.012

18.10. SELECT WIRE FROM GROUP

Time to select one wire equals the total number of wires in a bundle multipled by 0.006 minutes.

Therefore:

1 out of 5 wires	0.030 minutes
1 out of 10 wires	0.060 minutes
1 out of 20 wires	0.120 minutes
1 out of 30 wires	0.180 minutes
1 out of 40 wires	0.240 minutes

or

2 out of 5 wires	0.060 minutes

18.11. TOOL HANDLING VALUES

Reach for tool on bench, pick up, adjust grip if necessary, and set aside to bench.

SET-UP COMPONENTS OR WIRE AT WORK STATION

1. Single handled tools (screwdriver, spin wrench, etc.)   0.035 minutes
2. Two handled tools (pliers, cutters, wire stripper, etc.)   0.038 minutes
3. Pencil tip solder iron from bench stand and return   0.052 minutes
4. 150-watt solder iron from bench stand and return   0.054 minutes
5. Pneumatic wire wrap gun
   - Suspended   0.029 minutes
   - On Bench   0.410 minutes

18.12. DEVELOP WIRE LIST FROM SCHEMATIC

Line out schematic point-to-point connection with crayon. Determine physical point-to-point destinations and routing. Write on wire list:

       Wire Number _____
       To   _____
       From   _____
           Total per wire   2.50 minutes

18.13. SET-UP COMPONENTS OR WIRE AT WORK STATION

The parts bins used at the assembly position are filled and labeled with the schematic symbol number. The operator or set-up worker then takes a tote box filled with bins and sets them up in the prescribed sequence at the work station. The bin is identified, checked with the drawing or methods sheet for assembly sequence, and positioned to the parts rack. To tear down the position, the binds are disassembled from the racks and returned to storage. Labels are removed.

     Fill parts bins and label   2.02 minutes/bin
     Place empty bins in storage
     Move parts bins to work stations and position   0.70 minutes/bin
     Return bins at job end   2.72 minutes/bin

SECTION NINETEEN

SOLDERING

19.1.	Solder Wire to Terminals	248
19.2.	Wire Gauge and Diameter Data	249
19.3.	Dip and Wave Soldering of Circuit Boards	250
19.4.	Spot and Seam Soldering	253

Soldering in this section refers to soft soldering with a 63% tin, 37% lead alloy. Silver soldering or brazing is not applicable to soldering electronic wiring connections. The 63–37 alloy is the eutectic composition, meaning that the melting point is lower than either pure tin or pure lead or any other combination of the two metals. The completely molten and completely solid temperatures for eutectic solder are 365°F and 364°F.

The detail time values of this section are combined with those for wire preparation and crimping in Section 18 to obtain average wiring values for the Summary Table 18 in Section 3. The table values in parentheses have been used for this purpose.

19.1. SOLDER WIRE TO TERMINALS

The time values in Figure 19.1 are based on the following:

Solder—63% tin, 37% lead, 0.062 in. diameter, resin core, approximately ½ inch used per joint.
Solder Iron—100–150 watt copper tip.
Move Distance—Values include an average 4-in. move from joint to joint.

WIRE GAUGE AND DIAMETER DATA

Solder wire to terminals

Wire gage	Number wires on Terminal	Minutes per Wire					
		Solid terminal	Hole terminal	Hollow terminal		Solder to base/chassis	
				Prefill	Heat insert lead	Hold to cool	No cooling time
26 - 30	1	.039	.031	.072	.081	.140	.081
	2	.052	.037				
	3	.056	.048				
18 - 24	1	.054	.046	.081	.092	.155	.092
	2	.067	.052				
	3	.070	.063				
16	1	.069	.061	.092	.104	.170	.107
	2	.082	.068				
	3	.089	.082				
14	1	.084	.076	.104	.115	.185	.122
Spot solder	0						.081
Braid pigtail	1					.190	.115

FIGURE 19.1. Solder wire to terminals.

Hollow Terminals—These are the type used on cable connectors.

Tool Handling—The table values do not include the pick up and lay aside of the solder iron or solder.

Add per occurrence:

Pick up solder iron and solder.	0.030
Lay aside iron and solder.	0.024
Total	0.054 minutes
Install and remove heat sink from one position.	0.09 minutes

19.2. WIRE GAUGE AND DIAMETER DATA

The following gauges and diameters are typical of those used for electronic wiring. Soldering time is influenced significantly by the increased mass of the heavier gauge wires.

American Wire Gauge (A.W.G.)	Conductor Diameter for Solid Wire (in.)
30	0.0100
28	0.0126
26	0.0159
24	0.0201
22	0.0253
20	0.0320
18	0.0403
14	0.0641
12	0.0808

19.3. DIP AND WAVE SOLDERING OF CIRCUIT BOARDS

Semi-Automatic Operations:

Masking. Boards are hand loaded between feed belts which pass a tape dispenser. Tape is applied along the top and/or bottom of the board and is pressed in place by rollers. Tape can be applied only parallel to the direction of board travel. Boards are manually removed after masking, visually inspected to check that tape has adhered properly, and are then placed in a tote.

Masking of other areas is done by hand.

Dip Soldering. Boards to be dip soldered are loaded onto a carrier which transports the boards over a tank of molten solder. The boards are then lowered into the tank and raised again. The transport mechanism next moves the board to the rinse station and then to an unload station. Some installations do not have mechanized transport and boards are hand dipped. In such instances the depth of dip is controlled by mechanical stops. The mechanical stops are either located in the bath or engage the board carrier. The depth of solder in the tank must be held within close limits.

Wave Soldering. The process of wave soldering (Figure 19.2) involves passing a board with the components already inserted above a tank of molten solder. A generator causes a wave of molten solder to rise and pass down the length of the tank, thereby causing molten solder to contact the bottom of the board where it makes a sound electrical joint between conductors and components. Solder can be induced to flow up through holes drilled in the boards also. High flow velocities with their consequential rapid heat transfer rates are desirable.

DIP AND WAVE SOLDERING OF CIRCUIT BOARDS

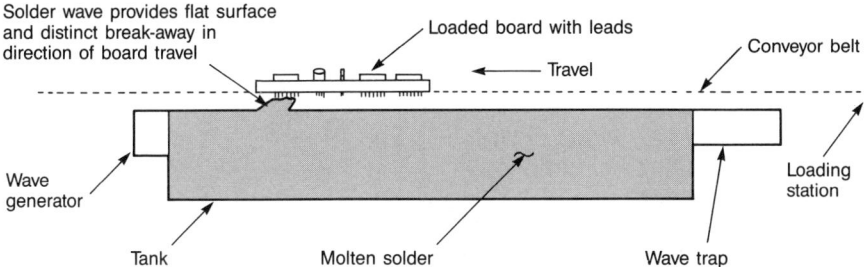

FIGURE 19.2. Wave solder tank and conveyor.

Rinsing. Boards are washed and rinsed, either manually or by automatic operation with conveyor line such as that shown in Figure 19.3.

Set-Up	Minutes	
	Dip Solder	Wave Solder
Set-Up Dip or Wave Solder Line		
Set-up tank, adjust conveyor, adjust line speed, add solder as required, adjust temperature, skim as needed. Does not include melt down time.	24	24
Set-Up Board Rinsing Line		
Adjust conveyor, adjust line speed, sample rinse tanks, make additions as needed		
If rinse is connected with solder line operation	8	8
If rinse operation is independent of solder line operation	10	10

Operating Time	Minutes Per Operation		
	Board Size		
	5 × 8 in.	8 × 10 in.	10 × 16 in.
Line Operation Times			
Dip solder—per board	0.37	0.37	0.37
Wave solder—per board	0.14	0.15	0.28

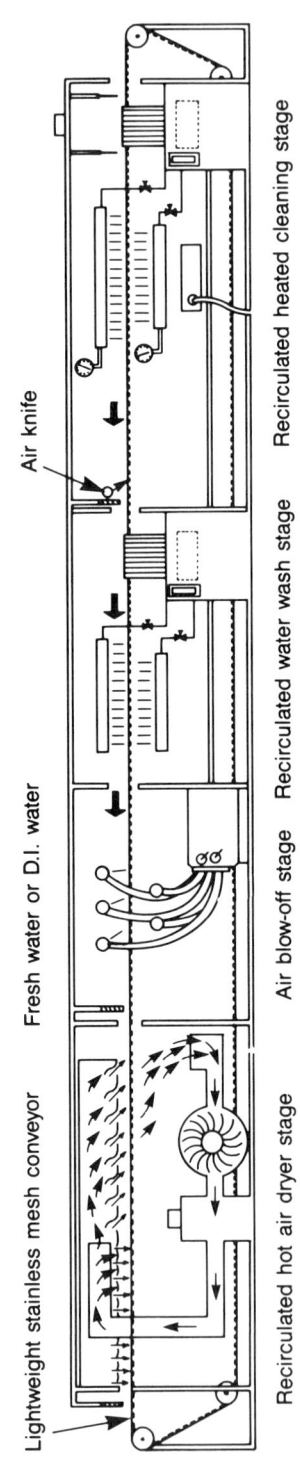

FIGURE 19.3. PCB In-Line Aqueous Spray Cleaning System.

SPOT AND SEAM SOLDER

	Minutes Per Operation		
	Board Size		
	5 × 8 in.	8 × 10 in.	10 × 16 in.
Optional Operation (Allow if Required)			
Manually mask additional areas at 0.0345 minutes per linear inch. Assume 7, 14, and 21 in. of tape respectively.	0.24	0.48	0.73
Dip into flux bath, brush flux liquid on board	0.10	0.20	0.30
Manually wash board & dry with air blast	0.45	0.75	1.05
Inspect board with magnifier	0.72	1.20	2.40
Touch up soldered circuit with hand solder iron. Average of 3, 6, and 9 points per board at 0.10 minutes/point.	0.30	0.60	0.90

19.4. SPOT AND SEAM SOLDER

	Minutes
150-watt solder iron—by hand	0.052
First inch	0.138
Per additional inch	0.092

These estimates apply to light-gauge work only, and are not applicable to galvanized sheet metal components.

SECTION TWENTY

FIBER OPTICS AND OPTOELECTRONICS

20.1.	Background	255
20.2.	Components	256
20.2.1.	Fibers and Cable	260
20.2.2.	Connectors and Splicing	263
20.2.3.	Transmitters	264
20.2.4.	Receivers	265
20.2.5.	Test Equipment	266
20.2.6.	Transducers	266
20.2.7.	Optoelectronics	267
20.3.	Applications	269
20.4.	Typical Costs	271

Fiber optic technology has become a legitimate alternative to wire and microwave communications systems. In addition, a spin-off of fiber optic communications research has been the discovery and development of fiber optic transducers for a wide variety of physical phenomena.

"Fiber optics" refers to the transmission of light through very thin glass or plastic fibers. In the extreme, the fiber may have a cross section of only one or two light wavelengths. There are three main consequences of using light in fibers rather than electrical signals in copper wires.

First, fiber optics have a greater information capacity per channel than any wire or microwave system. Channel capacity is proportional to the frequency of the carrier signal; light frequencies are extremely high. In practice, for instance, telephone transmissions can combine 24 two-way conversations on two pairs of twisted wire, one pair for each direction. By comparison, a single pair of fibers can accommodate 672 two-way conversations.

Secondly, the relatively shorter wavelength of light causes it not to interact with electromagnetic radiation in the more familiar ranges. Fiber optic systems neither generate nor are they disturbed by electromagnetic interference (EMI) or radio frequency interference (RFI). This means that fiber optic communication systems are immune to EMI/RFI, whether from natural or artificial sources, whether accidental or intentional. Also, it is impossible to tap or "bug" a fiber optic system without physically breaching the fiber, which would cause an immediately noticeable change in signal level and/or characteristics.

The lack of electrical currents means that ground loops and other problems with circulating currents are non-existent on fiber optic systems. Similarly, fiber optics are intrinsically safe to use in explosive or flammable environments; there is no electrical signal to generate a spark.

Third, the combination of very high capacity per channel and low weight per fiber means that for comparable capacity, fiber cables are much lighter and smaller. For instance, an optical cable of 144 fibers can carry 48,389 two-way conversations and yet be only 1 in. in diameter and weigh less than 6 ounces per ft. A coaxial copper cable to carry a comparable amount of information would be about 3 in. in diameter and weigh about 10 pounds per ft., or 27 times as much.

Any given application of fiber optics takes advantage of one or more of these characteristics. Long-distance telecommunication circuits make use of the very high information density and lower cost of fiber optics. As prices drop, the point at which fiber optics becomes more economical drops as well. In other applications, the raw cost of the fibers may be relatively unimportant, but the installation costs may be large. Some situations dictate the need for small cable size and weight. There is little space left today in the Manhattan underground cable systems. It means a lot to telecommunications installers to know that removing old copper cables leaves room for much higher capacity fiber optic systems. In other applications, immunity to EMI/RFI, or security may be an overriding consideration.

20.1. BACKGROUND

The development and rapid growth of fiber optics has depended upon the development of adequate cable and high-speed emitters. As will be shown, the losses in fibers have been dramatically reduced in recent years. Where once effective communications was possible over only short distances, now distances of up to 30 km between repeaters is possible. Some idea of the technology involved can be seen in the comparative clarity of glass used in long-distance fibers and ordinary window glass. The losses through 1 km of fiber are about equal to the losses through one pane of ordinary window glass.

256 FIBER OPTICS AND OPTOELECTRONICS

Fiber technology was closely followed by opto-electronic devices suitable for launching energy into the fibers and for receiving the light energy at the other end. Other areas that required development were connection and splicing techniques.

That fiber optic technology is coming of age is clear from Figure 20.1, which shows that by 1985 about 1% of the U.S. telecommunications cable is expected to be fiber, displacing 7.5 million pounds of copper. Translated in terms of cable production (Figure 20.2), annual growth rates as high as 35% have been predicted to 1990. The diversity of fiber optic applications is shown further in Figure 20.3 for a 1981 market of $134.5 million. Projections of these same market segments are shown in Figure 20.4 to reach $814 million in 1986 and $1868 million by 1990.

The following sections describe the components, applications, and typical costs of fiber optic systems.

20.2. COMPONENTS

The basic elements of a fiber optic communication link are presented in Figure 20.5. The transmitter converts an electrical signal into light by means of either a light emitting diode or a laser diode. The light energy is focused into the fiber cable where upon it is converted back into an electrical signal by either a p-i-n diode or an avalanche photo-diode at the received end.

The relative market share of these system elements is shown in Figure 20.6.

The key element, of course, is the fiber itself. Considerable support technology is required, from the cable that houses and protects the fiber to the

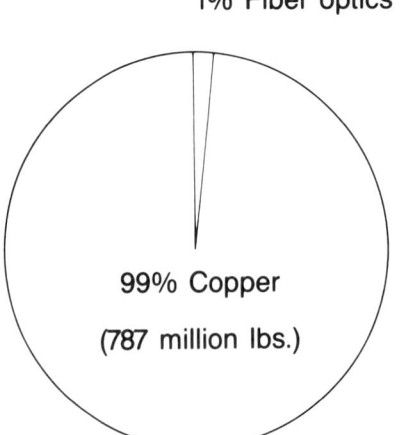

FIGURE 20.1. Fiber optic cable as a percentage of the 1985 U.S. copper telecommunications market.

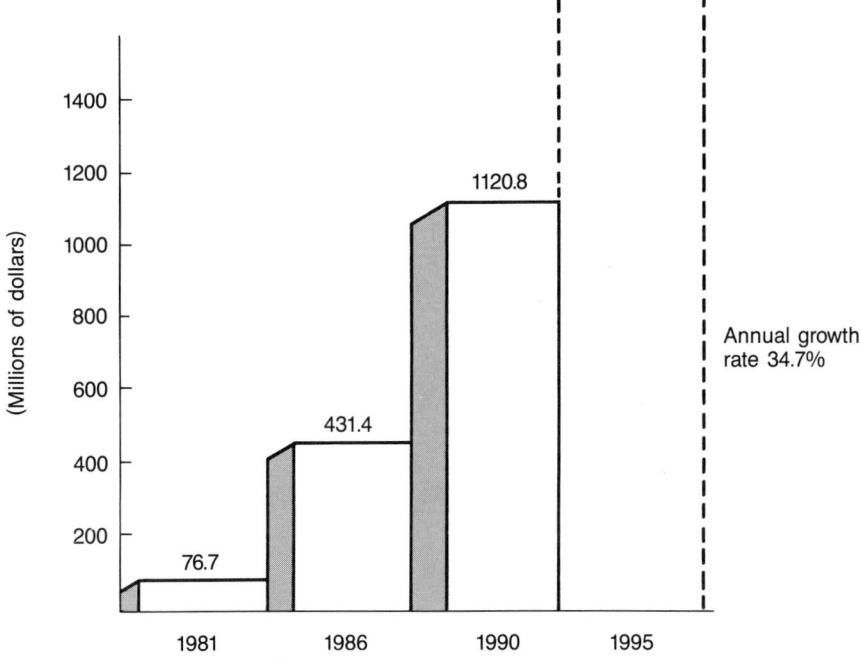

FIGURE 20.2. U.S. fiber optic cable production: 1981–1990.

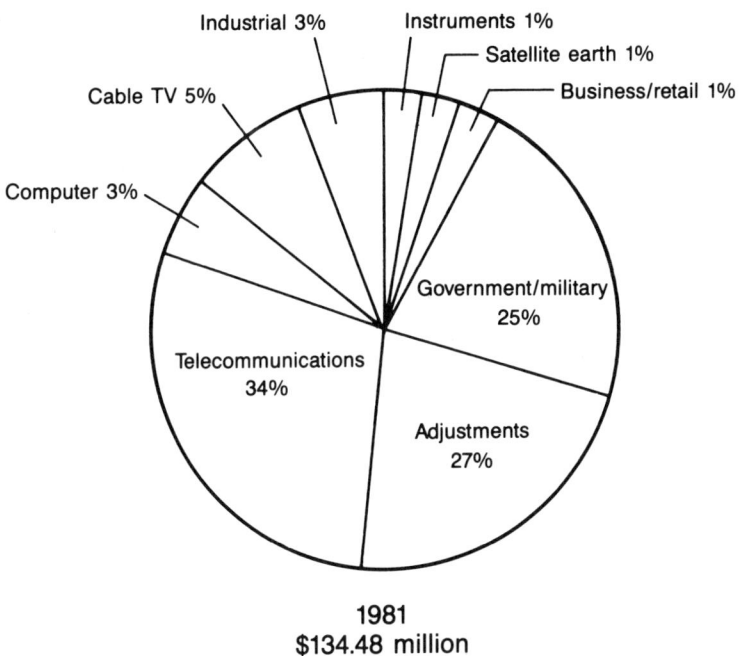

FIGURE 20.3. U.S. fiber optic component consumption by application.

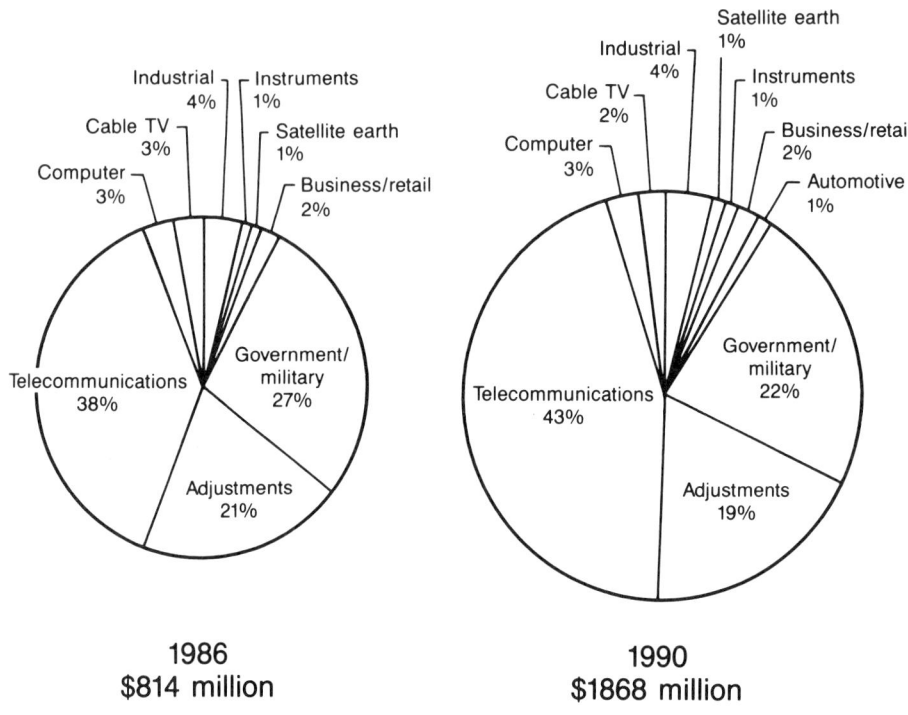

FIGURE 20.4. U.S. fiber optic component consumption by application: 1986–1990.

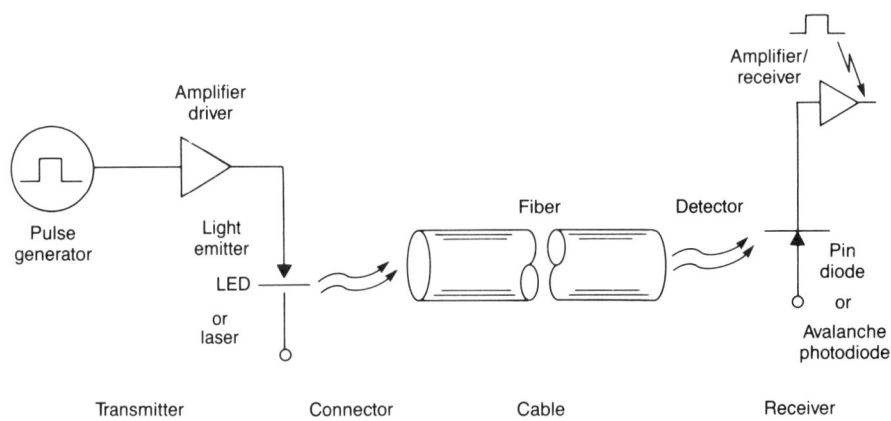

FIGURE 20.5. Basic elements of a fiber optic system.

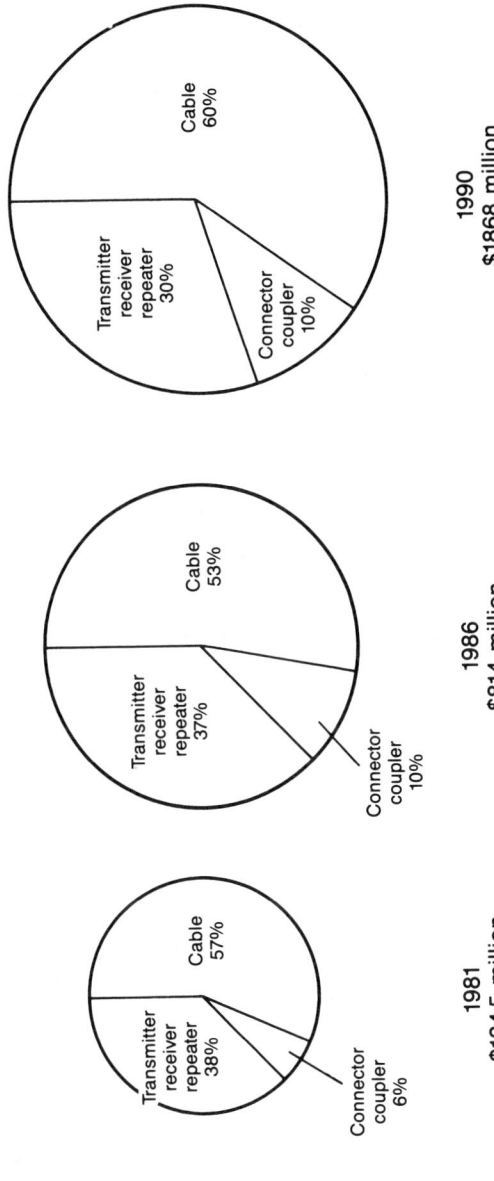

FIGURE 20.6. U.S. fiber optic production by component type: 1981–1990.

transmitters and receivers required to generate the light signal, launch it into the fiber, retrieve it at the far end, and convert it back into an electrical signal. In addition, special equipment is required to test optical systems. Each of these is covered in the following discussion.

20.2.1. Fibers and Cable

Fiber optics have been used for many years for very short-range transmission of light for various sensing and control tasks. The use of fibers for communications over longer distances was formerly impossible because of the attenuation of light in very long fibers and the lack of light emitters capable of high speeds. The obvious advantages of modulating light for high-density communication led to an intensive effort to develop low-loss glass. Figure 20.7 shows the results of that effort. By comparison, the attenuation of an electric signal on a twisted wire pair is about 0.21 dB per kilometer. In 1968, the lowest-loss fibers exhibited an attenuation of about 1000 dB/km. Eleven years later, in 1979, fibers were demonstrated with an attenuation of 0.2 dB/km, on the same order of attenuation as copper wire in twisted pairs.

Fibers are not uniform filaments of glass. The inner, or core glass is surrounded by a jacket of glass cladding with different index of refraction. The distribution of index of refraction across the fiber leads to different modes of signal propagation.

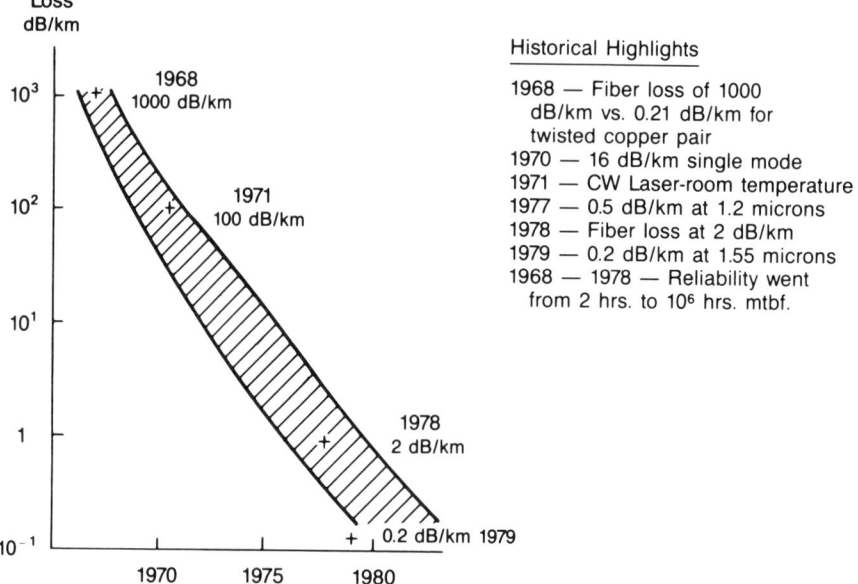

FIGURE 20.7. Advances in fiber technology. (*Source:* High reliability communications with fiber optic systems by Avery H. Hevesh, Boston IEEE Reliability Chapter Meeting, January 1982.)

COMPONENTS

As indicated in Figure 20.8, fibers are classified into three categories, depending on the mode of propagation of light.

Step-index fibers are made of two glasses which have different indices of refraction, with an abrupt change of index of refraction between them. Light is kept in the fiber better by reflection from a glass–glass interface than by reflection from an external surface. However, there are many different paths for light to take, as Figure 20.8 illustrates. Since each angle of reflectance leads to a different path length through the fiber, there is a large signal dispersion if the fiber is very long, with some parts arriving at the receiver before others. This blurs the edges of a signal. The only way to avoid loss of data is to limit the bandwidth of the modulation, which means limiting the data rate.

The step-index is the least expensive to manufacture, since only two types of glass are involved, and the tolerances are relatively loose. There are at present two ways of improving on its performance.

Graded-index fibers also have two layers of glass and are about the same in overall dimensions as the step fibers; the difference is in the core glass. Its index of refraction varies with distance from the centerline, as shown in Figure 20.8. The effect of this is that all light ray paths have the same propagation time. Since the signal is much less dispersed, the maximum data rate is increased.

	Single mode	Multi Mode	
		Step	Graded
Typical Characteristics			
Core diameter bandwidth Data content Coupling Velocities	2 to 10 um Large High Difficult Single	50 um Limited Low Easier Several (High dispersion)	50 to 100 um Moderate Medium Easiest Same for all modes
		Refraction index	

FIGURE 20.8. Fiber properties by type.

The graded fiber is also quite tolerant of angular alignment error in the coupling of light into fiber.

Single-mode fibers have a core of very clear glass that is only a few times larger than the wavelength of the light propagated. As a result, there is effectively only one path through the fiber, and very low signal dispersion. Single-mode fibers have the largest data capacity.

There are attendant problems and costs, however. Since the core is typically only 2 to 10 microns in diameter, manufacturing tolerances are very close. Considerable precision is required to insure integrity of fiber alignment at joints, couplings, and splices to minimize light losses.

Elliptical-cored fibers have recently been developed in Japan. The internal structure of the fiber is much more complex than that of cylindrical fibers. Its purpose is to preserve the polarization of light launched into the fiber. Maintaining polarization may lead to even higher data rates than are obtainable with single-mode fibers. But, more importantly, polarization allows more precise signal control, improving resolution of fiber optic sensors. (See Section 20.2.6.)

Although single fibers are very flexible and have yield strengths greater than that of steel, they are somewhat fragile due to small size. Cable is required to insure protection against forces of ordinary handling, pulling through cableways, and so on. As shown in Figure 20.9, the fibers are arranged around a strong central core, and surrounded with several layers of cushioning and protection. The fibers are spiraled around the central core, which allows displacement between fibers under bending and flexing.

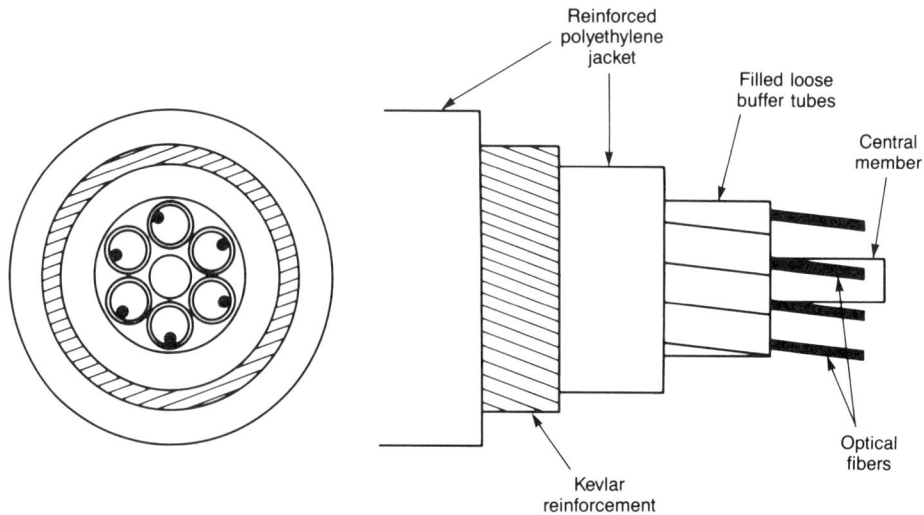

FIGURE 20.9. Multi-fiber cable.

COMPONENTS

20.2.2. Connectors and Splicing

Connections in fiber optic systems require great precision to avoid large signal losses, illustrated by the range of fiber diameters (2 to 100 microns, or 0.08 to 4 thousandths of an inch) in Figure 20.8.

Connectors are commonly available with insertion losses in the 0.3 to 2 dB range. Several manufacturers produce standard ranges of fittings. Typical configurations are shown in Figure 20.10.

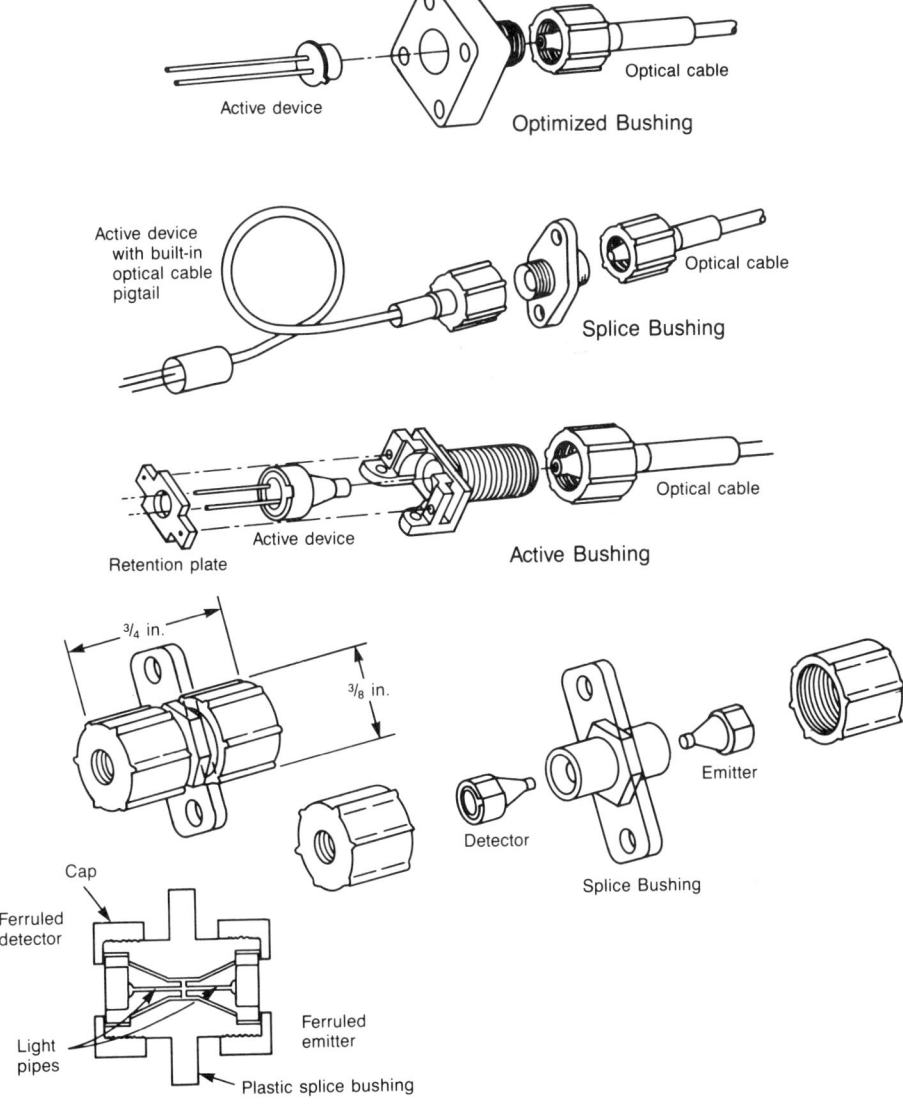

FIGURE 20.10. Typical connectors. (*Source:* GE Semiconductor, Power Electronics Semiconductor Department.)

Connectors have matured considerably in the past few years. Where once it was necessary to epoxy a fiber in place in a fitting, and then grind and lap the end of the finished assembly to obtain adequate optical characteristics, now there are proprietary systems which include a special tool for cleanly fracturing the end of the fiber, leaving a planar surface. That insures rapid and accurate positioning of fiber to connector.

Some connectors use thread sizes and rings that are standardized for coaxial cable. Others are completely new, such as the "hermaphroditic," where all fittings are alike, permitting any fitting to mate with another.

Permanent splices are also possible. A low-loss approach uses an electric arc to fuse fiber ends. The most important part of the procedure is obtaining and maintaining adequate alignment of the fibers.

20.2.3. Transmitters

The transmitter in a fiber optic system provides electrical-to-optical conversion. The transmitter is generally a single printed circuit card or hybrid component; its key element is the light emitter.

Obtaining the high bandwidths that are theoretically possible with light requires a very fast emitter. Although fast photo-detectors have been available for many years, fast emitters are relatively new. The two types of emitters in use are laser emitting diodes and light (or infrared) emitting diodies (LEDs or IREDs).

Laser diodes are fast, powerful, and expensive. Their optical rise time is typically less than one nanosecond. Certain types of laser diodes are capable of emitting as much as 10 watts. However, this type requires high drive current, which in turn leads to requirements for forced cooling. Lower-power laser diodes are more commonly used in fiber optic systems with outputs in the 10 to 20 mW range, which are usually adequate and are much easier to integrate into systems. Prices of $1000 or more are not unusual for a high-powered laser diode; this does not include supporting circuitry.

LEDs are much less expensive, on the order of fifty dollars or so. They can deliver a maximum of about 10 mW, but are capable of essentially continuous operation. For short- haul applications, often only 30 to 200 microwatts into the fiber is entirely adequate for reliable communications. At these outputs, the power and cooling requirements are very low. Where laser diodes are commonly guaranteed for 10,000 hour life, and "expected" to reach 100,000 hours, LEDs have operating lives in the range of 1 to 10 million hours.

Figure 20.11 shows the relative market shares of laser diodes and LEDs at present and as projected for the next few years.

Both laser diodes and LEDs can emit light at the wavelengths where low-attenuation windows are found in glass fibers. The visible red range (0.66 microns) is popular because of its visibility and workability, and falls in a relatively low-loss region of typical step index fibers. Generating near infrared (0.8 to 0.9 microns) is more efficient, and has the added benefits of matching

COMPONENTS

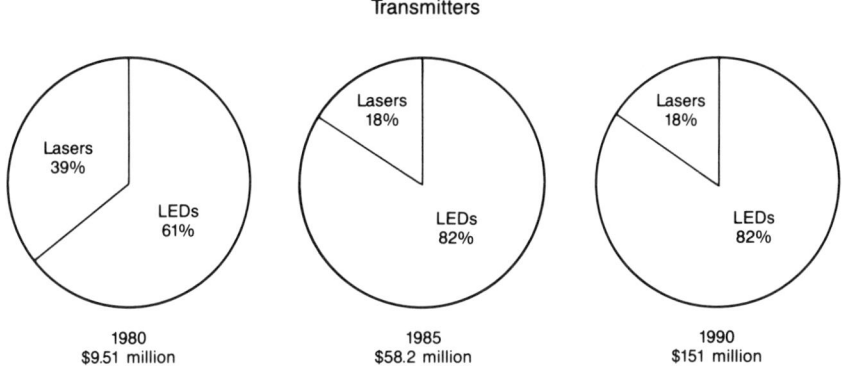

FIGURE 20.11. U.S. market for fiber optic transmitters.

both the high-sensitivity regions of typical receiver diodes and a low-attenuation region of typical graded index fibers. The far infrared (≥ 1.3 microns) holds promise for long-haul links, but emitters and detectors are currently quite expensive.

Transmitter circuitry is relatively simple. For digital transmission, a plain Transistor–Transistor Logic (TTL) gate can drive a transistor to modulate an LED for data rates to about 20 Mbit/second. Up to 50 Mbit/second transmitter circuits are excessively complex. For analog signals, a LED can be driven by a suitable general-purpose operational amplifier.

20.2.4. Receivers

Although sensitive, high-speed detectors have existed for many years in vacuum tube form. They have been replaced by solid state avalanche or p-i-n diodes for the usual reasons: reliability, size, and efficiency.

Avalanche diodes cost 5 to 10 times as much as p-i-n diodes, are less reliable, are temperature sensitive, and require high-bias voltages on the order of 400 volts. On the other hand, they offer 10 to 20 dB greater sensitivity and greater bandwidth. Avalanche diodes can function up to about 100 GHz, while p-i-n diodes are limited to about 1 GHz.

Figure 20.12 shows the relative market shares of avalanche and p-i-n diode detectors.

Because of the very low levels of received power, receiver circuitry is more complex than transmitter circuitry. Especially with the very high frequencies and low signal levels involved at the high-performance end of the spectrum, the detector circuitry must be carefully designed to minimize noise. Under these conditions, thermal random noise becomes a factor. Extremely low error rates are required by the very high speeds involved. Consider that at the 10^{-9}-bit error rates usually quoted, and at a relatively slow 10-Mbit/second data rate,

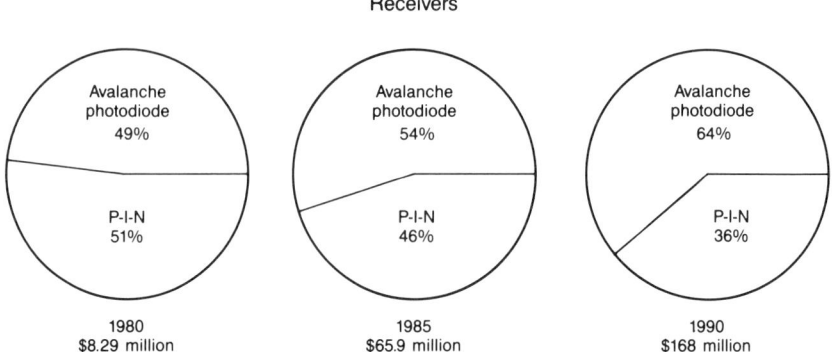

FIGURE 20.12. U.S. market for fiber optic receivers.

an error is likely every 100 seconds, on the average. Error rate is at least partly a function of input power to the detector. Shorter hauls and higher input powers both improve reliability.

Transmitters and receivers are commonly offered as matched sets, but may also be purchased separately.

20.2.5. Test Equipment

A fiber optic system is unique only in the optical portion: source, signal path, and detector. A calibrated optical detector and signal source are all that is required in the way of unique test equipment for fiber optic systems. Given the appropriate conversion devices, standard electronic test equipment can be used to make level measurements, display waveforms, and so on, although fiber optic "multimeters" are also available and are more convenient in some cases.

Converters in the visible to near-infrared (0.4 to 1.0 micron) range cost approximately $250. Far infrared detectors are considerably more expensive, about $500 for a comparable unit in the 1.3 to 1.6 micron range.

High-frequency signal sources are priced at about $350, plus or minus $25 depending upon choice of wavelength.

20.2.6. Transducers

A relatively new and undeveloped technology is the use of fiber optics as sensors. Changes in the fiber due to temperature, strain, and intense magnetic and electrostatic fields affect the optical signal to varying degrees. Although the effects are generally quite small, it is possible to wind a very long fiber into a small package and so increase the magnitude of any given effect. At present, researchers are confident that fiber optic sensors can be made at least as sensitive as current devices for a very wide range of phenomena, as listed in Table 20.1.

Various predictions indicate that size of the fiber optic sensor market lags the communication market by 5 to 10 years. Predicted market expansion rates range around 30% annually, leading to a $50 to $100 million market by 1990.

Fiber optic transducers have several features that make them competitive or uniquely applicable to some measuring tasks. For instance, the glass fibers can withstand temperatures well in excess of those tolerated by other kinds of sensors and solid state supporting circuitry. Where a fiber is used to communicate a light signal between transducers in a hostile environment, complete electrical isolation provides safety and signal advantages as well.

Research is initially focusing on military applications, in particular underwater sound and aircraft engine monitoring. Other possible applications include sensing and control systems for industrial robots, commercial aircraft, ships, buildings, and farm equipment.

20.2.7. Optoelectronics

The emitters and detectors used in fiber optic systems are also applied without the use of fiber light paths. The primary use of such devices is either to transmit signals without electrical connections or to sense the position of an object by monitoring changes in a light beam.

Opto-isolators, or photon-coupled isolators, are produced in a wide variety of capabilities, from emitter/detector sets mounted in close proximity within a single DIP or other package, to devices using a short (few millimeter) connecting fiber to reduce the isolation capacitance and increase the breakdown voltage, such as shown in Figure 20.13.

The evolution of the optoelectronic coupler has led to a completely solid

TABLE 20.1. Fiber Optic Sensor Applications

Linear position
Angular position
Temperature
Pressure
Strain
Acceleration
Rotational rate
Magnetic field intensity
Electrostatic field intensity
Radiometric dosage
Waterborne sound (SONAR)
Airborne sound

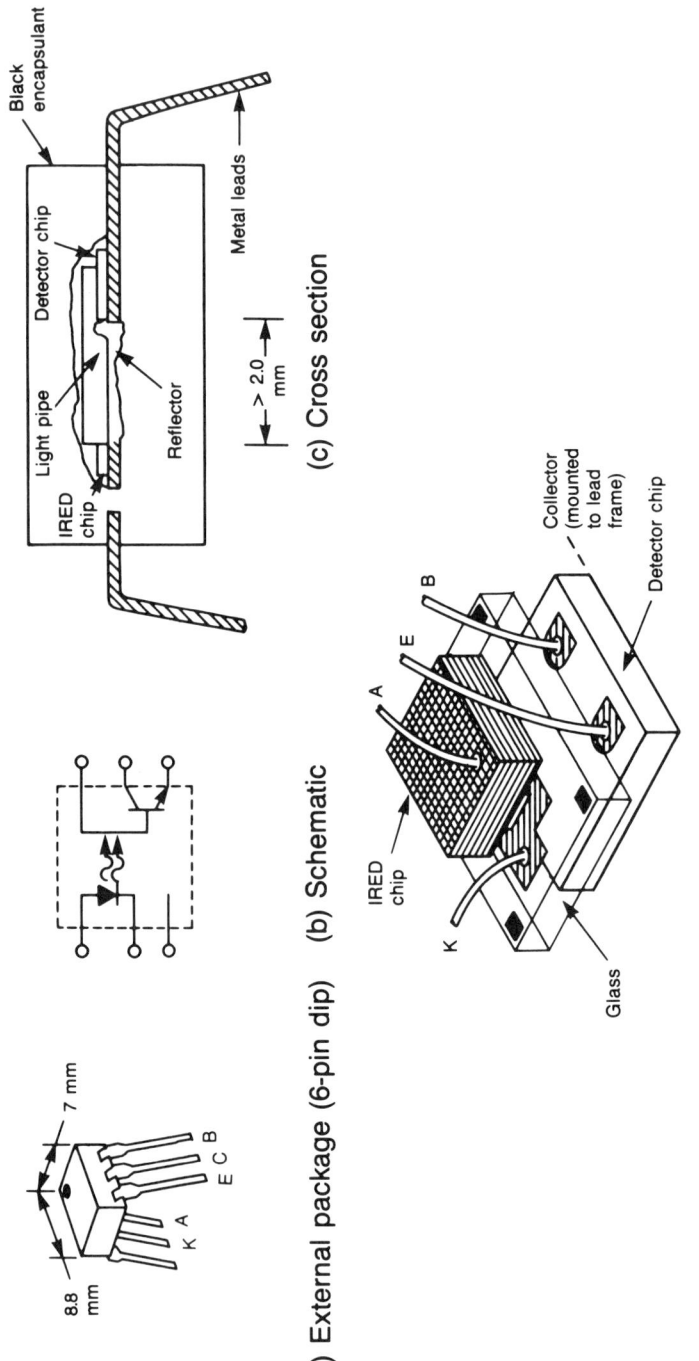

FIGURE 20.13. Opto-isolator. (*Source*: GE Semiconductor, Power Electronics Semiconductor Department.)

APPLICATIONS

state relay. In addition to the normal switching functions of a relay, this technology offers solid state reliability, zero voltage switching, and most importantly, a direct interface between logic-level signals and the electrical powerline. Figure 20.14 is a block diagram of a solid state relay.

Pairs of emitters and detectors are also used for sensing the position of an object. The devices can be bought separately, or in a single package, as illustrated in Figure 20.15.

20.3. APPLICATIONS

The list of applications of fiber optics is very long and growing rapidly. In addition to the well-known telecommunications uses, a few of the more recent applications include the following examples.

Fiber optic technology is used to transmit aircraft carrier radar data from the radar equipment room to the combat information center. A seven-fiber cable weighing a total of 15 pounds replaced 7 tons of wire carrying 375 separate signals 750 ft. The cable is ¼ in. in diameter and costs $30,000 to install, compared with $1,000,000 for the wires. A significant part of the cost of the wire system is due to the need for large parallel conductors to reduce the effects of ground loops; nothing comparable is required for the fiber system.

An operator-guided missile based on a fiber optic link is under development for the Army. Here a fiber provides a high-bandwidth channel for video, commands, and other information between the missile and operator. The intent is to make it feasible to launch from completely concealed positions, to reduce the launchers' exposure, whether ground- or helicopter-based. In the initial system, 10 km (6.2 miles) of fiber is contained in a canister with an overall length of 12 in. and a diameter of 6 in. Longer fibers are being developed.

The command, control, and communications network for the MX missile is being based on fiber optics, to obtain the benefits of high data rates, immunity to many of the effects of nuclear weapons, and security.

The YC-14 experimental air transport incorporated a "fly-by-light" system, where fiber optics carried control surface commands from the flight deck to the

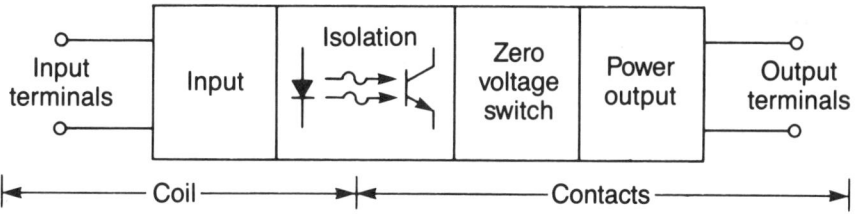

FIGURE 20.14. Block diagram of a solid state relay.

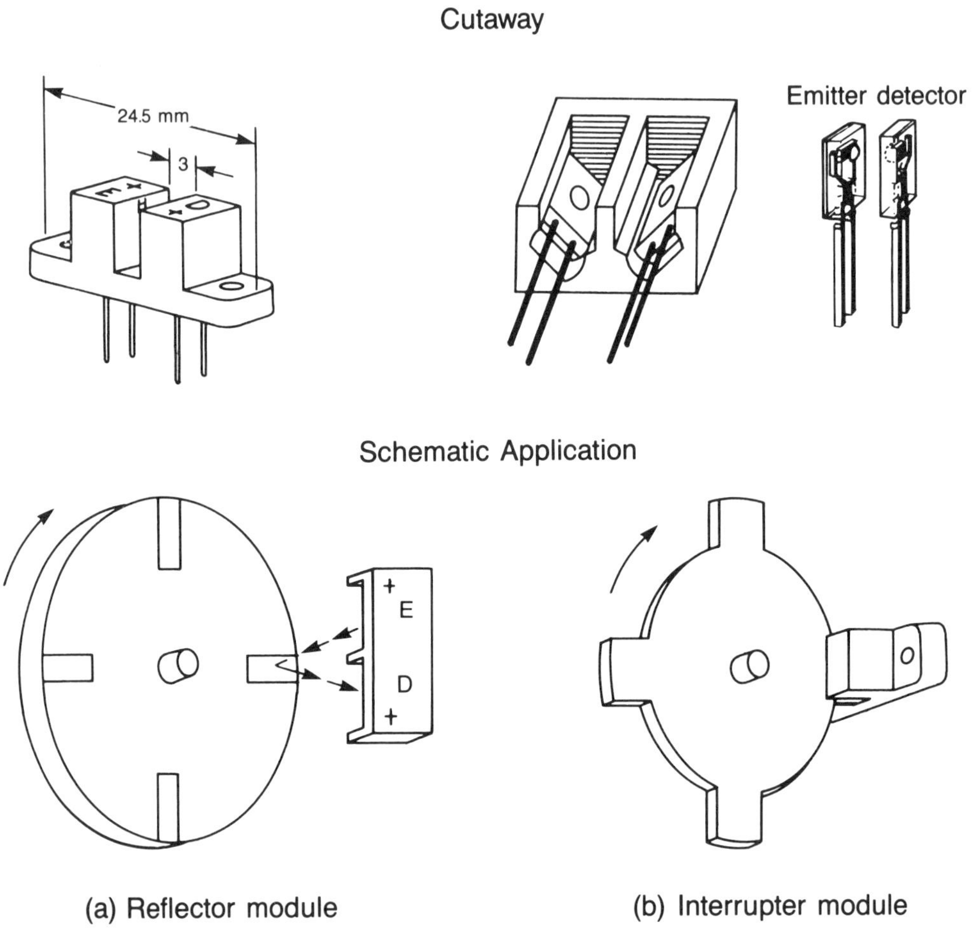

FIGURE 20.15. Optoelectronic sensors. (*Source:* GE Semiconductor, Power Electronics Semiconductor Department.)

actuators, with complete electrical isolation. Other systems are under consideration for the full range of aircraft, from the helicopters and V/STOL through supersonic fighters. In addition, at least one company is actively developing a fiber optic system to make it easier to add electronics to existing aircraft. The fiber optic system is lightweight, compact, and neither generates nor is sensitive to RFI.

A fiber system being developed for troop field communications is lightweight, compact, and totally secure, in addition to providing much greater bandwidth than a typical wire system.

20.4. TYPICAL COSTS

Much of the high-cost reputation of fiber optic systems is due to the very high performance goals of long-haul telecommunications. Systems operating at 250 Mbit/second, with 10 to 15 km between repeaters, require cooled laser diodes and temperature-stabilized avalanche diode detectors in addition to first-quality fiber cable. The emitters and detectors are expensive themselves, and the physical support systems are also expensive. However, short-haul systems are much less expensive.

Prices in this field are rapidly changing. A well-known manufacturer offered a fiber optic system including transmitter and receiver modules and 5 meters of cable capable of data rates from DC to 5 MHz for \$55 in late 1980. Two years later, the same system is advertised for \$27.50 (both prices for single quantities).

There are many off-the-shelf systems, as well as a variety of components that can be engineered into special purpose systems. As an indicator of the range of prices involved, consider the following systems:

\$148/pair (transmitter and receiver module), quantity 100, 40 Mbit/second, 500 meters, bit error rate of 10^{-8}.

\$183/pair, 10 Mbit/second, 150 meters, bit error rate of 10^{-9}.

\$325/pair, 25 MHz, 1 km, bit error rate not given.

\$350/pair, DC to 30 Mbit/second, 2 km, bit error rate of 10^{-11}.

SECTION TWENTY ONE

INSPECTION

21.1.	Inspection Estimating Ratios	272
21.2.	Visual Inspection and Record Keeping	273
21.3.	Machine Shop Gauging Times	274
21.4.	Gauging Frequencies	275
21.5.	AQL and Inspection by Random Sampling	276

Inspection costs may or may not have to be estimated as a separate direct labor cost according to specific company cost accounting policies. Some companies classify inspection labor as indirect, and therefore, the cost is included in the burden rate (which is then prorated over all direct labor).

Inspection labor can be estimated by using overall percentage ratios or by detailed time standards. The percentage technique is most practical for short-run production jobs. The elemental time standards should be used where the production quantities are large, continuous, and completely defined. Under these conditions, inspection labor can be measured in the same manner as other types of labor.

21.1. INSPECTION ESTIMATING RATIOS

A common method of estimating inspection labor is to use a percentage of the productive labor hours. Actual head counts of inspectors vs. production personnel result in ratios similar to this breakdown:

	Percentage of Production Labor		
Type of Inspection	Minimum	Average	Maximum
Receiving and source inspection	2	5	7
Production inspection	5	10	15
Total inspection	7	15	22

The ratios are based on the following definitions:

Productive labor. Total machining, process, assembly, and wiring direct labor hours.

Receiving and Source Inspection. Inspection of material purchased from vendors; inspection of materials at the vendors' facility. Source inspection is most prevalent with MIL SPEC. procurements.

Production Inspection. Inspection of all fabricated parts, subassemblies, assemblies, units, and systems both during and at completion of the production process.

21.2. VISUAL INSPECTION AND RECORD KEEPING

Time to perform visual inspections for surface coating, damage, general condition, and so on is based on the number of eye fixations required to scan the object plus the required handling time. For normal inspection severity, 0.06 minute per eye fixation is appropriate.

A large percentage of visual inspection is handling time. Normal handling times for objects on a bench can be estimated as 0.10 minutes per simple single hand movement, plus 0.03 minutes for each of: second hand required, moderately careful alignment required, and upper body movement required.

With a hand-held magnifier, additional time is required to put the device into use, and an eye fixation naturally covers a smaller area. Approximately the same time is required to use an ultraviolet or infrared illuminator.

Microscopic inspection has become common in the electronics industry. The normal eye fixation time is appropriate within the microscope field of view, but positioning the work piece and adjusting the microscope require additional time.

Tables 21.1 and 21.2 give elemental times for visual inspection and record keeping.

TABLE 21.1. Visual Inspection

Task	Minutes
Per eye fixation	0.06
Handling	
Simple hand movements, pick and place	0.10
Additional for: second hand, torso movement, or accurate placement	0.03
Select, turn on, turn off, and replace hand-held magnifier of UV/IR illuminator	0.32
Place item on microscope stage	0.16
Change objective, focus microscope	0.23
Adjust vernier stage, each axis, initial	0.28
subsequent	0.17

TABLE 21.2. Record Keeping

Task	Minutes
Read and verify 12 character part number	0.22/12 characters
Record data	0.07/character
Calculate (includes writing):	
5 digits ± 5 digits	0.90
3 digits × 3 digits	0.82
File form in three-ring notebook on shelf	1.03
Retrieve form from three-ring notebook	0.59

21.3. MACHINE SHOP GAUGING TIMES

The elemental time values on Table 21.3 apply to both the machinist while making the part, and the inspector for checking the part. The times include: pick up measuring device, adjust to dimension, read dimension, and lay aside. The times do *not* include handling the part to be measured or travel to obtain tools, and so on. (See Section 15.)

Times which include more than this basic content are noted below:

Micrometers—Includes checking the dimension at several points to ascertain accuracy.

Plug Go/No Go—Includes checking both Go and No Go sides of gauge.

Thread Gauge—The time for an average of ten threads.

TABLE 21.3. Machine Shop Gauging Times (Minutes)

Instrument	Tolerance (In.)	Set-Up	First Reading	Successive Reading
OD Micrometer, 0–4 in.	±0.001	—	0.28	0.17
4–12 in.	±0.001	—	0.41	0.20
Additional to obtain high precision	±0.0005	—	0.12	0.12
Additional for awkward position		—	0.10	0.10
ID Micrometer 2–4 in.	±0.001	0.42	0.71	0.61
Snap gauge		—	0.08	0.08
Telescoping or ball gauge	±0.001	0.74	0.56	0.33
Depth micrometer to 3 in.	±0.001	0.31	0.37	0.32
Profile		—	0.16	0.16
Go/No Go Gauge to 2 in.		—	0.20	0.15
Spring Calipers to 8 in.				
Outside		0.59	0.57	0.52
Inside		0.59	0.60	0.55
Optical Comparator		—	2.34	2.25
Scale, 12 in.	±1/16	—	0.17	0.11

21.4. GAUGING FREQUENCY

Gauging frequency is inversely proportional to the part tolerance. In other words a dimension held to ±0.0002 in. may be checked five times during machining while a part with a ±0.030 in. tolerance is checked on only 1 out of 100 parts. The accompanying table serves as a guide to gauging frequency. Individual parts, operation sequences, and production quantities will vary in actual cases.

Maximum Tolerance (in.)	Gauging Frequency, Number of Parts per Measurement
0.0002	1/5
0.0005	1/2
0.001	1
0.002	5
0.005	15
0.010	30
0.015	50
0.030	100

TABLE 21.4. Typical Sampling Plan (AQL = 1.0%)

Lot Size	Sample Size	Allowable Defectives Without Rejecting Lot
1–13	All	None
14–150	13	None
151–500	50	1
501–1200	80	2

Note: An AQL of 1.0% means that an average rate of defectives of 1 per 100 is acceptable. The sampling plan is statistically designed to specify sample size and rejection criteria which assure that accepted lots will, on the average, have less than the AQL rate of defectives. This sampling plan is taken from MIL-STD-105.

21.5. AQL AND INSPECTION BY RANDOM SAMPLING

Random sampling can be used to economically determine whether or not a large number of units have an acceptable quality level. Although statistical analysis cannot tell for certain the quality of each and every unit, it is frequently more economical to establish an AQL and accept a known risk of defects than to perform 100% inspection.

AQL is defined as the maximum percent defectives that can be considered satisfactory as a long-term process average. A sampling plan for a given AQL specifies the numbers of samples to be taken and the criteria used to accept or reject a given lot in order to maintain that AQL. There are a variety of sampling plans available. The most common sampling plans used for military hardware are given in MIL-STD-105, "Sampling Procedures and Tables for Inspection by Attributes" and MIL-STD-414, "Sampling Procedures and Tables for Inspection by Variables for Percent Defective."

The AQL for which a plan is designed applies with high reliability only to a long-term average of many lots. Any sampling plan includes some risk of accepting lots with many defectives and of rejecting lots with few defectives. The operating characteristics (OC) curve for a given plan shows the probability of accepting or rejecting a single lot with a given percent defective. Both cited MIL-STDs include OC curves for their sampling plans.

They also provide three severity levels of inspection. In addition to normal inspection, reduced inspection plans may be used where the quality of production has been consistently better than required, and tightened inspection plans may be used when quality has been consistently lower than required.

Details of the various plans are given in the MIL-STDs. Inspection requirements are normally specified by citing the governing MIL-STD and designating an AQL.

Table 21.4 shows a typical normal level sampling plan.

SECTION TWENTY TWO

TEST

22.1.	Test Estimating Ratios	277
22.2.	Elemental Test Time Standards	279
22.3.	Troubleshoot and Retest Allowances	280

Test ratios and standards in this section cover receiving and production tests. They do *not* include environmental or life tests. The receiving test effort may or may not have to be included as a direct labor cost, according to specific company accounting policies. Estimating ratios for both receiving and production tests are given separately and may be applied as required. The experience ratios are for approximate budgeting of total test effort while the elemental time standards are for accurate measurement of individual test operations.

Unique tests are discussed in the appropriate process section. For instance, electrical testing of printed circuit boards for shorts and opens and testing of LSI chips are covered in their respective sections.

22.1. TEST ESTIMATING RATIOS

A technique for estimating total unit test labor is to use a percentage of the total productive labor hours. Experience has shown that, within a range, test labor varies directly with the amount of both fabrication and assembly labor required to build a system. More circuits mean more components, more wires, and more assembly labor. More components and wiring mean larger chassis, more etched circuit boards, and more fabrication labor. The additional machine

TABLE 22.1. Test Estimating Ratios

Labor Base	Percentage of Base Labor		
	Simple	Average	Complex
Fabrication and Assembly			
Receiving test	1	2	4
Production test	9	18	36
Total	10%	20%	40%
Assembly			
Receiving test	2	3	7
Production test	15	32	63
Total	17%	35%	70%

shop and mechanical assembly required for precision electro-mechanical units is accompanied by an increase in electro-mechanical check-out time. Experience ratios in Table 22.1 are given for a total productive labor base, and for an assembly labor base only. The total productive labor base can be used to estimate total system test time, but to estimate test time by subassembly, the assembly labor base should be used. (The total test effort is thus allocated over the assemblies requiring test, and *not* over fabricated piece parts.)

One note of caution, the ratios are an estimating shortcut. They are *not* meant as a replacement for a detailed test estimate. They should be used accordingly.

The ratios are based on the following definitions:

Total Productive Labor—Includes machining, process, assembly, and wiring direct labor hours.

Assembly Labor—Includes precision mechanical assembly as well as hardware assembly and wiring direct labor hours.

Receiving Test—Electrical test performed on purchased components and subassemblies prior to their acceptance by the receiving department.

Production Test—Electrical test of subassemblies, units, subsystems, and systems. This includes all phases from circuit checkout through alignment and functional test of the system. Environmental and life test are *not* included.

Simple, Average, Complex—The complexity of the electronic circuitry will vary the ratio of test labor to other productive labor. For example, a small receiver of standard circuit design will result in about a 10% ratio, while a space probe system which is advancing the state of the art may require a 30 to 70% ratio.

22.2. ELEMENTAL TEST TIME STANDARDS

The following standards are for use with detailed test procedures. Test labor can be measured with reasonable accuracy by standardizing the operation, applying elemental time standards, adding normal time study allowances, and adding an additional troubleshoot and retest allowance. Troubleshoot allowances are discussed in the next paragraph, and general labor allowances, multipliers, and learning curves are discussed in Sections 26 and 25 respectively.

Preliminary to Test:
Pick up and lay aside unit (average)	0.38
Cover or panel-disassembly, reassemble (force fit)	0.52
(with screws)	1.04
Ground unit and technician, where required for CMOS devices	0.28
Hook up and unhook:	
Clip small cable, plug, jack.	0.14
Attach microprobe to integrated circuit on dense P.C. board	0.22
Large cable and connector	0.32
Tube or shield	0.22
Crystal	0.58
Dummy load (clip on)	0.27
Seal part with cement, glyptol, and so on	0.27
Convert voltage to decibels via formula	1.50
Read and write serial number	0.32
Add or subtract (6) digit number and write on data sheet	0.84

Test Equipment Adjustments:
Toggle or rotary switch: off and on. (Simple)	0.06
Adjust Variac or loose tolerance knob (Medium)	0.13
Adjust knob to given frequency (Difficult)	0.17
Adjust to zero beat (sound or scope)	0.34
Adjust scope for volts, cycles, phasing, etc.	0.34
Observe scope and analyze pattern	0.23
Inspect meter position	0.02
Accurately read meter	0.10

Adjustments on Unit Being Tested:
Circuit check with probe and volt–ohm meter (per point) (add meter-reading time)	0.06
Add to hook up ground clip and probe	0.14

Adjust potentiometer with tool	0.17
Adjust coil core to tune	0.34
Adjust capacitor (trimmer)	0.34

22.3. TROUBLESHOOT AND RETEST ALLOWANCES

Troubleshooting time will vary greatly from day to day on the same test, but over a period of weeks or months an average allowance can be developed. These allowances are over and above the normal personal, fatigue, delay, and learning allowances, which are used with elemental time standards.

For example:

Elemental times plus PFD allowance	1.0 standard hour.
Add trouble shoot allowance at 50%	1.5 standard hours.
Add learning allowance at 2.0	3.0 realized hours.
Expected shop performance	3.0 hours.

Average trouble shoot and retest ratios are given below:

Type of Test	Percentage of Work Day	Percentage of Straight Test Time
Subassembly or Unit Tests		
One overall test per unit	33	50
Preliminary and final test:		
Preliminary	40	67
Final	20	25
Quality control surveillance test	5	5
Subsystem test	10	10
System test	10	10
Composite system average	33%	50%

SECTION TWENTY THREE

AUTOMATIC TEST EQUIPMENT

23.1.	Products	282
23.2.	Market and Suppliers	284
23.3.	Processes	289
23.4.	Cost	293

Automatic testing has become a standard process in today's manufacturing operations due to the influx of high-density electronic packaging and the increased use of integrated circuit chips. Automatic test equipment (ATE) is found in pre-screening and incoming inspection of components, bare board testing prior to component insertion, and final PC board testing after assembly. In more specialized applications it is used in subsystem- and systems-level testing as well as field testing. It is also used in burn-in and environmental qualification of products.

Market distribution of ATE applications is shown in Figure 23.1. As indicated production was the dominate user with 76% of the market, followed by incoming inspection, engineering, and service, respectively.

Within the broad spectrum of ATE, different techniques and technologies are employed, yielding varying results. Basic to all this, two fundamental premises govern its economics. The first is that testing must add more value to the product being tested than the overhead cost of the test itself. Secondly, whatever testing is employed it must prevent future loss of the product, the occurrence of which would be greater than the cost of testing. Stated more simply, each level of test must justify its existence by correcting faults that

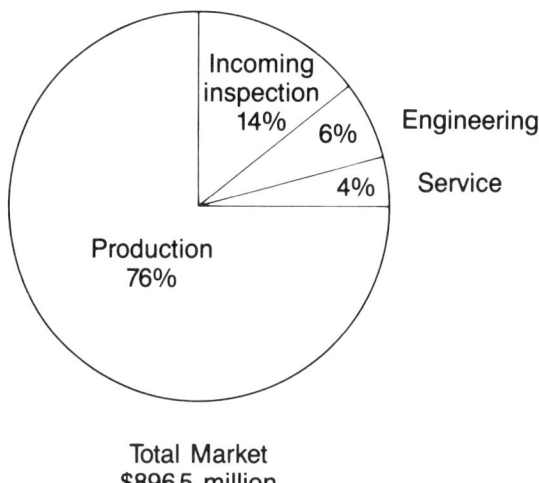

FIGURE 23.1. ATE applications—1981. (*Source:* Dataquest, Inc.—*Electronics Business*, October 1982.)

would have otherwise been corrected at a later, more-expensive stage of the manufacturing cycle. It has been said that the least-expensive place to find and correct a fault is in the first test operation after the occurrence of the failure. This is illustrated in Figure 23.2, where the relative cost of test and repair increases rapidly through the manufacturing cycle and can become extraordinarily expensive, on a life cycle basis, once the product enters the field. At the lowest level, smaller circuit packages permit a well-structured process for detecting and diagnosing faults using automation and ATE. Once programmed, this process requires very little skill. After accounting for the investment of the test system, it imposes the lowest possible cost.

Failures that are detected at the subsystem or higher levels require considerably greater skill to diagnose and correct. The product is more complex at this stage of manufacture. Economically, ATE is not as viable a solution as at the board level because of the comprehensive nature of the test programs and specialized instrumentation requirements. An unstructured approach to diagnostics invariably results. The skills of highly trained technicians must be employed due to the impracticality of automating the detection and diagnosis of all possible fault combinations. Test and troubleshooting time sharply increases, intensifying labor content and cost. This condition continues well after the product reaches the field.

23.1. PRODUCTS

In general, there are four classes of commercially available ATE that are widely used in manufacturing. These are component, integrated circuit, inter-

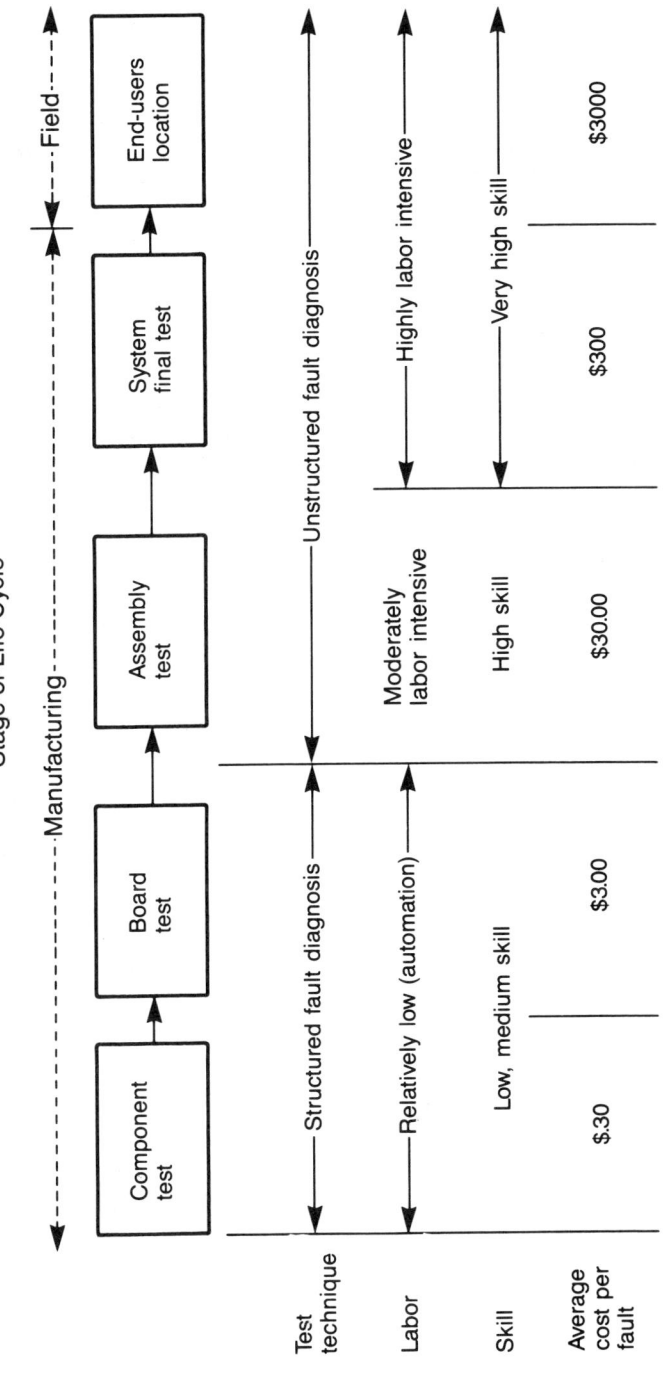

FIGURE 23.2. Relative cost and repair per fault. (*Source:* How should a manufacturer test its printed circuit boards by George Mabry, Teradyne Incorporated.)

connect/bare board, and loaded PC board test systems. A fifth category covers a wide variety of automated field-service testers.

The component testers concentrate primarily on linear and discrete semiconductors, although some passive components are included in this category. Integrated circuit test systems include VLSI/LSI, memory, and microprocessor chips. Continuity testers of bare printed circuit boards represent the majority of applications in the interconnect field.

The loaded/fully assembled PC test systems are the most elaborate and are generally identified as either in-circuit or functional testers or a combination of both. In-circuit testers are designed to verify correct PC board assembly by assessing proper component placement and value. Test connection is accomplished by contacting the internal circuit nodes by use of a bed-of-nails fixture. Functional board testers, on the other hand, verify board performance by duplicating the electrical system environment under which the board operates. Access to the board is via the normal edge connector where the test system provides stimulus and evaluates responses.

Both test systems have a number of advantages and disadvantages. In-circuit testers are notably efficient in detecting and isolating both single and multiple faults during a single test cycle. Their shortcoming is that the time it takes to test a good board is only slightly less than that required to isolate a failed one. Functional test systems on the contrary are highly efficient at sorting out good and bad boards. Their disadvantage is that they are much slower in diagnosing and isolating failures. A number of ATE manufacturers have attempted to overcome the shortcomings of in-circuit and functional testing by combining the best features of both. The choice of a test system is based on the peculiar advantages and disadvantages each has to offer for a given application. The decision is heavily influenced by the cost of fault detection and isolation at one level versus the cost of allowing the failure to go undetected until the next higher level and paying the penalty of more expensive diagnosis and repair operations.

23.2. MARKET AND SUPPLIERS

ATE is the fastest growing segment of the test equipment market with a 10.5% growth in 1981 to $989.5 million, followed by 21.4% for 1982. Based on anticipated economic improvements over the longer term, ATE is predicted to grow 68.2% between 1982 and 1985. Within the total test equipment market, ATE share will increase moderately from 23% in 1981 to 28% by 1985 as specialized and general test equipment segments lose some ground. As indicated in Figure 23.3, this share is part of a total test equipment market base of $4,277 million in 1981 and rising to as much as $7,174 million by 1985.

Segmentation of the ATE market is shown in Figure 23.4. Fully assembled PC board and integrated circuit testers represent the bulk (74.2%) of the total market in 1981. By 1985 the combined share of these two segments are not

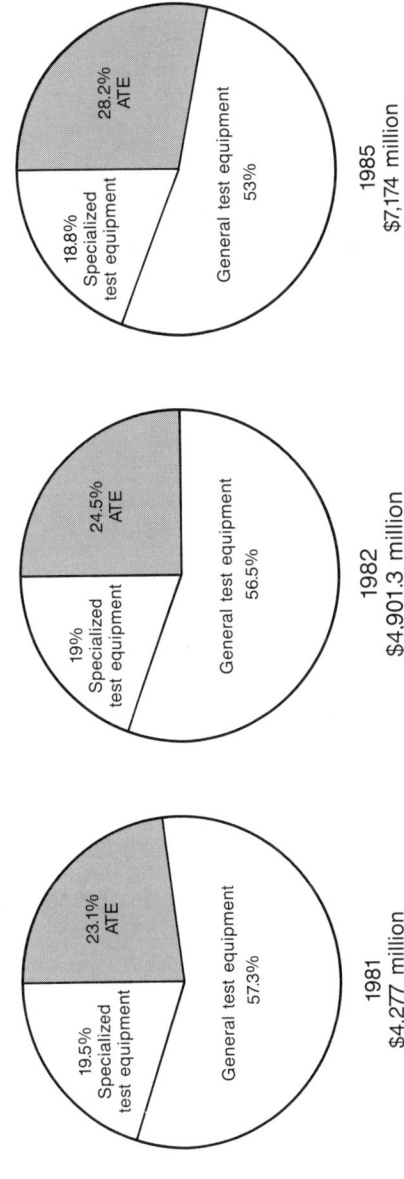

FIGURE 23.3. ATE market share of the U.S. test equipment market: 1981–1985.

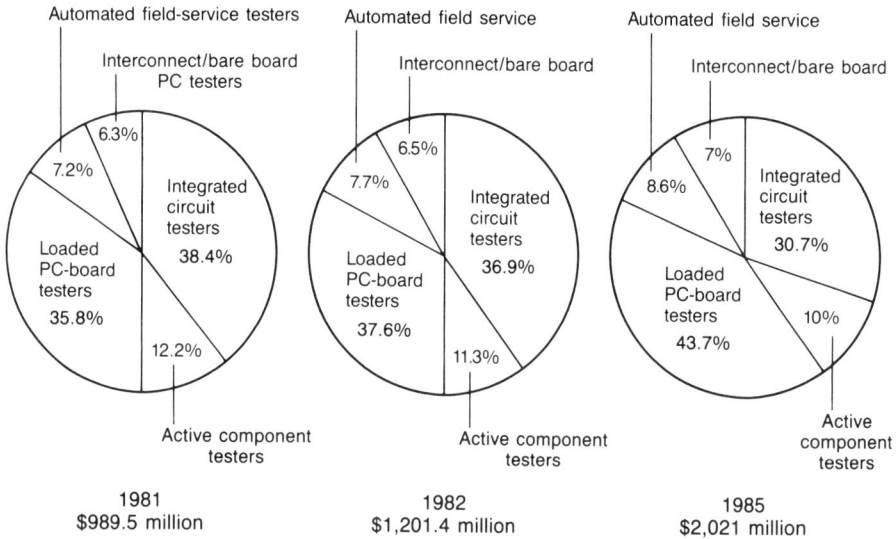

FIGURE 23.4. U.S. ATE sales by component type: 1981–1985.

expected to change significantly; however, the loaded PC board testers will rise to 43.7% of the market while the integrated circuit share will decline some 8%. This is due to the higher expected growth of the PC board market resulting in greater demand for automated testing. The total ATE market is expected to almost double from 1982 to 1985 reaching sales of $2,021 million.

The leading automatic test equipment manufacturers in 1982 are presented in Figure 23.5. Of these, the top three are all in the PC board tester market, an area where the largest growth is projected (see Figure 23.4 and 23.6). Fairchild and Teradyne are also leaders in the semi-conductor (integrated circuit) and interconnect-continuity fields.

Figure 23.6 illustrates the problems manufacturers face in matching their products against the changing demands of tester capability. A lot of this is attributed to market strength and weaknesses of products to be tested as well as constant technological changes of both the products under test as well as the testers themselves. In the semi-conductor area, 1984 showed significant increase in demand over the previous year for testers capable of testing linear ICs and discrete components. Requirements for large scale integrated (LSI) circuits and memories for the same period declined. However, for 1984 and beyond, the biggest demand for tester capability was supposed to emerge in the area of LSI circuits and memories.

In-circuit testing of PC boards continue to out run demand for functional testing, although the latter still do well in dollar sales. The strong demand for in-circuit capability is due to recognition of the benefits of fault diagnosis and correction at the PC board level.

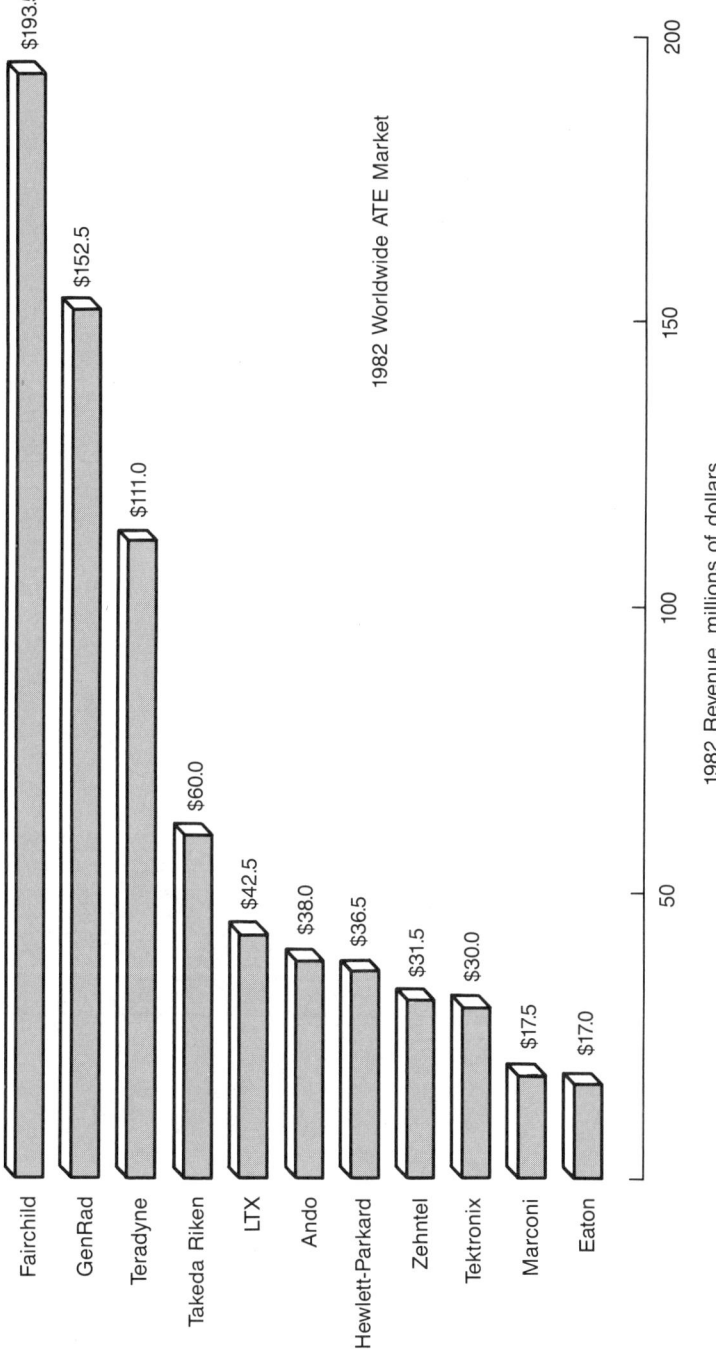

FIGURE 23.5. Leading manufacturers of automatic test equipment. (*Source*: Prime Data.)

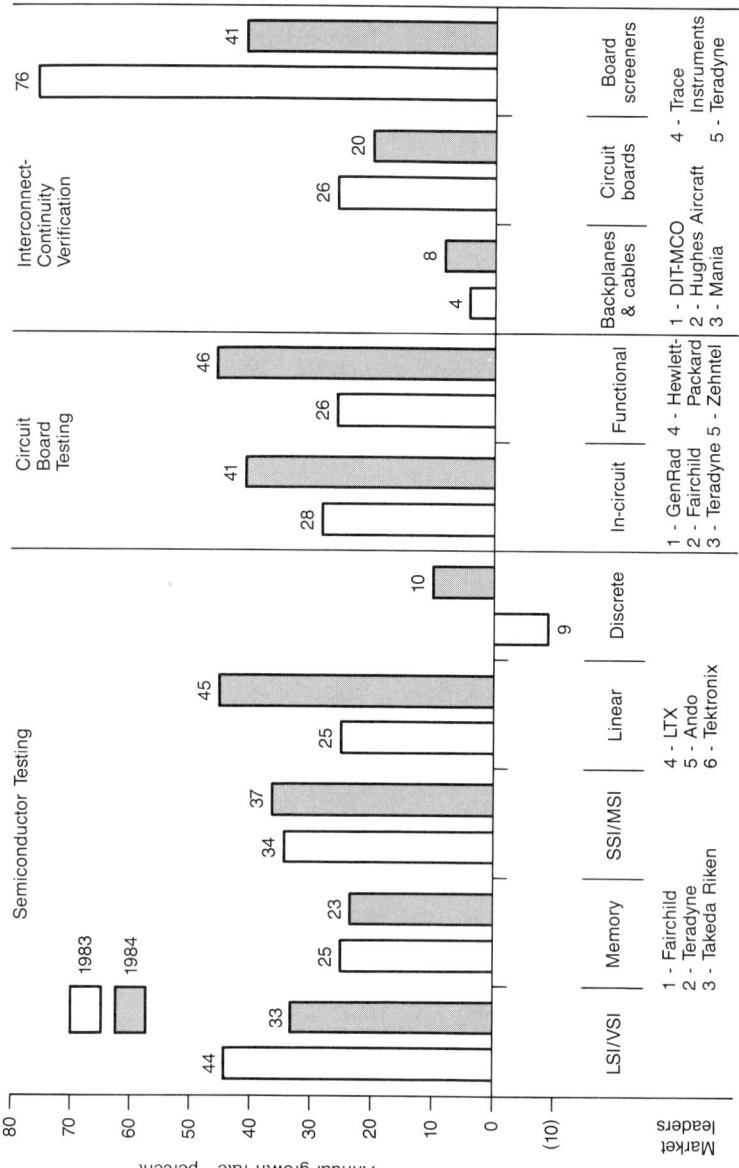

FIGURE 23.6. ATE market segments. (*Source:* Prime Data.)

23.3. PROCESSES

Various kinds of processes are employed by ATE in manufacturing testing, depending upon the application and volume of products produced. These processes include prescreeners of semi-conductors, passive components, and other discrete devices at incoming inspection; in-circuit and functional testing at the board level; and specialized computer-controlled simulators at higher assembly levels. These processes generally employ one of four testing techniques that are illustrated in Figure 23.7 along with their pros and cons.

In-system testing is one of the least expensive to perform from a test equipment standpoint, but its capability is severely restricted to failure detection with little to no capacity for fault diagnosis. It requires using one of the products being manufactured as a working test system. The device (component or board) under test is inserted into the working system in its normal operating position using a mother extender board, which provides edge pin accessibility. Known operational inputs to the system are then compared against desired outputs by the ATE. A device failing this comparison test is either rejected outright or fault isolation is instituted by making manual signal measurements at the pin connections.

Comparison testing is commonly used in the functional approach to testing PC boards. The test system compares the device under test against a known set of criteria, usually a reference board or a standard device. The test system is completely standalone, not requiring any product out of inventory other than the device being tested. It has the advantage of being able to stress the tested unit by varying the inputs over a prescribed range. It is the quickest method for detecting failures, but is much slower and less versatile for diagnosis than other methods.

In-circuit testing generally employs either a stored response technique or a series of algorithm-based exercises. Both techniques draw on the resources of the ATE's computer to a much greater degree than the other techniques. A wide range of stimuli and measurement can be programmed along with dynamic analog and digital test patterns at clock speeds up to 40 MHz. Fault isolation is accomplished through bed-of-nails contacts at circuit nodal points, where discrete components and integrated circuits can be electrically isolated and evaluated as individual elements. The stored-response method is widely used for component test in addition to circuit boards. Test application programs are virtually unlimited, restricted only by cost. A number of ATE manufacturers offer a complete library of software routines containing the operating characteristics of many of the commonly used TTL, CMOS, and VLSI semi-conductor devices.

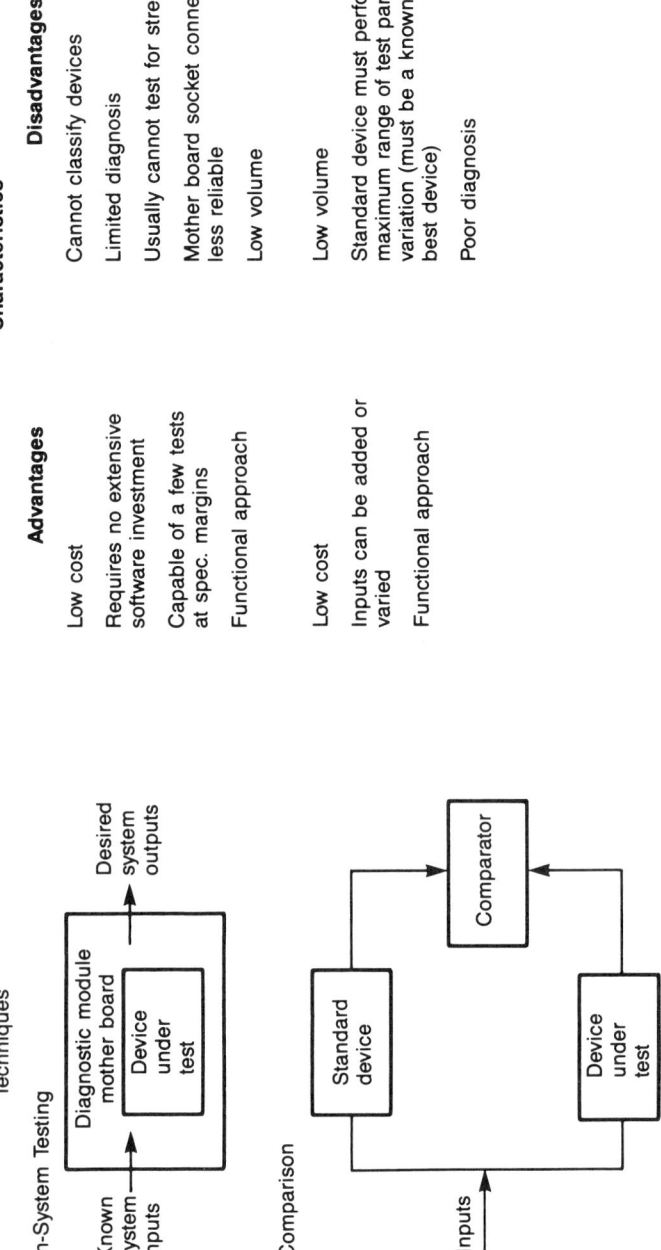

Techniques

In-System Testing

	Characteristics	
Advantages		Disadvantages
Low cost		Cannot classify devices
Requires no extensive software investment		Limited diagnosis
Capable of a few tests at spec. margins		Usually cannot test for stress
Functional approach		Mother board socket connection less reliable
		Low volume

Comparison

Advantages	Disadvantages
Low cost	Low volume
Inputs can be added or varied	Standard device must perform over maximum range of test parameter variation (must be a known best device)
Functional approach	Poor diagnosis

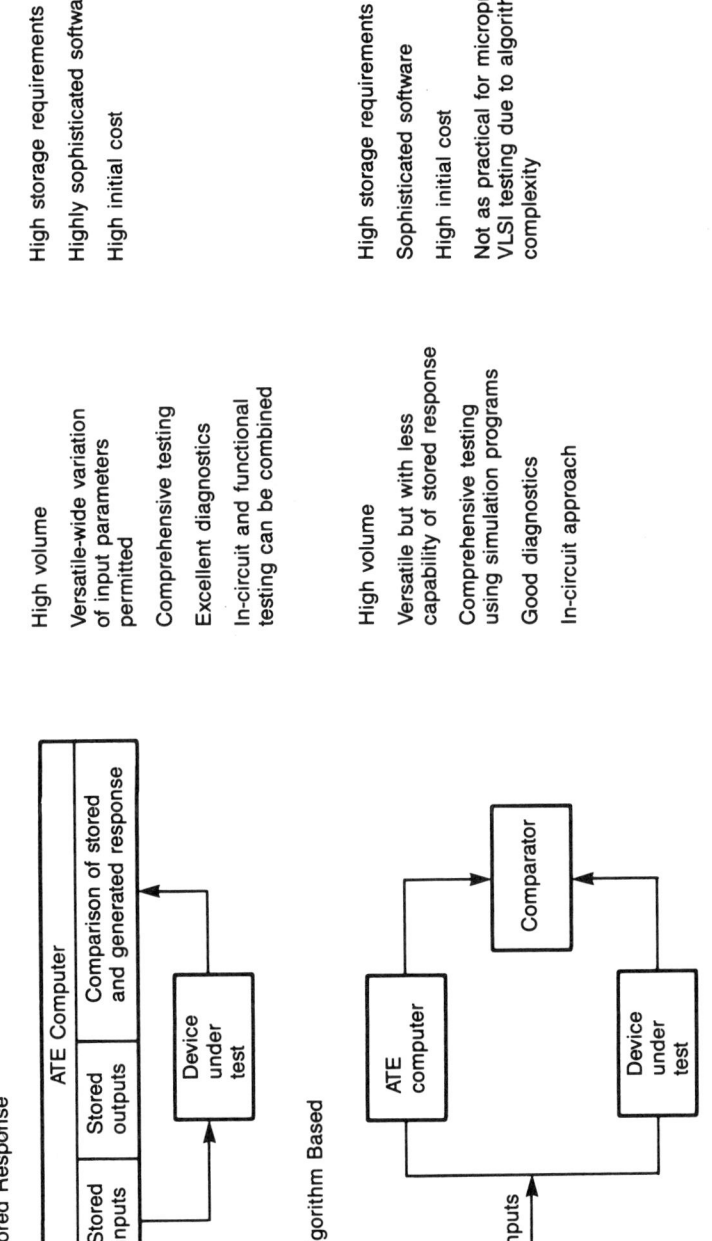

FIGURE 23.7. ATE techniques.

AUTOMATIC TEST EQUIPMENT

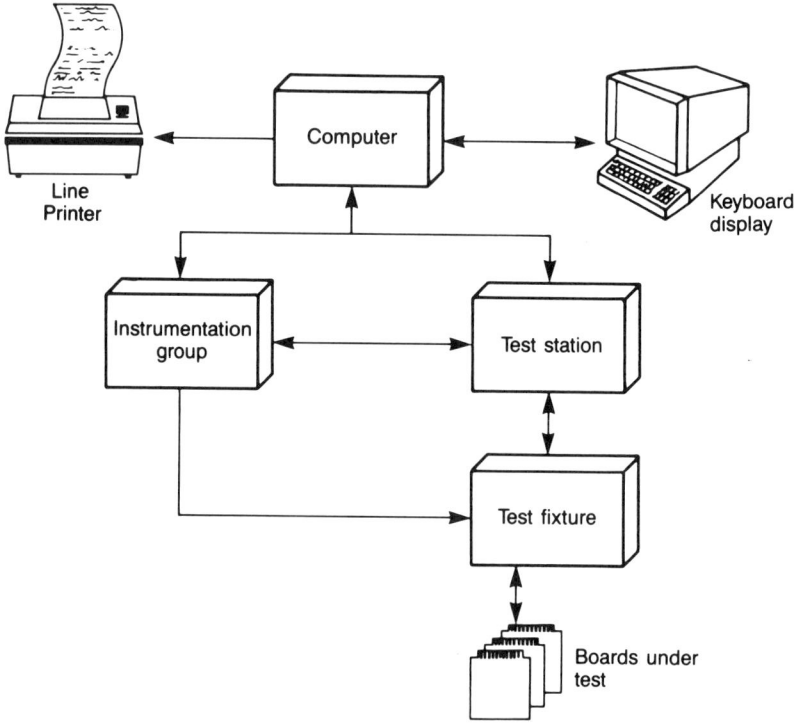

The algorithm method concentrates on simulating patterns that provide a more complete overall dynamic exercise of the unit being tested. This approach is not as well suited at the component level involving VLSIs or microprocessors due to the complexity of providing the large quantity of logic combinations that a test of these devices require.

The basic elements of a typical ATE board tester are shown above. The computer group under test program instruction sets up the appropriate stimulus and measurement functions of the instrumentation group. These functions are then routed to the proper pin connections of the unit under test via the test station and test fixture. The test fixture acts as an adaptor to match the particular edge connector of any unit to be tested to the ATE switching network.

Test results can be displayed at either the keyboard terminal or in hard copy from a line printer.

Most of the advanced systems have software generation capability using a high-level symbolic language that is captive to a manufacturer's brand. This language is often used in conjunction with existing libraries of device operating characteristics and bussing protocols such as the IEEE-488 interface standard.

The functions of a typical test program generator are illustrated by the following:

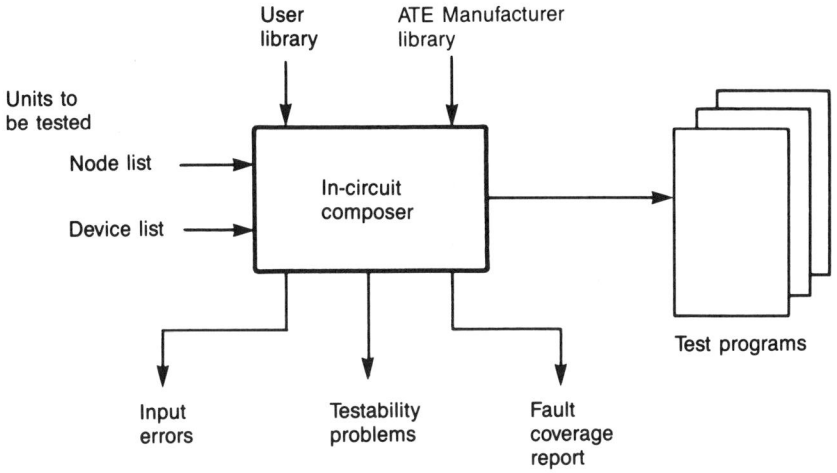

An important feature of ATE at the higher end of the scale is an ability to generate new test programs on-line without interfering with the tester's day-to-day manufacturing test schedule. This enters into the overhead burden of the total cost of manufacturing test and affects the return on an ATE investment.

A list of representative ATE systems and some of their features for component and board testing is presented in Table 23.1.

23.4. COST

Factors affecting the cost of testing include a majority of the following:

Quantity of units tested.
Characteristics such as size, and number and types of components.
Quality of materials and fabrication.
Environmental operating requirements.
Test methods and thoroughness of test.
Test equipment characteristics and cost.
Labor content.

Technology advances over the years have done much to alter the cost of testing. Higher circuit packaging densities, faster clock speeds, and broadening of the frequency spectrum have contributed towards obsoleting the manual approach to instrumentation and measurement. The Figure 23.8 shows the impact on the testing caused by one segment of the electronics field, semi-

TABLE 23.1. Representative ATE Systems

Component Testers

Model	Manufacturer	Features
Sentry VII	Fairchild	Performs functional tests to 10 MHz. 60 programmable PINS.
M110A	Micro Control Inc.	Functional tests patterns up 10 MHz. Not completely programmable—requires changing performance boards for each device tested.
MX–17	ADAR Associates	Uses reference device to generate test patterns one clock cycle ahead of device under test. Cassette device for program storage.
T8048	Systems Sales Inc.	Test all functions of Intel 8048 family only.
More-Recent Systems		
GR-16 & 18	General Radio (Gen Rad)	VLSI tester with 288 PIN capacity—tests microprocessors, microcomputers, chips, RAMs up to 40 MHz.
System 7900	Accutest	Precision VLSI tester; 50 MHz rate, 144 PIN capacity.
Sentry 21	Fairchild	VLSI tester at 40 MHz rate, 120 PIN I/O.

PC Board Testers

Model	Manufacturer	Features
3060A	Hewlett-Packard	Performs advanced in-circuit component measurements and board level functional stimulus and response tests.
5800	Data Test Corp.	Fixed pattern and programmed pattern stimulus. Programmed response and transitional redundancy check.
2225	General Radio (Gen Rad)	Functional field tests and fault diagnosis.
"Herbie"	Bendix Corp. Test Systems Div.	Go/no go testing of digital board, modules and devices with a limited debugging capability.
More-Recent Systems		
GR 2271, 2272	General Radio (Gen Rad)	Tests: ECL, VLSI, CPUs, Memory and I/O boards; 3584 hybrid PIN count; in-circuit test, has extensive device library of TTL, CMOS, VLSI, etc.
30/333	Fairchild	Combination of in-circuit and functional test modes (capacity of 2207 digital and 959 analog test points).
System 80	Marconi	In-circuit tester with emphasis on fault isolation and reduced repair time.
3050A	Fluke	Analog/digital functional tester with diagnostic capabilities measured against known good board tests; LSIs up to 10 MHz with 240 PINS.

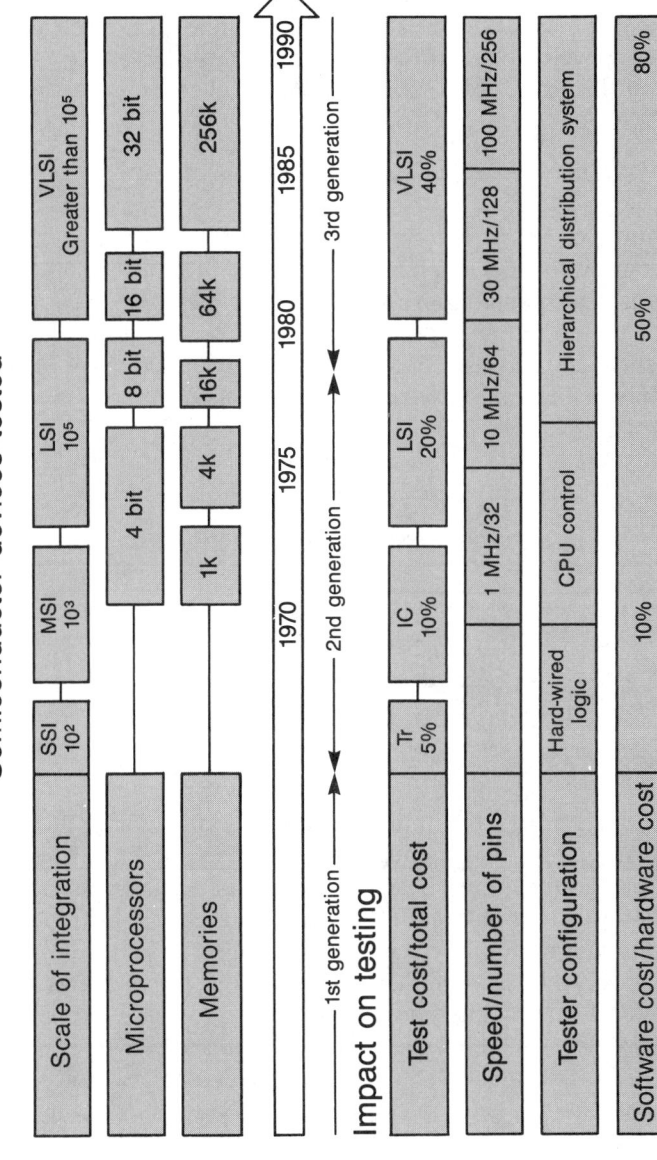

FIGURE 23.8. Impact of semi-conductor technology on testing. (*Source*: Japan Electric Measurement Instruments Manufacturer's Association—*Electronic Business*, October 1982.)

conductor technology. The scale of integration is projected to grow substantially over the next decade, with 32-bit microprocessors and 256K memories as common as 16-bit and 64K chips. The higher logic speeds and rising pin counts of these devices forces corresponding increases in tester versatility and complexity. Testing will absorb an increasingly larger share of overall device cost, rising to 40% of the cost of VLSI from 20% of today's LSI. More significantly, test program development costs will increase even faster, with software costs approaching the cost of the ATE equipment.

23.4.1. Test Economics

The economics of testing involves the assessment of the characteristic of the device being tested (its complexity, function, type, and quantity of components); some understanding of rate of failures per batch; test and diagnosis times based on test method and equipment employed; labor content in work-minutes per device; and cost. The test-and-repair analogy basic to all manufacturing testing is illustrated by the following diagram:

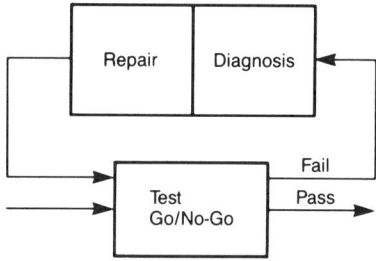

The total cost of testing is the summation of this analogy across all elements and levels over the complete manufacturing cycle of the product. The test method employed has a large bearing on the quality of testing and the cost in the diagnosis and repair of failures. Table 23.2 compares the attributes of various testing techniques at different levels of test. These techniques are illustrated diagrammatically by the test models in Figure 23.9. The in-circuit approach can be either single or multiple phase depending upon tester capability or test program strategy. In both cases, once the failure is detected its location (diagnosis) is known, due to the in-circuit tester's ability of accessing circuit nodal points. The test process stops at that point while repair is instituted, usually off-line. Testing then resumes if all faults have not been detected on the first pass. Although taking longer to complete, in-circuit testing accomplishes failure detection and diagnosis in one operation.

Functional testing identifies failures in a much shorter time than in-circuit

TABLE 23.2. Test Method Times and Labor Rates[a]

Test Method	On-Line						Off-Line			
	Test Time	Handling Time	Diagnosis Time	Repair Time	Utilization Factor	Test Labor Rate	Diagnosis Labor Rate	Repair Labor Rate	Diagnosis Time/Fault	Repair Time/Fault
In-Circuit Test (Single Phase)	50 seconds	—	—	—	0.80	$6.00	$7.00	$5.00	2 minutes	7 minutes
In-Circuit Test (3-Phase)	t_1 = 18 seconds t_2 = 12 seconds t_3 = 20 seconds	—	—	—	0.80	$6.00	$7.00	$5.00	2 minutes	7 minutes
Functional Board Test	2 seconds	8 seconds	60 seconds	—	0.80	$6.00	$7.00	$5.00	2 minutes	7 minutes
Analog Functional Test	15 minutes	2 minutes	15 minutes	7 minutes	0.80	$9.00	—	—	—	—
Systems Test	50 minutes	5 minutes	—	—	0.80	$13.00	$7.00	$5.00	30 minutes	7 minutes

[a]Direct labor without overhead.
[b]From Applying quality curves for economic comparisons of alternative test strategies, by K. P. Taschioglou, 1981 IEEE Test Conference © 1981 IEEE.

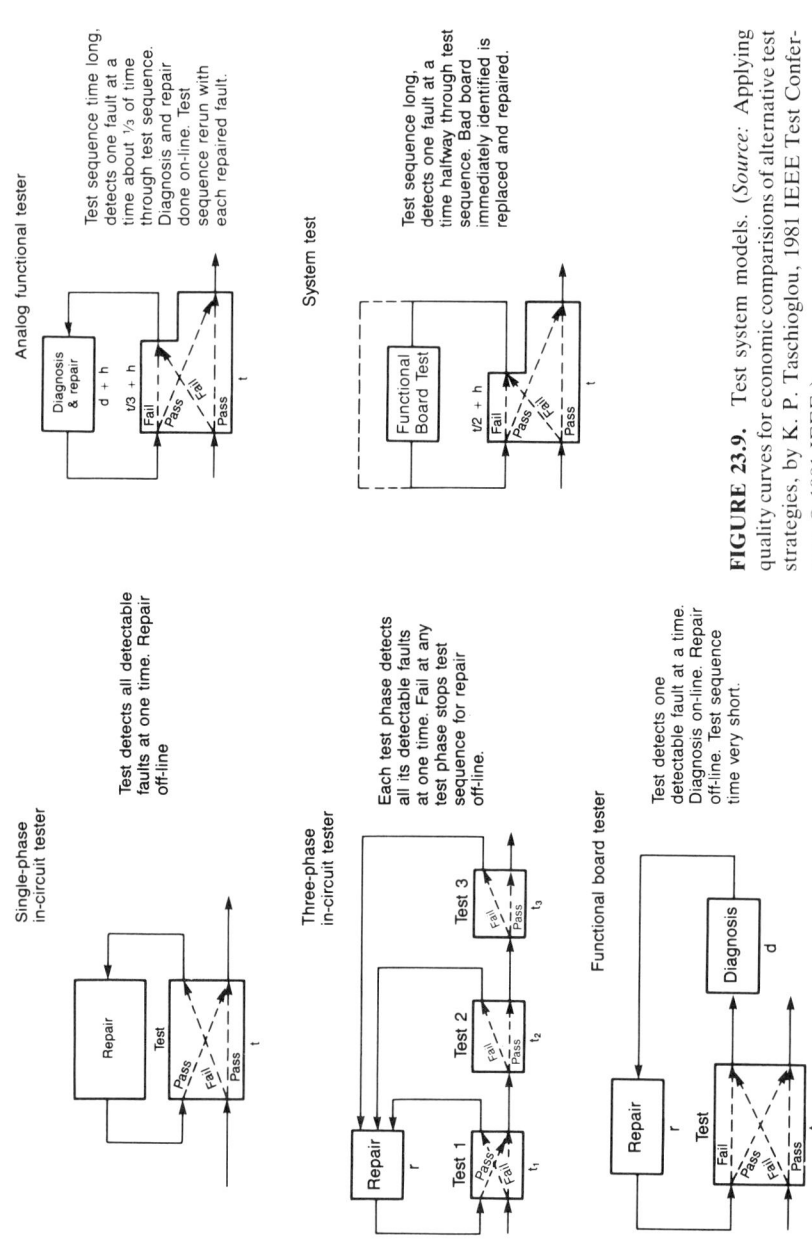

FIGURE 23.9. Test system models. (*Source:* Applying quality curves for economic comparisions of alternative test strategies, by K. P. Taschioglou, 1981 IEEE Test Conference © 1981 IEEE.)

testing; however, once identified, failure diagnosis must be done on-line, often requiring manual probing of circuit paths, tying up the tester from testing of other devices.

System test has the disadvantage of requiring long test sequences due to multi-fold complexity of the unit at the higher assembly stages. Failure detection requires that the faulty board or module be replaced before the remainder of a test can be completed. Replaced units have to be returned to functional and in-circuit levels for diagnosis and repair, necessitating that system test have a complete stock of replaceable boards and modules in order to insure test sequence completion.

Table 23.2 provides test-sequence times and labor time and rates for the process steps involved in fault detection, diagnosis, and repair. As a general rule of thumb for large quantities of units to be tested, it becomes economically attractive to functionally test all devices first, which sorts out bad from good units in the shortest possible time. Those units identified as failed can then be retested and diagnosed on the slower but more thorough in-circuit tester. Table 23.3 compares the test times and ATE system cost for the in-circuit and functional test methods.

Note that functional testing has significantly shorter test time as well as more complete fault-detection coverage. However, the same method requires more than double the time to diagnose the source. Equipment cost (another factor) trades-off programming and fixture cost between the two methods. These cost of quality trade-offs can also be expressed graphically, in terms of labor cost per unit tested and number of detected faults. Figure 23.10 shows this for PC boards where, for a given number of detected faults, the cost of labor can be found for the different test models described in Figure 23.9. The least labor

TABLE 23.3. Test Times and ATE System Cost for Circuit Boards

Test Time (Per Failure)	Test Method	
	In-Circuit	Functional
Fault detection	3.0 minutes	1.3 Minutes
Troubleshooting	3.3 minutes	8.0 Minutes
Fault coverage	92%	98%
System Cost		
Equipment	$100K–$175K	$150K–$250K
Test programs	1K– 3K	2K– 5K
Test fixtures	1K– 2.5K	50– 100

Source: "The Economics of Testing," by John Cline. *Evaluation Engineering,* September–October 1981.

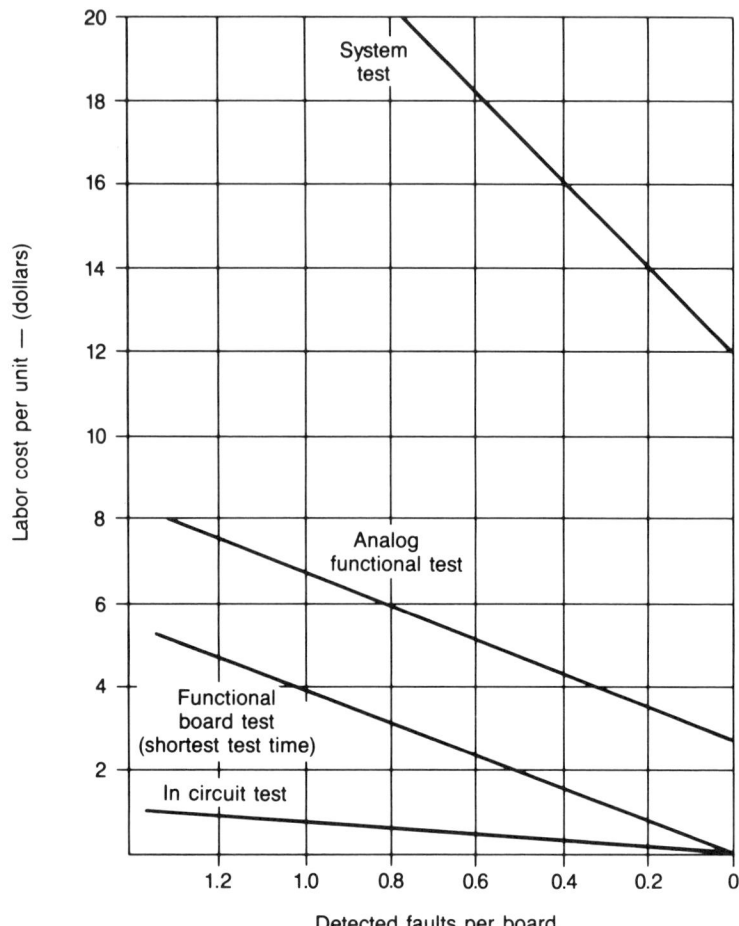

FIGURE 23.10. Cost of quality for failures detected at different test stages. (*Source:* Applying quality curves for economic comparisons of alternative test strategies, by K. P. Taschioglou, 1981, IEEE Test Conference © 1981 IEEE.)

cost is for functional and in-circuit testing where the former has the edge for boards with low fault indices.

An important cost element is the quality level of the product being tested. Devices experiencing higher levels of defects will necessitate greater emphasis on diagnostics and repairs. This would favor in-circuit over functional testing. Filling this category are VLSIs and very complex circuit boards which have higher incidence of manufacturing shrinkage due to circuit density. Distribution of good and bad yield data for a typical PC boards containing 60 integrated circuits and 50 passive components (resistor, capacitors, etc.) is shown as follows:

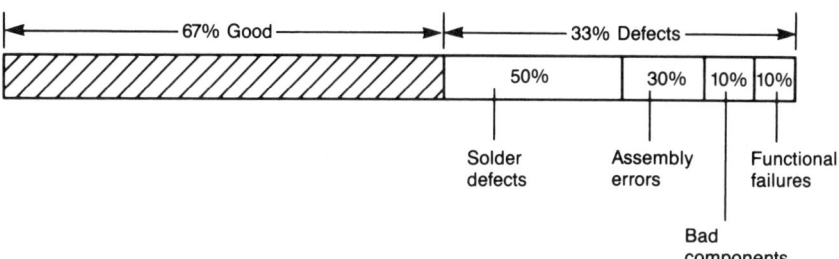

(*Source:* "The Economics of Testing," by John Cline, *Evaluation Engineering*, September–October 1981.)

23.4.2. Cost Calculations

Approximate costs of testing can be calculated by selecting the appropriate test method (from Figure 23.9, System Models) and then applying the test-and-repair times from either Table 23.2 or Table 23.4 and multiplying by the labor rates. The following examples are typical:

Printed Circuit Board
    (Number of good boards) × (Test time) × (Labor rate) = $ _____
    (Number of failed boards) × (Diagnosis-and-repair time) × (Labor rate) = $ _____
    (Number of failed boards) × (Retest loop factor) × (Test time) × (Labor rate) = $ _____
        Total cost = $ _____

The number of good and failed boards out of a tested batch may be calculated from expected or given failure-rate data or by assuming such values from a description of board complexity and component characteristics. For lack of better data, the failure occurrences listed in Table 23.4 may be used, with suitable adjustments for products that are new and have not established a quality curve.

Software test program development costs can be estimated from the example worked out in Table 23.5. The example is based on testing large-scale integration and microprocessor devices. The total labor content can also be used for most PC boards.

An important cost factor is whether or not the generation of test programs involves taking the test system off the production line. Appropriate allowances should be made if this is the case. Although a $50/hour fully loaded labor rate is used in the example, a different rate may be employed as appropriate.

TABLE 23.4. Test and Repair Data for Typical Circuit Boards

	Time (Minutes)
Test and Failure Detection	
Shorts test on assembled board	0.33
In-circuit test	3.00
Functional test	1.30
System test	4.00
Failure diagnosis on functional test	4.00
Diagnosis and Repair	
1 Failure after shorts test	2.00
1 Failure after in-circuit test	3.30
1 Failure after functional test	8.00
Manual diagnosis and repair	
1 Failure after system test	120.00

	× (Test + Repair)
Times Around Repair and Retest Loop	
After shorts test	1.0
After in-circuit test	1.2
After functional test	1.2
After systems test	1.5

	Percentage Rate of Occurrence/Board
Fault Categories	
Defective passive components	1.0
Defective active components	3.0
Etched shorts per bare board	0.04
Etched opens per bare board	0.02
Solder shorts per assembled board	0.1
Solder opens per assembled board	0.05
Component insertion errors	0.05
Functional faults/board	0.3
Undetected faults—functional test	0.02
Undetected faults—any ATE	0.01

Source: Selection of test strategy for best return on investment, James J. Faran, Jr., 1980 IEEE Test Conference © 1980 IEEE.

TABLE 23.5. ATE Test Program Development Costs[a]
(Large Scale Integration, Microprocessors)

Process Steps	Using On-Line Production Test System			Using Off-Line Test System Simulator		
	Labor Hours	Cost at $50/Hour[b]	Percentage	Labor Hours	Cost at $50/Hour[b]	Percentage
Program conception	40	$ 2,000	20	40	$ 2,000	22
Generation, compilation and editing	100	5,000	46	70	3,500	39
Integration and debugging	60	3,000	30	60	3,000	33
Clean up and documentation	8	400	4	8	400	˙6
Total labor	208 hours	$10,400	100	178 hours	$ 8,900	100
Lost productivity while test system is removed from service	133 hours	$39,900 at $300/hour		40 hours	$12,000 at $300/hour	
Simulator cost	—	—		8 hours	$ 752 at $8/hour	
Total cost		$50,300			$21,652	

[a] *Source*: "Cost Reduction Method of LSI Test System Programming," by Ron Griffin and David Mees, *Electronics Test*, April 1982.
[b] Fully burdened rate.

SECTION TWENTY FOUR

PACKAGING AND PACKING

24.1.	Packaging and Packing Methods	305
24.2.	Time and Material Requirements	306
24.3.	Calculation Method	306
24.4.	Packing Material Price List	311
24.5.	Reusable Shipping Containers	316

Packing, packaging, and preservation is a very complex and detailed field best left in the hands of a specialist. It is not easy to devise a packing system that will adequately protect equipment in shipping and storage, even with the help of the many available standards and specifications. In fact, the set of standards and specifications is so complex that establishing compliance is on occasion almost as difficult as protecting the equipment.

Only a few levels of protection are commonly used to pack and package electronics equipment. Within each level, additional variables must be specified in detail, but these will in general have negligible effects on the overall labor and material requirements, for the accuracy of this level of estimating. This section describes five methods of packing that cover the range of protection normally required for electronics equipment. The methods used here are derived from MIL-P-116, "Methods of Preservation."

The terms "packing" and "packaging" are used loosely to mean the same thing, but they have distinct meanings. "Packaging" refers to the material immediately surrounding an item to protect it from shock, vibration, and moisture; the packaging is generally not exposed to handling or the weather. "Packing" refers to the material that contains one or more packages to protect them from handling and the weather. Thus, items are first packaged, then one or more packages are packed in a shipping container.

24.1. PACKAGING AND PACKING METHODS

Table 24.1 shows the packing and packaging methods covered in this section. Method III is the lowest level of protection considered here. It includes a layer of cushioning wrap and a close-fitting weatherproof container, which is a cardboard carton for smaller sizes and a wooden box for larger ones.

Four levels of protection are included beyond Method III, two of physical protection and two of moisture protection. "Moderate" physical protection is provided by Methods IA8, IC1, and IIC. These include wrapping, cushioning, and a moisture-proof bag to package each individual item, plus packing equivalent to Method III for one or many packages. "Maximum" physical protection is provided by Methods IA14, and IIB, which include an individual carton or box for each package, in addition to the provisions of "moderate" protection.

Moisture protection is provided by water-vapor-proof barrier bags in each of the above methods except Method III. Additional protection in Methods IIC and IIB is provided by placing desiccant within the moisture barrier; a humidity indicator card is required as well.

Electronic components and equipment fall into three size ranges. Small components and PC boards are generally smaller than about 200 in.^{3} volume and 200 in.^{2} of surface area. Hardware is usually packed with Method III; PC boards and small subassemblies are typically packaged with sub-method IA8 or IC1.

Rack-mount modules, individual pieces of test equipment, and so on fall into an intermediate size range of 0.5 to 3 ft.^{3} or about 500 to 2000 in.^{2} of surface area. Any of the five packing and packaging methods given may be required for items in this size range. Other, similar methods may be specified in actual practice, but the costs will closely correspond to the comparable level of protection included here.

Equipment in racks, larger cabinets, major modules, and so forth seldom

TABLE 24.1. Packaging and Packing Methods

| Physical Protection | Moisture Protection | Typical MIL-P-116 Sub-Methods | Packaging | | | Packing |
			Inner Container	Barrier Bag	Desiccant	Outer Container
Minimal	—	III				X
Moderate	Sealed bag	IA8, IC1		X		X
Moderate	Sealed bag and desiccant	IIC		X	X	X
Maximum	Sealed bag	IA14	X	X		X
Maximum	Sealed bag and desiccant	IIB	X	X	X	X

run much larger than about 500 ft.^{3} or 10,000 in.^{2}. These are usually packed for maximum physical protection, Method IA14 or IIB. Larger packing jobs require consideration that is beyond the scope of this section, and should be estimated by other means.

24.2. TIME AND MATERIAL REQUIREMENTS

The time and material described in Section 24.3 are sufficient to perform all direct charge activities required to package an item. Package design, or packaging engineering time is not included. Labor is given in standard minutes and includes a 15% allowance for personal, fatigue, and delay, but not time for rework or other variances.

Custom low-volume packing is much more expensive than standardized high-volume operations, but is typical of the requirements of high-quality electronic equipment manufacturing.

The basic data in Section 24.3 are for single quantities. Time and cost decrease rapidly when more than one unit is to be packed.

Custom packaging times follow approximately a 65% curve. That is, the time required for each package goes down quickly as the number of identical packages goes up. Section 26 discusses learning curves in general, but the 65% curve and instructions necessary to use it in estimating packing and packaging are included in Section 24.3.

Industry experience is that packaging and packing material costs follow approximately a 75% material discount curve. This is quite steep, but is due primarily to the initial costs of custom containers. Bulk materials, such as wrapping and cushioning material, labels, tags, tape, and so forth contribute little to the material costs for low-volume packaging. The major cost component is custom cartons or boxes. Although the stock from which they are made is relatively inexpensive, set-up costs and high initial scrap rates make small quantities quite expensive.

Section 29 discusses material discount curves in general, but the 75% curve and instructions for using it to estimate packing and packaging material costs are included in Section 24.3.

The material costs given in this book reflect 1982 prices. A representative list of material prices is given in Section 24.4; escalation of materials can be made if necessary. As a practical matter, packing materials have escalated relatively slowly, averaging about 3½% from 1964 to 1982. For additional information see Section 27, "Material Costs and Allowances."

24.3. CALCULATION METHOD

The general procedure is to first estimate the cost of packaging the individual items, then add the cost of packing the packages in shipping containers,

CALCULATION METHOD

and finally add them together to arrive at a total cost in minutes of labor and dollars of material cost.

Packing and packaging cost data are supplied in chart form, plotted against the total surface area of the package in square inches. Figure 24.1 gives the labor required in minutes as a function of package surface area for each of the five methods described above. The data for each of the four packaging methods (that is, all methods except Method III) do not include packing requirements, as often several packages are combined into a single pack. Figure 24.2 gives the material cost data in a similar format.

Figure 24.7 is a sample cost estimating worksheet and Figure 24.8 shows a sample calculation. Following Figure 24.8, the first step is to record the size and number of items to be packed, and the method to be used. If packaging is required, as is nearly always the case for electronics equipment, use the appropriate lines from Figure 24.1 and 24.2 to determine the labor and material required. Packaging the units of the example, which have surface areas of 80

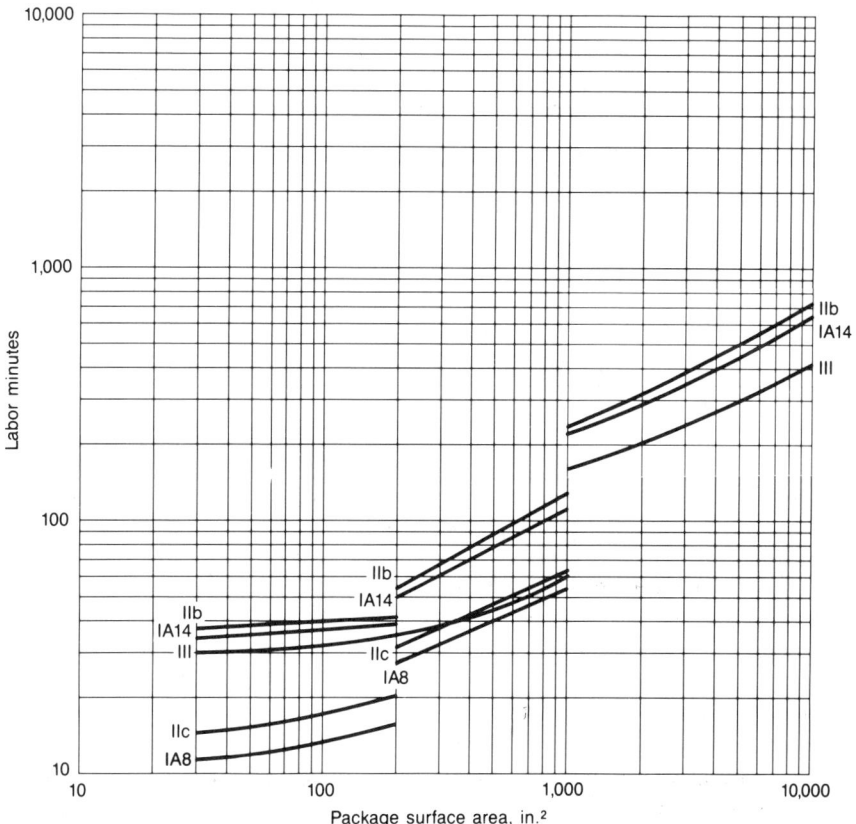

FIGURE 24.1. Packing and packaging time.

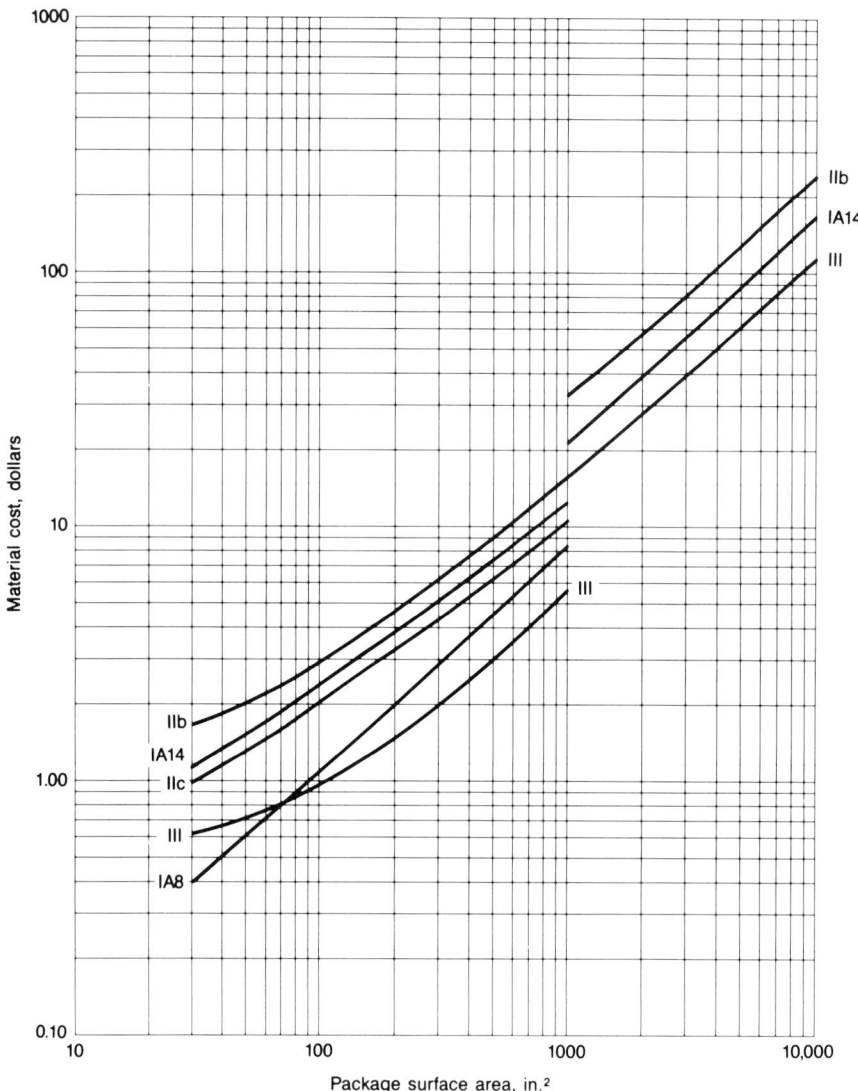

FIGURE 24.2. Packing and packaging material cost.

square inches, requires 13.2 minutes of labor, by Figure 24.1 and $0.86 of material by Figure 24.2.

The values of time and material cost from these figures are for single units. Use the cumulative average cost factors from the 65% learning curve, Figure 24.3 for time, and the 75% material discount curve, Figure 24.4 to determine the average costs for the given quantities. Multiplying the single unit costs

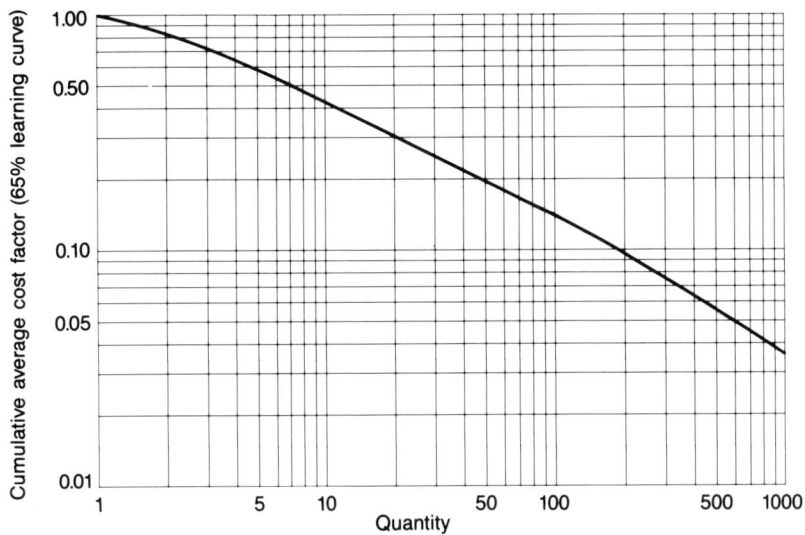

FIGURE 24.3.

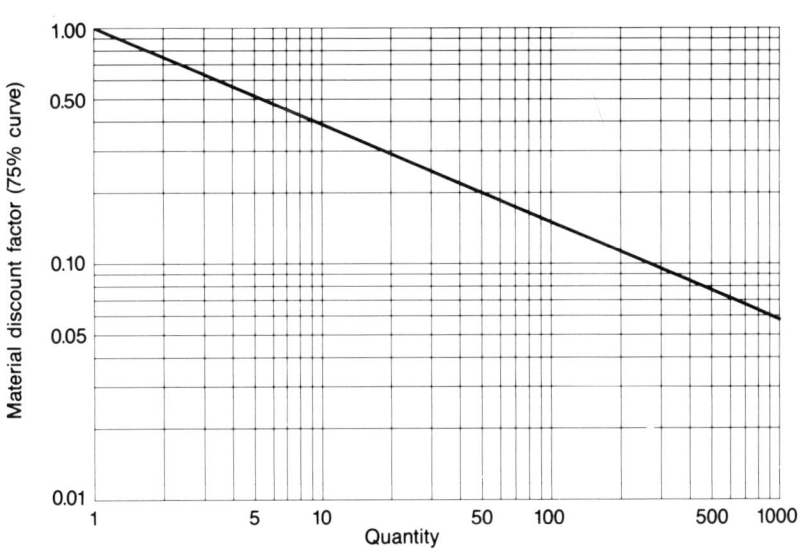

FIGURE 24.4.

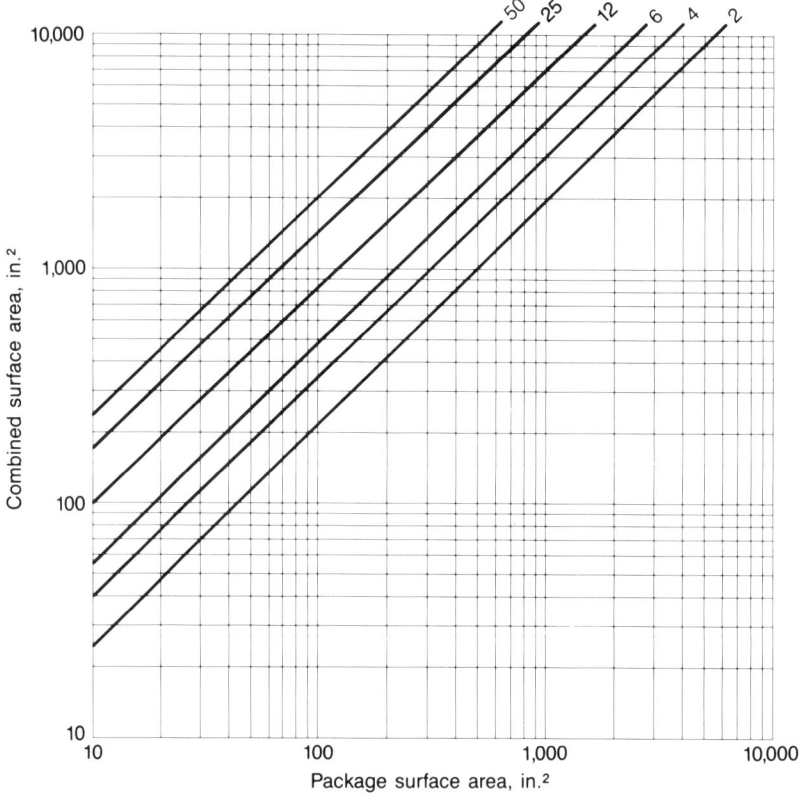

FIGURE 24.5. Combined surface area of a group of several packages assembled into one shipping container.

times the cumulative average cost factors and times the number of units involved gives the total time and material required to package the lot.

To determine the costs of packing the lot, first find the number of units to be shipped together in each shipping container, and divide into the lot size to find the number of shipping containers required.

To find the surface area of each shipping container, enter Figure 24.6 with the surface area of the individual units and the number in each container. Enter Figure 24.6 with the same information to find the time required to assemble the packages into a group for loading into the shipping container. In the example, Figure 24.5 shows that the surface area of a group of ten unit packages is 550 in.^{2}; 0.28 minutes are required to assemble them into a shipping container.

Next, use the surface area of the group to enter Figures 24.1 and 24.2, this time to find the Method III packing requirements. Adding the labor to assemble the packages into a group and that to pack the group in accordance with

PACKING MATERIAL PRICE LIST

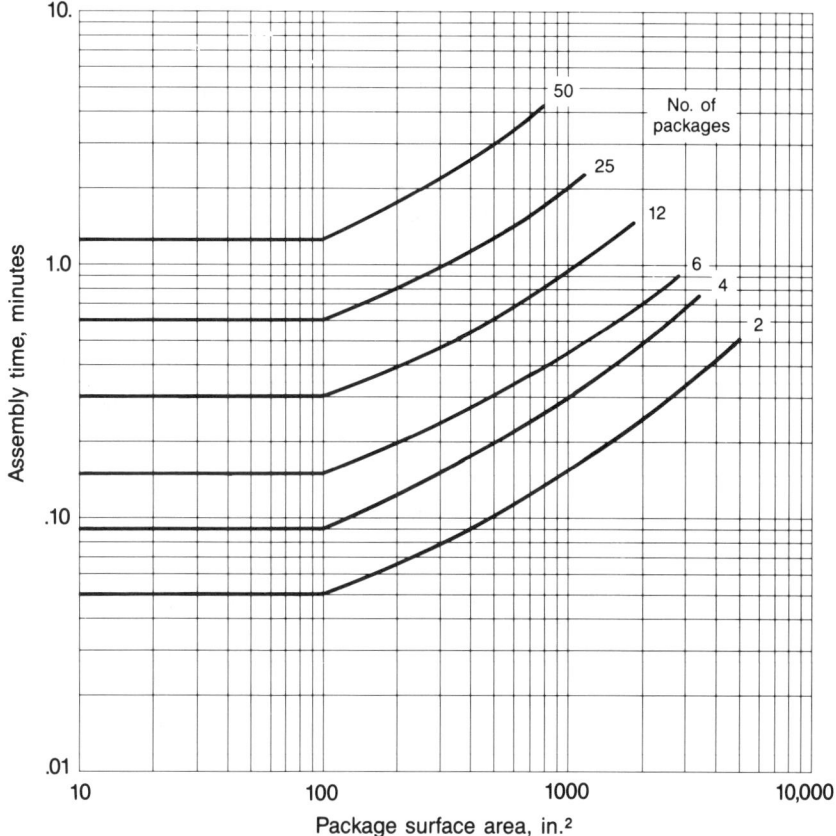

FIGURE 24.6. Time to assemble a group of several packages into one shipping container.

Method III, find the time for one pack, which is then multiplied by the number of packs required to complete the lot. Adding the packaging and packing requirements gives the totals for the entire lot.

24.4. PACKING MATERIAL PRICE LIST

Table 24.2 presents a representative list of the packing and packaging materials, and their costs. Escalation of these values can be accomplished with the methods of Section 21 as required. As a practical matter, packing materials costs tend to escalate slowly, on the order of 3½% per year between 1964 and 1982.

DATE: _____   BY: _____

BASIC DATA:
Items in Lot _____ Packaging method _____
Item Surface Area _____ Sq. In.
Items in Each Shipping Container _____
Shipping Containers in Lot _____

PACKAGING:
TIME _____ Min. (Figure 24.1)
   × _____ Learning Cost Factor (Figure 24.3)
   × _____ Items
1.   = _____ Min.

Cost $ _____ (Figure 24.2)
   × _____ Material Discount Factor (Figure 24.4)
   × _____ Items
2.   = $ _____

PACKING:
Shipping Container Surface Area = _____ Sq. In. (Figure 24.5)
TIME _____ Min. (Figure 24.1, Method III)
   × _____ Learning Cost Factor (Figure 24.3) for # of Shipping Containers
   = _____ Min.
   + _____ Min. to Assemble Packages (Figure 24.6)
   = _____ Min.
3. × _____ Shipping Containers in Lot   = _____ Min.
COST $ _____ (Figure 24.2, Method III)
   × _____ Material Qty. Discount Factor (Figure 24.4)
   × _____ Shipping Containers
4.   = $ _____

TOTALS 1 + 3   = _____ Min.
        2 + 4   + $ _____

FIGURE 24.7. Packing and packaging cost estimating worksheet.

DATE: 5/10/82   BY: RKJ

```
BASIC DATA:
    # Items in Lot ____10____     Packaging method __IA8__
    Item Surface Area __80__ Sq. In.
    Items in Each Shipping Container __10__
    # Shipping Containers in Lot ____1____
```

PACKAGING:
  TIME __13.2__ Min. (Figure 24.1)
   × __.47__ Learning Cost Factor (Figure 24.3)
   × __10__ Items
1   = __62.0__ Min.

  Cost $ __0.86__ (Figure 24.2
   × __.37__ Material Discount Factor (Figure 24.4)
   × __10__ Items
2.   = $ __3.18__

PACKING:
  Shipping Container Surface Area = __550__ Sq. In. (Figure 24.5)
  TIME __45.__ Min. (Figure 24.1, Method III)
   × __1.0__ Learning Cost Factor (Figure 24.3) for # of Shipping Containers
   = __45__ Min.
   + __.28__ Min. to Assemble Packages (Figure 24.6)
   = __45.28__ Min.
3. × __1__ Shipping Containers in Lot   = __45.28__ Min.
  COST $ __3.70__ (Figure 24.2, Method III)
   × __1.0__ Material Qty. Discount Factor (Figure 24.4)
   × __1.0__ Shipping Containers
4.   = $ __3.70__

TOTALS 1 + 3   = __107.3__ Min.
       2 + 4   + $ __6.88__

FIGURE 24.7. Packing and packaging cost estimating worksheet.

TABLE 24.2. Packaging Material Price List

Material	Mil-Spec Number	Stock Size	Cost($)[a]
Wraps			
Kraft, waterproof, greaseproof	Mil-B-121 grade A, type II, class 2	36" × 200 yds	$0.046/ft.^2$
Kraft, greaseproof, waterproof moldable	Mil-B-121 grade C, type II, class 2	36" × 200 yds	$0.086/ft.^2$
Volatile corrosion inhibitor	Mil-I-3420		$0.06/ft.^2$
Non-Electrostatic	Mil-B-81705	36" × 200 yds	$0.15/ft.^2$
Transparent, waterproof, greaseproof	Mil-B-22191 Type II	36" × 200 yds	$0.088/ft.^2$
Polyethylene film	L-P-378	Varies	$0.015/ft.^2$
Dunnage			
Flexible corrugated	PPP-C-291	Varies	$0.04/ft.^2$
Cellulosic cushioning	PPP-C-843		$0.05/ft.^2$
Bubble wrap	PPP-C-795		$0.07/ft.^2$
Rubberized hair	PPP-C-1120		$0.22/ft.^2$
Polyurethane foam, 2 in. thick, 1.25 density	PPP-C-26514	2 × 8 ft.	$0.32/ft.^2$
Polyethylene foam, 2 in. thick, 2.2 density	PPP-C-1752	2 × 9 ft.	$1.20/ft.^2$
Barrier Bags			
Waterproof, greaseproof	Mil-B-121 grade A, type II, class 2	12 × 15 in.	$0.12/ft.^2$
Water-vapor-proof	Mil-B-131, Class 1	12 × 15 in.	$0.26/ft.^2$
Water-vapor-proof	Mil-B-131, Class 2 (used for less than 10 lbs.)	5 × 6½ in.	$0.04/ft.^2$
Non-electrostatic	Mil-B-81705	8 × 10 in.	$0.22/ft.^2$
Transparent, waterproof greasepoof	Mil-B-22191, type II	6 × 8 in.	$0.07/ft.^2$
Containers			
Corrugated fibreboard	PPP-B-636		
200 lb. test		6 in. cube	0.21 each
		12 in. cube	0.39 each
275 lb. test		24 × 24 × 18 in.	2.15 each
Miscellaneous			
Tape			0.04/ linear ft.

[a] Based on supplier costs, Boston area, April 1982.

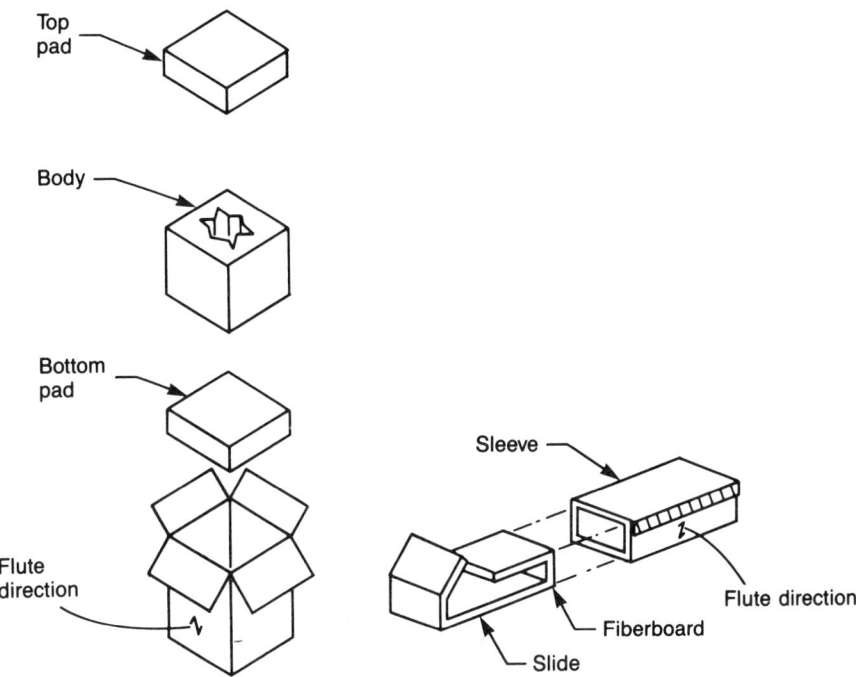

FIGURE 24.9. Star pack.

FIGURE 24.10. Slide pack.

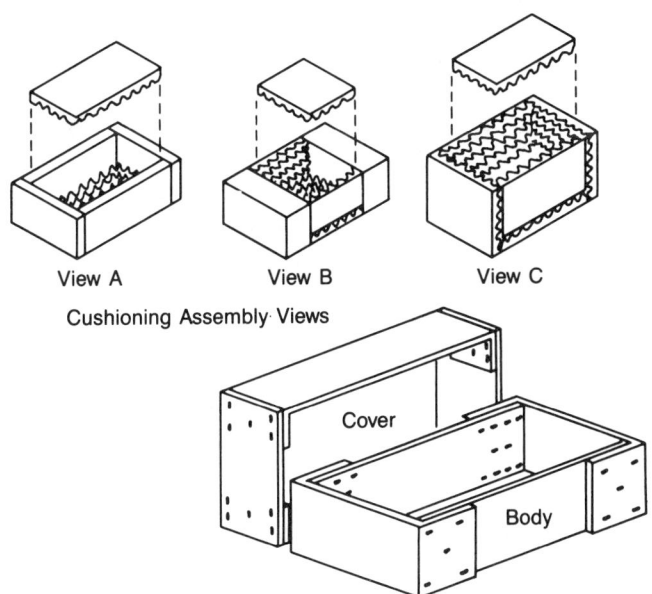

FIGURE 24.11. Full telescoping cover reusable carton.

TABLE 24.3. Reusable Shipping Container Prices

Type	Size (In.)	Unit Price (Quantity 25–50)
I	6 × 6 × 10	$ 9.63
Star pack	12 × 12 × 14	18.32
	14 × 14 × 16	28.27
II	6 × 5 × 2½	5.23
Slide pack	9 × 6 × 3½	5.70
	12 × 8 × 3½	5.94
	18 × 12 × 3½	6.00
III	25 × 14 × 14	38.15
Full scope box		

24.5. REUSABLE SHIPPING CONTAINERS

Reusable shipping containers, such as cardboard cartons with fitted foam cushioning, offer advantages over one-time-use packaging methods. First, the container is reusable. Even though the initial cost is higher, there is a long-term benefit to the user for logistically recycled equipment. A container used to ship a replacement printed circuit board, for instance, can be used to return the defective one for repair.

Moreover, due to the integral construction of the cushioning, less labor is required to pack items in these containers. These are often referred to as fast packs for this reason. The labor required is about equal to that for Method III (minium physical protection), whereas the protection is comparable to Method IA8 or IIC (moderate physical protection).

Figures 24.9 through 24.11 illustrate typical configurations of reusable containers. A list of prices for various sizes and configurations is presented in Table 24.3. These prices were effective in mid-1982, and apply to quantities of 25–50.

SECTION TWENTY FIVE

LEARNING CURVES

25.1.	Basic Principles	318
25.2.	Industrial "Learning"	318
25.3.	Applications of Learning Curves	320
25.4.	Mechanics of the Learning Curve	321
25.5.	Learning Curve Selection	324
25.6.	Uses of Learning Curves	329
25.7.	Completion vs. Expenditure, Considering Learning	332

The learning curve is a way of analytically describing something everybody knows: the time needed to do a job goes down with experience, but it goes down more rapidly at first. The first published analysis of this phenomenon was made by aircraft manufacturers in the late 1920s and early 1930s. Managers noticed that the direct-labor hours required to manufacture airframes decreased rapidly with experience. In fact, the direct labor required for the tenth plane of a series tended to be less than half that required for the first. Investigation showed that the decrease in labor for successive units was adequately described by an exponential function, and the learning curve was born.

The learning curve is a powerful estimating tool. It makes possible the use of standard time data for estimating, regardless of the production quantity involved. Standard times apply to well-established, well-planned work carried out by experienced workers. In practice, it takes many production units to achieve this situation; standard hours are normally well below what is actually required to do the work. As will be demonstrated later, the proper way to use standard data is to determine how many units would have to be produced to achieve standard performance, and then use the learning-curve techniques to properly account for the additional time that will inevitably be required for low-volume runs or for the first portion of high volume runs.

25.1. BASIC PRINCIPLES

The basic principles of the learning curve are straightforward. Experience has shown that the graph of production labor against number of units produced is very nearly a straight line when plotted on log-log paper. This means that labor goes down by the same amount with every doubling of the quantity produced.

A particular learning curve is characterized by a percentage, such as "the 80% learning curve." On the 80% learning curve, each doubling of the production quantity brings the labor required to 80% of its former value. For example:

Unit Number	Labor Required (80% Learning Curve)
1	100
2	$80 = 100 \times 0.80$
3	70.2
4	$64 = 100 \times 0.80 \times 0.80$
5	59.6
6	56.2
7	53.4
8	$51.2 = 100 \times 0.80 \times 0.80 \times 0.80$

On the 90% learning curve, each doubling of the production quantity brings the labor cost to 90% of its former value.

Practical application of the curve is somewhat more complicated. In particular, the estimator is normally interested in what the average unit cost will be, or in the total cost for a given production run, rather than in the cost of a specific unit. Methods of calculating these are presented in Section 25.4. The intervening sections describe further the basis and application of learning curves.

25.2. INDUSTRIAL "LEARNING"

The "learning" in "learning curve" encompasses much more than the improvement of the skill of the individual worker at a single task over time. Taking "learning" to refer to the decrease in unit labor requirements with experience in producing these units, learning is not the result of faster soldering, or hole drilling, or bolt tightening. Rather, it is the collective effect of many people in the production organization working to improve the process. The individual worker learns to plan his job better, and spends more time doing and less time getting ready. The planners learn to coordinate work better, so that less time is spent waiting for material or for the next prior step to be finished. In all, learning is of four major types:

(a) Operator Learning. The operator learns the sequence and specific techniques required by his operation.

(b) Improved Methods, Processes, Tooling, Machines, and Designs for Producibility. Improvements in any of these areas will reduce the direct unit labor required. Some improvements of this type will happen simply as a result of people looking for better ways to do the work. Others will result from technological advances in other fields. Most, however, are because large quantities make more specialized tooling, machines, and so on possible economically. These have high initial costs which can be offset by the reduced labor per unit across large quantities.

(c) Management Learning. Every person in the manufacturing process from the machine operator to the plant manager experiences learning the idiosyncrasies of a new production unit as it applies to the management of his portion of the process. The net result is overall improvement in the management system. The causes for material shortages are eliminated. Plant layouts are improved. Paperwork systems and procedures are improved. All of the activities supporting the direct labor worker undergo an improvement as the job progresses. This contributes to the reduction of indirect labor.

(d) Debugging of Engineering Data. Minor dimensional and material callout errors on engineering drawings can cause large additions to the time required to produce the first units of a new design. Identifying and correcting drawing and data errors is a major task, which is properly done before production begins. This is normally not possible, causing high initial costs, but rapid improvement toward lower costs.

A major factor in the adoption of computer-aided design technology has been the reduction of engineering data errors, with the attendant reduction in direct labor costs.

Although these elements are necessary to the best realization of learning through experience of the production plant as a whole, they are not always sufficient. The learning curve is a powerful tool, but to be of value it must be used, and it must be used carefully.

Labor reduction per the learning curve does not just happen, it must be made to happen. Many companies have bid quantity production jobs on the learning-curve theory and then sat back, waiting for the labor reduction to occur. To their dismay, this has not always happened. Here are the additional time-reduction ingredients predicted by the learning curve.

1. Detailed production planning and manufacturing engineering.
2. Suitable special tooling, fixtures, and test equipment.
3. A stable design. The next best thing is a means of segregating the costs of engineering changes, whether or not they increase the complexity of the unit.
4. Production goals which incorporate the time reduction of the learning curve. It is important that these goals be both realistic and specific.

A shop that is left to devise its own methods and tooling and sets its own standards of output may well show little or no improvement across a whole production run. On the other hand, a shop that has help in the form of planning, methods analysis, tooling, cost reporting systems, and attainable goals will find it normal to keep itself within the projected cost-reduction rate.

25.3. APPLICATIONS OF LEARNING CURVES

Learning curves were first used in the aircraft industry, but they have been successfully applied in fields from oil refining to electronics manufacture.

The shipbuilding industry in World War II was found to follow an 80–85% learning curve. Producers of complex machine tools have found that the labor required for new models progresses downward at rates between 75 and 85%. Detailed work measurement of repetitive clerical and bookkeeping operations have resulted in curves in the 75–85% range. Repetitive machining and punch press operations fall into the 90–95% range. Ranges for the electronics industry are given in Section 25.5, but vary from about 75%–90%, depending on the nature of the work. Machine work shows comparatively little learning (high learning-curve percentages) as opposed to manual work, which shows high learning (low learning-curve percentages).

Within an industry, learning curves are useful for more than just forecasting total labor cost. Some of the important applications are:

(a) Internal Cost Control. Learning curves were originally developed from historical data as an after-the-fact observation that could be expected to recur in similar situations. Plans based on the learning curve generally are closer to what the true output will be than simple averages, and so can be used more reliably to set consistent, attainable production goals.

(b) Delivery Rate. The rate of improvement of a group of workers is important in setting the delivery schedule to be expected. Note that it is important to take into account changes in the workforce. New worker's performance on a given job will back-up the learning curve compared with experienced ones.

(c) Labor. The learning curve effect should be taken into account when forecasting the manpower requirements of a program. To a large extent, schedule duration and labor quantity are related. The effect of learning will be to increase the rate of delivery or to reduce the number of workers required to maintain delivery rate. In practice, it is of course usually far preferable to maintain a level workforce. When labor adjustments are made, the learning curve can be used to predict the rate at which the new adjusted workforce will produce; it will be different than that of experienced personnel, although not as different for inexperienced personnel in an inexperienced production environment.

The learning curve is also useful for taking into account the effects of breaks in the work flow. A process that is well down the learning curve will exhibit a step backward if it is stopped for a period of time and restarted.

(d) Facilities. The learning curve is useful in predicting the amount of facilities and equipment necessary to accommodate a given production rate by relating the facilities needed to the size of the workforce.

Finally, it is important to note that the addition of more tooling is accompanied by an increased cost per unit until personnel experience and debugging have been manifested. The temporary jump in unit cost which accompanies the second set of tooling and the additional crew is especially significant.

25.4. MECHANICS OF THE LEARNING CURVE

The characteristic curve of unit cost vs. quantity drops steeply at first and then levels off in the general manner of Figure 25.1. It was found empirically that the same values follow essentially a straight line when plotted on a log-log scale, as in Figure 25.2. The following general formula describes such a relationship

$$Y = A X^{-k}$$

where

Y = Cost per unit
A = Theoretical cost of the first unit
X = Cumulative number of units produced
K = Exponential of the slope (See Table 25.1.)

The logarithmic form of the expression is

$$\log Y = -K \log X + \log A$$

which shows analytically how the relationship plots as a straight line on log-log paper.

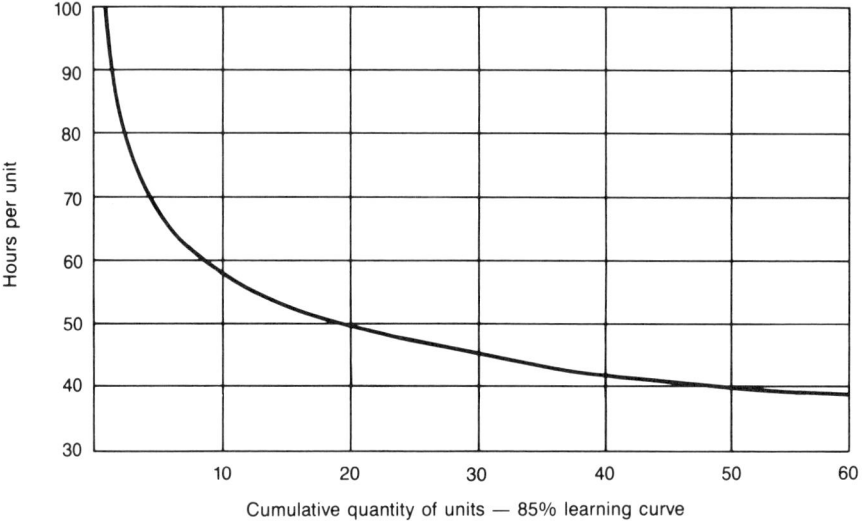

FIGURE 25.1. Characteristic learning curve.

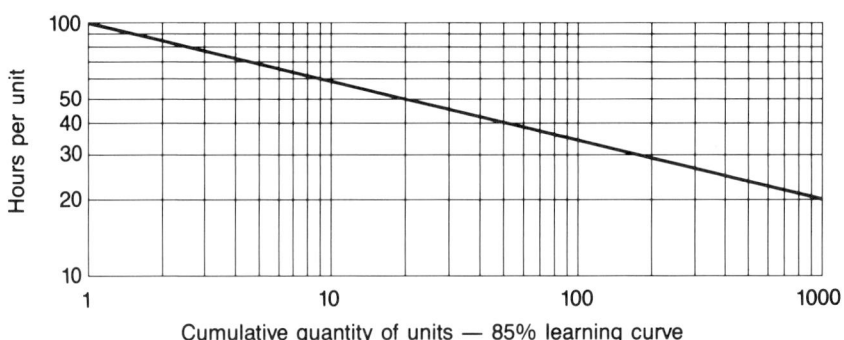

FIGURE 25.2. Learning curve (log-log scale).

Both individual unit cost and cumulative average unit cost are required. The use of estimating standards such as those given elsewhere in this book leads to unit costs, but the estimator is generally seeking the cost of the entire production lot, which is based on the cumulative average.

For production quantities above 50 or 100, the unit cost and the cumulative average cost plot as straight lines that are parallel to each other. Parallel straight lines on a log-log scale differ only be a constant factor. Table 25.1 gives the cumulative average factors for a range of learning-curve percentages. The cumulative average factor is the ratio between cumulative average cost and unit cost.

TABLE 25.1. Learning-Curve Factors

Percentage of Curve Slope	Exponent $\left(K = -3.32 \log \dfrac{\%}{100}\right)$	Cumulative Average Factor $\left(\dfrac{1}{1-K}\right)$
60	0.737	3.80
61	0.713	3.48
62	0.689	3.22
63	0.666	2.99
64	0.643	2.80
65	0.621	2.64
66	0.599	2.49
67	0.577	2.36
68	0.556	2.25
69	0.535	2.15
70	0.514	2.06
71	0.494	1.98
72	0.474	1.90
73	0.454	1.83
74	0.434	1.77
75	0.415	1.71
76	0.395	1.65
77	0.376	1.60
78	0.358	1.56
79	0.340	1.52
80	0.322	1.47
81	0.304	1.44
82	0.286	1.40
83	0.269	1.37
84	0.251	1.34
85	0.234	1.31
86	0.217	1.28
87	0.201	1.25
88	0.184	1.23
89	0.168	1.20
90	0.152	1.18
91	0.136	1.16
92	0.120	1.14
93	0.105	1.12
94	0.089	1.10
95	0.074	1.08

For small production quantities the unit-cost and cumulative-average lines approach each other. Obviously the cumulative average and unit cost are the same for the first unit. Therefore, only one of these can plot as a straight line. The two major methods of applying the basic learning-curve theory are based on the two possible choices.

The Crawford method holds that the unit cost follows a straight line on log-log paper, and that the cumulative average is thus a derivative function that bends outward from the first unit to eventually become parallel to the unit-cost curve. The Wright method holds that the cumulative average follows a straight line, and that the unit cost is the derivative function. The Crawford method describes a situation where the unit cost decreases by a constant amount with each doubling of the production quantity; the Wright method describes a situation where the cumulative average cost decreases by a constant amount with each doubling of quantity.

Both methods find widespread use, and either can be used to obtain satisfactory results. It is important not to mix methods in application, however, especially when dealing with relatively small production quantities. At quantities above 50 to 100, where the unit cost and cumulative average lines have become parallel for all practical purposes, the methods are essentially the same.

The Crawford method is given here, where the unit cost follows the straight-line learning-curve mathematics, and the cumulative average is a curve that bends outward from the first unit to become parallel to the unit cost curve.

As shown in Figure 25.4, the percentage of difference between the true cumulative average curve and the straight line approximation decreases rapidly. The difference at 10 units is only about 10%, dropping to about 6% at 30 units, and to less than 1% at 100 units for the 85% learning curve.

Curves for a number of commonly used learning curve percentages are given in Section 25.5. If values for other percentages are required, they can be obtained graphically or analytically. Table 25.1 gives the required values of the exponent K and the ratio of cumulative average to unit cost for the range of common learning-curve percentages. Remember that the cumulative average factor is only accurate for large quantities. For smaller quantities either prepare a curve, based on calculating a few values and fairing a smooth curve through the resulting points, or use a programmable calculator to calculate the values exactly.

25.5. LEARNING CURVE SELECTION

The learning curve has been discussed so far in generalities. The crucial question is of course, "what curve for what job?" The variables are many. One situation calls for projecting total direct labor effort on a plant-wide basis. Here, the set-up as well as the recurring labor is included. Rework of inspection rejects and engineering changes are also included. Another situation calls for recurring labor only while another calls for the projection of a specific operation. Others call for a projection of machining labor versus assembly and test labor.

Like any line, the learning curve is completely determined by two points or by a point and a slope. In practice, the estimator uses point and slope. He or she has available the standard time and some knowledge of the job, the production facility, and the workforce. To pick a learning curve the estimator must decide on the slope of the learning curve and on the number of units required to attain standard production performance. He or she can then use the learning curve, or the associated cumulative average curve to determine the cost of the early production units. On short runs, the standard time may never be achieved, but the principles still apply.

Since it is not possible to use one learning curve in all situations, it is the purpose here to give a basic family of curves and some general guidelines for choosing among them. The user can then choose a curve based on knowledge of the company and the job. For instance, one of the greatest variables of all is at what level has the estimate or standard been set? One company reaches their standard in 100 units. Another company in 1000 units, and still another company in the mass production field in 10,000 or 100,000 units.

Table 25.2 presents an overview of the major variables controlling the rate of progress and the number of units required to reach the standards given in this book. Figures 25.3 through 25.5 are learning curves for typical percentages.

Complexity of Product. This variable could just as well be stated "complexity of operation." It is given here in terms of average job cycle in standard hours and in terms of the total labor content of the unit. The typical units' names and total labor content are given only as approximate points of reference. The basic idea here is that learning is based on the number of times an operation is performed, not on the number of complete units produced. The standards in this book are generally attainable in 1000 units. If there are 10 similar subassemblies in an equipment rack, then performance on the subsassemblies will reach standard in about 100 racks.

The operation cycle time gives a useful way of estimating how many units must be built to reach standard. In general, the smaller the average cycle time, the more times that cycle will be performed per complete unit, and the smaller the number of complete units required before that portion of the labor is performed at standard.

Newness of Production (Maturity). The rate of progress varies with the "newness," or amount of innovation possible, which in turn depends in part on characteristics of the work and in part on the experience of the production facility with that general type of work.

A useful rule of thumb is that operations that are made up of only machining or other processes show essentially no reduction of time from unit to unit, which would correspond to a 100% learning curve. Operations that are purely

TABLE 25.2. Learning Curve Selection Chart. Curve Slope and Number of Units Required to Meet the Standard Hour Estimate.

Newness of Product—Opportunity for Innovation
←———— Large   Small ————→

Complexity of Production by Fabrication Assembly Standard Hour	New Product New Manufacturing Methods Standard Hour Based on 1–1,000 Quantity		New Product Standard Manufacturing Methods Standard Hour Based on 1–1,000 Quantity		Variation of Basic Product Standard Manufacturing Methods Standard Hour Based on 1–1,000 Quantity		Mass-Produced Product Standard Manufacturing Methods (Punch Press Components, and so on) Standard Hour Based on 1,000–10,000 Quantity	
	Individual Operation Progress: Recurring Work only	Total Program: Start-Up, Recurring, and All Variances	Individual Operation Progress: Recurring Work only	Total Program: Start-Up, Recurring, and All Variances	Individual Operation Progress: Recurring Work only	Total Program: Start-Up, Recurring, and All Variances	Individual Operation Progress: Recurring Work only	Total Program: Start-Up, Recurring, and All Variances
Type unit								
Ave. job cycle								
Total hour/unit								
Components								
0–0.05	Negligible	Negligible	Negligible	Negligible	Negligible	Negligible	95% 1,000	90% 1,000
0–1.0	80% 100	75% 100	85% 100	80% 100	90% 100	85% 100	95% 10,000	90% 10,000
Subassemblies								
.06–0.20								
1.1–10.0								
Rack chassis								
0.21–0.50	80% 300	75% 300	85% 300	80% 300	90% 300	85% 300	—	—
11–100								
Full racks								
0.51–2.00	80% 1,000	75% 1,000	85% 1,000	80% 1,000	90% 1,000	85% 1,000	—	—
101–1,000								

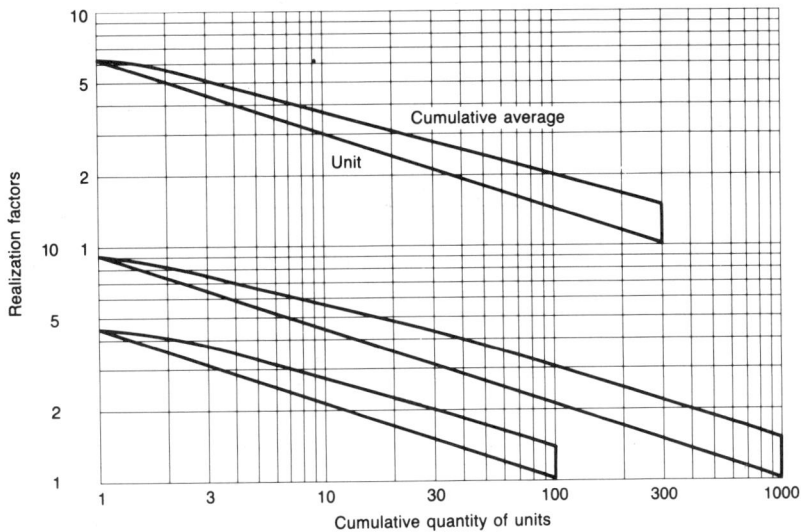

FIGURE 25.3. 80% learning curve.

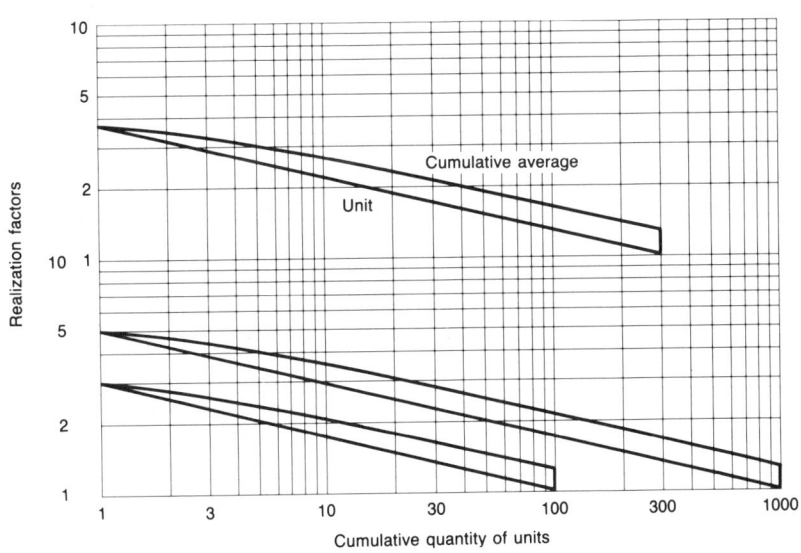

FIGURE 25.4. 85% learning curve.

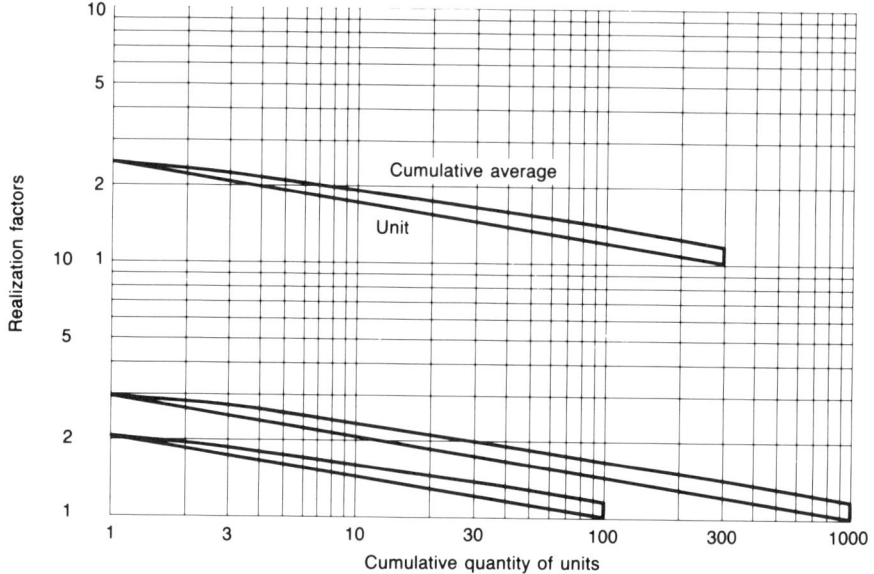

FIGURE 25.5. 90% learning curve.

manual in nature, and contain no machine or process time, follow about a 60% learning curve. Since most operations are mixed, a useful rule of thumb is that machining operations as a group follow a 90% learning curve and assembly operations follow an 80% curve.

This rule should be used with care; some machining operations, particularly those involving new alloys, require the development of new methods. The learning curves for these fall in the 70 to 80% range. Similarly, some assembly operations show little or no improvement on subsequent operations. This is usually because the operation is made up of simple combinations of very small operations, which are quickly learned to standard and which improve very little thereafter. For instance, assembling 100 identical resistors to a PC board will show very little learning from board to board, because the learning took place from resistor to resistor. Assembling each of 100 different resistors to 100 different points will show 80% to 85% learning.

It should be noted also that a product or type of work that is new to one company may be merely a variation of an established product line for another. There is inherently less innovation possible in the experienced workforce in a facility that has been producing similar products for a number of years and exhibits a 90% rate, whereas a company building a type of unit for the first time will perform at about the 85% rate.

The 85% rate is representative of the average electronics unit. A newly developed unit that is advancing the state-of-the-art will show more rapid learning, at the 75 to 80% rate. Note that the rates that apply to the individual operations do not include other costs of a total program.

Individual vs. Total Program Progress. Another variable in the rate of progress is the amount of start-up labor that is prorated in the cost per unit. An 85% curve represents the time reduction of an individual operator doing the recurring work on a new product using standard manufacturing methods. On the other hand, if the total labor effort of set-up cost, repair of inspection rejects, and miscellaneous non-recurring activities are added to the strict recurring cost, it will be found that the total labor effort is more closely approximated by an 80% curve. The difference of individual versus total program curves holds true for each category of product newness.

Job Shop vs. Mass Production. The time standards of this book are achievable in the 1–1000 quantity range. They are geared to job shop electronics equipment manufacturing. The mass-production learning curve is given as a comparison where the standard is normally set with much greater precision and based on the methods used for the 1000 to 100,000 production range. As a result the standard is much tighter and a mild learning process is experienced before the standard is met. The learning curve for this type production is not steep, but it is of a much longer duration in terms of number of units.

25.6. USES OF LEARNING CURVES

The usual use of learning curves in estimating is to determine time requirements for production quantities less than that required to reach standard time. Figures 25.3 through 25.5 are curves for 80 to 85%, and 90% learning rates.

The curves are given in terms of realization factors, which equal one at standard and some higher value for each lesser production quantity. The following examples show how to use them to estimate a first production lot and succeeding lot.

Given:
Initial production lot.
Quantity—100 units.
Description—a new electronic system requiring standard methods of manufacture.
Labor Content—200 standard hours of fabrication
300 standard hours of assembly.
Labor Progress—recurring labor only at the individual operation level.

Determine:
Realization fabrication and assembly hours per unit for a 100 lot.
From the Learning Curve Selection Chart, Table 25.2 it is determined that the progress rate for this system is 85% and that 1000 units must be produced to reach standard.

From the 85% Learning Curve, Figure 25.4, it is determined that the cumulative average realization factor for the first 100 systems is 2.25.

Therefore:

	Fabrication	Assembly
Standard hours/unit	200	300
Realization factor	2.25	2.25
Realized hours/unit	450	675

The basic fabrication and assembly labor can now have the appropriate set-up, rework, and variance allowances added to make up the total bid hours.

Given:
Same system.
Quantity—next 100 units.

Determine:
Realized hours per unit for the next 100 lot.
Follow on lots are determined from the unit cost line:
100th unit realization factor   1.78
200th unit realization factor   1.52
Average for the 100th to the 200th unit:
   Measure ½ the arithmetic distance between the above two points. This is the average for the second lot of 100 units.

Therefore:

	Fabrication	Assembly
Standard hours per unit	200	300
Realization factor	1.65	1.65
Realized hours/unit	330	495

Another application of the learning curve is to project future time requirements based on return cost data. Suppose that 100 units of the type described in the previous example had actually taken 110,000 hours to manufacture. The cumulative average would have been 1100 hours. From the 85% cumulative average curve, the realization factor for these conditions is 2.25. Dividing 1100 by 2.25 gives the standard time, which can then be used with the 85% curve to project the cost for any other quantity.

TABLE 25.3. Completion vs. Expenditures for Log-Log Learning Curves

Percentage Complete	Percentage Expended			Percentage Complete	Percentage Expended			Percentage Complete	Percentage Expended		
	90%	85%	80%		90%	85%	80%		90%	85%	80%
1	2.0	3.2	4.4	36	42.1	46.1	50.0	71	74.8	77.1	79.3
2	3.6	5.3	7.0	37	43.0	47.0	51.0	72	75.7	77.9	80.1
3	5.1	7.2	9.3	38	44.0	48.0	51.9	73	76.6	78.7	80.8
4	6.5	8.9	11.3	39	45.0	48.9	52.8	74	77.5	79.6	81.6
5	7.9	10.5	13.1	40	46.0	49.9	53.7	75	78.4	80.6	82.8
6	9.2	12.0	14.8	41	47.0	50.8	54.6	76	79.3	81.2	83.0
7	10.5	13.5	16.5	42	47.9	51.7	55.5	77	80.2	82.0	83.8
8	11.7	14.9	18.1	43	48.9	52.7	56.4	78	81.0	82.8	84.5
9	13.0	16.3	19.5	44	49.9	53.6	57.3	79	81.9	83.6	85.2
10	14.2	17.6	21.0	45	50.8	54.5	58.2	80	82.8	84.4	85.9
11	15.4	18.9	22.4	46	51.8	55.5	59.1	81	83.7	85.2	86.7
12	16.6	20.2	23.8	47	52.7	56.3	59.9	82	84.5	86.0	87.5
13	17.7	21.4	25.1	48	53.7	57.3	60.8	83	85.4	86.8	88.2
14	18.9	22.7	26.4	49	54.6	58.1	61.6	84	86.3	87.6	88.9
15	20.0	23.8	27.6	50	55.6	59.1	62.5	85	87.1	88.4	89.6
16	21.1	25.2	28.9	51	56.5	59.9	63.3	86	88.0	89.2	90.3
17	22.3	26.2	30.1	52	57.5	60.8	64.2	87	88.9	90.0	91.0
18	23.4	27.4	31.3	53	58.4	61.7	65.0	88	89.8	90.8	91.7
19	24.5	28.5	32.4	54	59.3	62.6	65.9	89	90.6	91.5	92.4
20	25.5	29.6	33.6	55	60.2	63.5	66.7	90	91.4	92.3	93.1
21	26.6	30.7	34.7	56	61.2	64.4	67.5	91	92.4	93.1	93.8
22	27.7	31.8	35.8	57	62.1	65.2	68.3	92	93.2	93.9	94.5
23	28.8	32.9	36.9	58	63.1	66.1	69.1	93	94.0	94.6	95.2
24	29.8	33.9	38.0	59	64.0	67.0	69.9	94	94.9	95.4	95.9
25	30.8	35.0	39.1	60	64.9	67.8	70.0	95	95.8	96.2	96.6
26	31.9	36.0	40.1	61	65.8	68.7	71.6	96	96.6	97.0	97.3
27	32.9	37.1	41.2	62	66.7	69.6	72.4	97	97.5	97.8	98.0
28	34.0	38.1	42.2	63	67.6	70.4	73.1	98	98.3	98.5	98.7
29	35.0	39.1	43.2	64	68.5	71.2	73.9	99	99.2	99.3	99.3
30	36.0	40.1	44.2	65	69.4	72.1	74.7	100	100.0	100.0	100.0
31	37.1	41.2	45.2	66	70.3	72.9	75.4				
32	38.1	42.2	46.2	67	71.2	73.8	76.3				
33	39.1	43.2	47.2	68	72.2	74.6	77.0				
34	40.1	44.1	48.1	69	73.0	75.4	77.8				
35	41.1	45.1	49.1	70	73.9	76.2	78.5				

25.7. COMPLETION VS. EXPENDITURE, CONSIDERING LEARNING

If the expenditure per unit decreases with experience than the percentage of budget expended will not be directly related to the expected percentage complete. For example, consider a program performing to the 80% learning curve. If performance is as planned, then when the program is 10% complete physically, 21% of the allotted manhours will have been expended. This ratio of 21% expenditure for 10% progress is alarming, but if the 80% learning rate is maintained the total expenditures will be on target.

Table 25.3 shows the percent of work-hours expended at any given percentage of physical progress for programs following the 80%, 85%, and 90% learning curves.

SECTION TWENTY SIX

LABOR ALLOWANCES AND MULTIPLIERS

26.1. Standard Hour Allowances   334
26.2. Standard Hour Multipliers   339

The application of the proper allowances greatly increases the number of uses for time standards. Engineered time standards reflect the best time expected of an hourly paid worker who has reasonable skill and training. Incentive-paid workers tend to perform the work in 65 to 75% of the standard time.

The time values in this book already include allowances for paid personal time of 24 minutes per day. A fatigue allowance is also included, which averages about 5% of the work time. In some cases additional fatigue has been allowed to compensate for lifting heavy loads. Another 5% has been allowed for work that was not observed and for delays that are beyond the control of the supervisor.

The time values are based on the premise that workers who have adequate experience will be well down the learning curve. As pointed out in Section 25, a learning curve reflects improvements in:

Operator Learning
Improved Methods and Processes
Management Learning
Debugging of Engineered Data

Adjustments to the time standards should be made in accordance with Section 25.

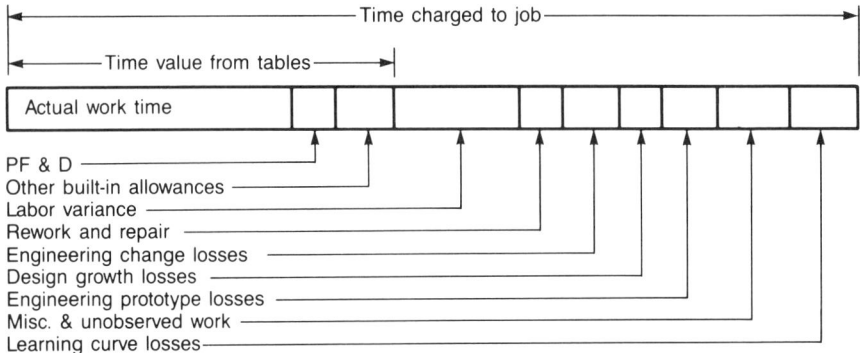

FIGURE 26.1. Standard vs. actual time elements.

26.1. STANDARD HOUR ALLOWANCES

These allowances make up the difference between internal shop control hours and total bid hours. The standard hour plus learning allowance represents a measured standard of performance for the shop. The standard is expected to be attained by the operator when the prescribed machine, tooling, and material are available, but rework of inspection rejects, engineering changes, temporary tooling, and so on require additional labor inputs before the job is completed. The original operator may or may not perform these functions, but nevertheless, the work-hours are part of the total job time. The standard is set without allowance for these variables, and it is expected that the actual labor charges will exceed the standard by a small but controlled amount. A performance standard plus a variance allowance is tailored to the standard cost type of accounting system. The "variance from standard" replaces the nebulous "contingency" factor of other costing techniques.

Figure 26.1 shows the relationship between standard (table) values and actual job time. The allowances in this section are intended to allow the estimator to switch from the book values to the actual values. Each of the allowance categories is explained below. The allowance values give a range and the estimator should use judgement in selecting a value within the range. Average values are summarized, but these values should not be used if the conditions being dealt with are not average.

	Minimum Percentage	Average Percentage	Maximum Percentage

Variance from Measured Labor
    Variation from standard time.
    Variation from standard method.
    Faulty tools, fixtures, and
        machines.

STANDARD HOUR ALLOWANCES

	Minimum Percentage	Average Percentage	Maximum Percentage
Material shortages.			
Re-set-ups.			
Total	5%	10%	20%
Normal Rework and Repair Allowance			
Rework of purchased material.			
Rework inspection rejects.			
Rework test rejects.			
Rework minor engineering changes.			
Repair units damaged in handling.			
Total	10%	15%	20%
Engineered Change Allowance This allowance is for projects where the design stability is very poor. This includes units designed and put into production on a crash program basis, and program phasing where field testing is done concurrent with production. Applicable only when design changes are required. To be used with discretion.			
Total	0%	to	15%
Design Growth Allowance Estimates based on a design concept or an early bread board model are subject to design growth. Experience has shown that as a complex system passes through the design stages of concept to final tested unit, the overall content will increase by:	0%	to	30%
Engineered Prototype Allowances The construction labor to build an engineered prototype is greater than that required to build the first production			

	Minimum Percentage	Average Percentage	Maximum Percentage
model. Reworks are higher and work is done from schematics and sketches rather than production drawings. The increase over first unit production labor ranges as follows:	15%	to	25%

Miscellaneous and Unobserved Work Allowances

When all the work steps are not readily identifiable, one allowance is often used to adjust for the omitted steps. However, a bid based on many small job elements is preferable to one composed of a larger (unknown) lump sum estimate.

Typical elements are material movement from work station to work station, the cost of first line supervision when such costs are direct charges (DL), production down time due to equipment maintenance, and tool and die upkeep.

Material Handling.	5	9	14
First line supervision.	5	7	9
Maintenance downtime.	4	6	9
Tool and die upkeep.	3	4	5
Average*	4	7	9

Another element is *Assembly Cycle Allowance* caused by loss of operator efficiency as the job cycle increases. An operator installing 100 different components will take longer per component than one installing 10 different components. The standard time values may be increased by *5%* if the cycle exceeds *9 minutes* and *10%* if the cycle is greater than *19 minutes*.

---

*Does not add due to rounding.

TABLE 26.1. Standard Allowance Selector

			Suggested Allowances				
Type of Operation	Labor Variance (%)	Rework and Repair (%)	Engineering Changes (%)	Design Growth (%)	Prototype (%)	Miscellaneous and Unobserved (%)	Suggested Total Allowance	Actual Total Allowance
Machining	5	10	5	5	15	5	45	
Sheet metal and fabrication	10	10	5	5	15	5	50	
Assembly by machine	5	10	5	5	15	5	50	
Assembly by hand	20	15	5	5	15	9	69	
Test	10	20	5	0	15	7	57	
Packing	5	10	0	0	0	4	19	

338   LABOR ALLOWANCES AND MULTIPLIERS

Type of Operation	① Actual Total Allowance	② Allowance as Per Cost of Workday	X	③ Minute to Hour Conversion	=	④ Multiplier
Machining		100 / 100-①*	x	0.0167	=	
Sheet metal & fabrication		100 / 100-①*	x	0.0167	=	
Assembly by machine		100 / 100-①*	x	0.0167	=	
Assembly by hand		100 / 100-①*	x	0.0167	=	
Test		100 / 100-①*	x	0.0167	=	
Packing		100 / 100-①*	x	0.0167	=	

*Value of column ①

FIGURE 26.2. Bid allowance work sheet.

Type of Operation	① Actual Total Allowance	② Allowance as Per Cost of Workday	X	③ Minute to Hour Conversion	=	④ Multiplier
Machining	45%	100 / 100- 45	x	0.0167	=	.0304
Sheet metal & fabrication	50%	100 / 100- 50	x	0.0167	=	
Assembly by machine	35%	100 / 100- 35	x			

FIGURE 26.3. Bid allowance work sheet example.

26.2. STANDARD HOUR MULTIPLIERS

A multiplier is used as an arithmetic shortcut. It converts minutes to hours and at the same time adds the personal, fatigue, and other allowances which make up the standard hour. Multipliers for typical operational classifications are

STANDARD HOUR MULTIPLIERS

given in Table 26.1. Suggested allowances are shown, but the estimator should select allowances which reflect his or her understanding of the uncertainties in the estimate.

To use the Standard Allowance Selector, Table 26.1, enter the sum of the allowances selected in the actual total allowance column for each type of operation. This actual allowance may differ from the suggested allowance. The actual total allowance is then transferred to the Bid Allowance Work Sheet, Figure 26.2.

To use the Standard Hour Multiplier, transfer the allowance selected in Table 26.1 to the actual total allowance, column 1 in Figure 26.2. In column 2 convert this percent into a fraction which is multiplied by column 3, the ratio of minutes to hours, that is,

$$\frac{1 \text{ hour}}{60 \text{ minutes}} = 0.0167 \text{ hours/minute}$$

to find column 4, the multiplier. This multiplier is then used to convert standard minutes to actual hours.

An example is shown in Figure 26.3.

This multiplier is then used to modify the minutes from Part 1 or Part 3 of this book. For instance, if the sum of all the machinery times including set-up came to 4820 minutes per part, the multiplier would be used to convert this to hours including allowances

$$4820 \text{ minutes} \times 0.0304 \, \frac{\text{hours}}{\text{minutes}} = 146 \text{ hours}$$

The 146 hours would then be ready for use in the cost estimating forms of Section 2.

In summary, the basic formula for expressing fully adjusted labor hours becomes

Labor Hours = Standard time values (in minutes) from tables
  × Learning curve losses (factors)
  × $\dfrac{100}{(100 - \text{Sum of allowances, Table 26.1})}$
  × 0.0167

SECTION TWENTY SEVEN

MATERIAL COSTS AND ALLOWANCES

27.1.	Raw Material Prices	340
27.2.	Line Stock and Raw Material Ratios	342
27.3.	Total Bid Material Allowances	343

The direct cost of a production program consists of two basic elements:

1. Direct Labor
2. Purchase Materials and Services

Although standard pricing data are given in this section for some raw materials, the balance of these items must be priced for the specific unit being estimated. Current rates of inflation and very volatile market conditions cause prices to vary widely from time to time. The change in cost is frequently very different from one material to another. Consequently, a general escalating factor for all material is not satisfactory. Ratios and allowances are given to convert basic parts list material dollars to total anticipated material expenditures.

27.1. RAW MATERIAL PRICES

The prices in Table 27.1 were representative for the Boston area in April 1982 and were obtained from raw material suppliers. Prices can also be obtained

TABLE 27.1. Raw Material Prices

Plastics

Sheets, 500 lbs. Type	Purchase Quantity Typical Size	Price/ft.^{2}
ABS	2 × 4 ft. × ¼ in.	$ 5.76
	× ½ in.	12.70
Acrylic	4 × 8 ft. × ¼ in.	3.01
	× ½ in.	7.39
Phenolic G10	3 × 4 ft. × ⅛ in.	5.00
Glass epoxy	× ⅜ in.	10.92
Polycarbonate	4 × 8 ft. × ¼ in.	3.95
	× ½ in.	10.90
	2 × 4 ft. × 1 in.	41.18
Teflon	2 × 4 ft. × 1 in.	146.50

Aluminum

Note: Aluminum is sold at a base price, plus "extras" for small quantities, unusual size, finish, and so forth.

6061-T6 Sheet, Bar and Shape: Base Price = $1.43/lb.

Quantity Extra		Sheet Thickness Extra		Note:
< 100#	$1.60/lb.	.032	$0.21	Negligible Size
300#	.68	.040	0.13	Extra for ⅜−6 in.
500#	.41	.063 and up	—	Bar and Shape
1000#	.20	Typical Sizes		
2000#	.14	.032 × 48 × 144, 22lb./sht		
4000#	.07	0.063 × 48 × 144, 42lb./sht		
		0.090 × 48 × 144, 61lb./sht		

6061-T4 Sheet

Quantity Extra		Typical Sizes	Base Price
100−199#	$0.41/lb.	0.032 × 48 × 144, 22-lb. sheet	$1.72/lb.
200−299#	0.31/lb.	0.063 × 48 × 144, 42-lb. sheet	$1.56/lb.
300−499#	0.21/lb.	0.090 × 48 × 144, 61-lb. sheet	$1.56/lb.
500−999#	0.11/lb.		
1000−1999#	0.06/lb.		
2000 > #	Base Price		

Steel

500 lb. Purchase Quantity — Avg $/lb.

	Size	lb/Sheet	Mild Steel (Cold Rolled)	302/304 Stainless	Note:
Sheet	22GA × 48 × 120	53	—	1.50	Negligible
	18GA × 48 × 120	80	0.37	1.44	Size Extra
	16GA × 48 × 120	100	0.37	1.41	for ⅜−6 in.
	14GA × 48 × 120	125	0.37	—	Bar and Shape
Bar and Shape			0.37	2.25	

342 MATERIAL COSTS AND ALLOWANCES

from the Bureau of Labor Statistics (BLS) publication *Producer Prices and Price Indexes*.* Updated monthly pricing information is provided for selected materials through a price index and, in some cases, the price per pound. However, caution should be used in basing material prices on BLS-supplied data. First, the most recent BLS data is usually 3 months out of date, and second, the price per pound of material does not specify material dimensions or the quantity purchased, two variables which have a direct impact on material costs. In many cases, suppliers report prices differing widely from BLS data since the quantity purchased is factored into their pricing structure. Actual supplier prices are still the best estimating basis when the raw material cost becomes a significant portion of the total unit price.

27.2. LINE STOCK AND RAW MATERIAL RATIOS

27.2.1 Line Stock

The screws, washers, solder, and so on, which are physically identifiable with the product, are low-value line stock items, and therefore do not warrant detailed cost estimating. Average percentage ratios are again the common method used by both estimating and cost accounting. The line stock ratio will vary from company to company, according to individual definitions of line stock vs. direct charge materials. For this reason, an example is given below for one particular situation. If all the line stock items except nuts, screws, and so on in this example were transferred to the direct material classification, the line stock ratio would drop to about ¼% of the material dollars. As the line stock category is expanded, the percentage will increase to about 4% of the dollar value, which may represent as much as 30% of the total items purchased.

Line Stock Material

Screws, nuts, washers, rivets, terminal lugs.
Solder, stand-offs, grommets, and so on.
Wire and cable.
Insulating sleeving.
Standard composition resistors, ¼ to 1 watt.
Electroplating, common and precious metals.
Paint and finishing compounds.
Roll and sheet stock packing materials.
Ratio to material dollars   3–4%
Ratio to direct labor dollars 9–12%

---

*Obtained either from the Superintendent of Documents, or the Bureau of Labor Statistics.

Direct Material

Machine shop and sheet metal raw material (bar, sheet, etc.).

Sheet laminates—copper clad and unclad.

All other material on a parts list except the above line stock items.

Overhead Material

Material used in the production process, which does not become a part of the end product. For example:
- Cleaning solvents
- Chemical process catalysts
- Machining fluids
- Tape, tags, rags

27.2.2. Miscellaneous Raw Material Ratios

	Percentage	Base
Process Material		
Electroplating		
Photographic		
Etched circuit processing		
Encapsulating		
Average recovery ratio:	3	Direct labor cost
Machine Shop and Sheet Metal		
Production quantities, 1–50		
Aluminum, steel and so on	15	Direct labor cost
Tooling Raw Material		
Tool steel, tool bits	2–5	Direct labor cost
Standard bussings, and so on		

27.3. TOTAL BID MATERIAL ALLOWANCES

These allowances make up the difference of what the estimated bill of materials adds up to and what is actually *expended* at the completion of the order. Some benchmarks for the required judgement factors are given here.

Attrition. An attrition allowance is required to finance the rebuying or original overbuying of material due to loss, breakage, solder burns, floor shortages, etc. Range 1–5%.

Engineering Change Allowance. A separate evaluation is made of material scrappage anticipated because of engineering design changes. Changes affecting units in production area are caused by engineering design tests being conducted during the initial stages of the production run and by failure of units in the field resulting in remedial design changes. Apply only when engineering change has been authorized. Range 0–10%.

Design Growth Allowance. Estimates based on a design concept or an early breadboard model are subject to design growth. Experience has shown that as a complex system passes through the design stages of concept to the final tested unit, the overall content will increase. Apply only when design change has been authorized. Range 0–30%.

Inflationary Price Increases. If work is bid at this year's price level but is to be performed in 1, 2, or 3 years, the economic trend must be taken into account to forecast actual costs during the year of performance. (See BLS data.)

SECTION TWENTY EIGHT

INTEGRATED CIRCUIT PACKAGING

28.1.	Background	345
28.2.	Integrated Circuit Packages	349
28.3.	Integrated Circuit Packaging Process	353
28.4.	Typical Costs	359

Because there are thousands of different IC chips and dozens of IC packages, it is not economical to produce or stock each chip in each package. Low to moderate volumes of integrated circuits are customarily packaged to meet the specific requirements of each application. As a result, the cost of packaging ICs often appears as a separate item in a cost estimate.

This section describes the types of packages available, outlines the chip packaging process, and gives approximate pricing information for the packaging.

28.1. BACKGROUND

The information-handling capability of modern electronics is based upon the integrated circuit (IC) chip. Chips now exist with half a million transistors on them. At modern speeds and circuit densities it is often practical to process analog signals in digital form, such as directly synthesizing complex frequency waveforms point by point through a combination of calculation and digital-to-analog output conversion. Integrated circuits have their limitations; higher frequencies and powers require specialized signal transistors and diodes. But as shown in Figure 28.1, the bulk of the semiconductor market is ICs. Within this segment, memories, standard logic, and microprocessors represent over 70% of all integrated circuits, a mix that remains fairly constant despite doubling of sales from 1982 to 1985, as indicated in Figure 28.2.

The advantages of ICs over discrete devices are those of size and power consumption, as well as cost. Further, the cost advantage of the ICs themselves is far outweighed by the effects of smaller space and power requirements on the total system. As shown by Figure 28.3, about 65% of the cost of an electronic system is directly related to the size of its components. Printed circuit boards, card racks, back planes, connectors, external enclosures, cooling systems, and power supplies all contribute heavily to the cost of a system. Far more of each of these is required for system based on discrete components than for one based on ICs.

There are other benefits as well. Consider the opportunity for error and failure in the interconnections among 500,000 individual transistors and their supporting circuitry. Clearly the IC offers much better reliability for a given capability.

Table 28.1. lists some of the characteristics of integrated circuits, and shows expected future values. Higher voltage capability, power dissipation, and

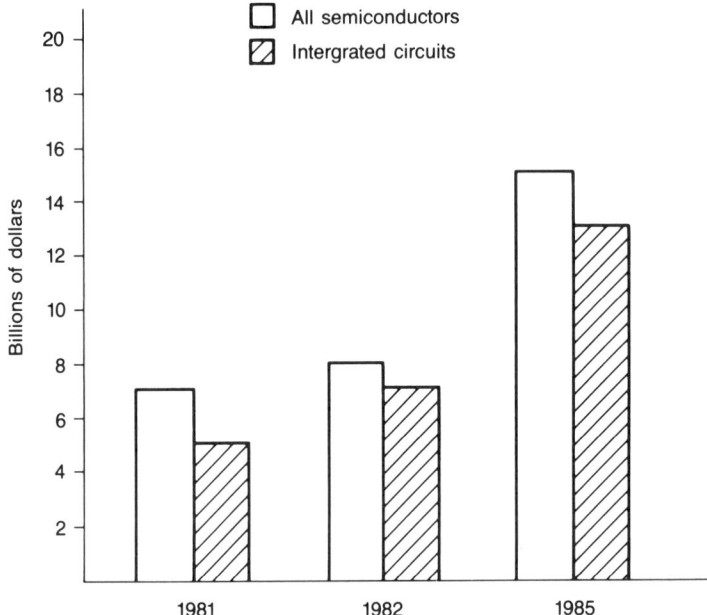

FIGURE 28.1. U.S. integrated circuits market as a share of total semi-conductor consumption: 1981–1985.

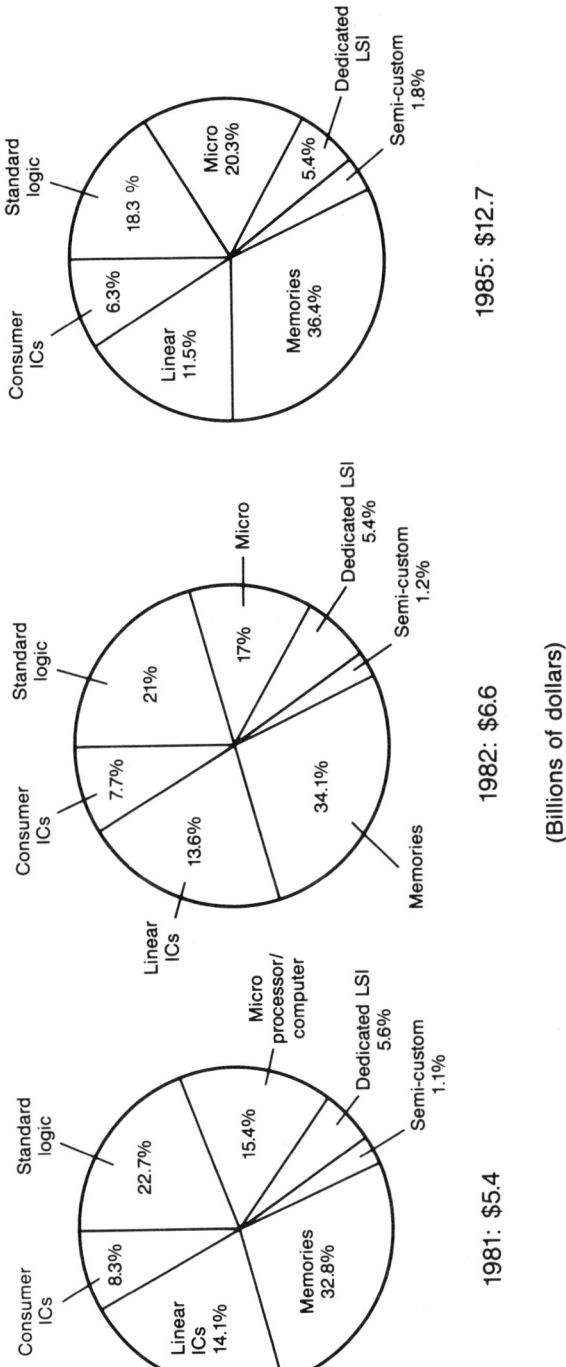

FIGURE 28.2. U.S. integrated circuit market share: 1981–1985.

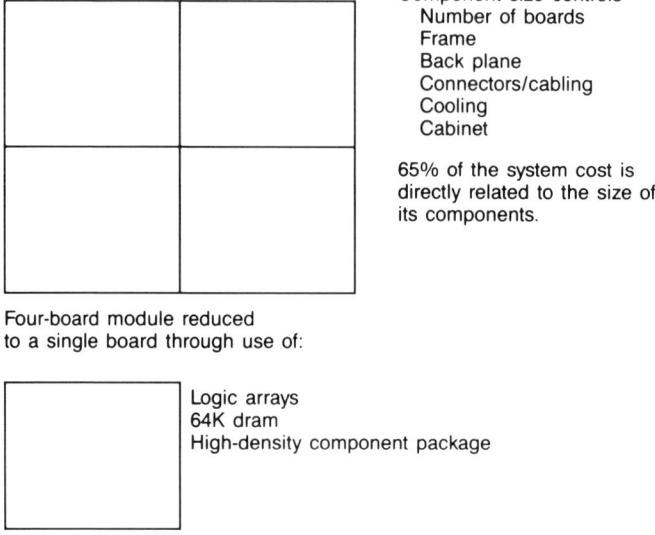

FIGURE 28.3. Semi-conductor package engineering—components vs. system cost. (*Source:* Plastic chip carriers, their advantages and applications by John W. Orcutt, Texas Instruments Semiconductor Group.)

impedance mean that fewer external elements are required to condition signals to suitable levels for processing on chip.

In addition, the feature size is being cut roughly in half every five years, while the maximum chip size is about doubling in the same time. The combined effect is that the number of active elements per chip is going up by a factor of between four and eight times every five years.

However, greater capability per chip means greater need for connections from the chip to external circuitry. As shown in the table, pin counts have not kept up with the other characteristics.

TABLE 28.1. IC Package Engineering Packaging Trends

	Quantification		
Advance	1975	1980	1985
Higher voltage (volts)	40	150	250–300
Higher power (watts)	1	2	6–10
Higher impedance (ohms)	10^9	10^{12}	10^{15}
Scaling and feature size reduction (micron)	5	3	1–2
Larger chip sizes (square mils)	40K	90K	200K
Higher pin count	40	64	256

Source: Plastic chip carriers, their advantages and applications by John W. Orcutt, Texas Instruments Semiconductor Group.

Where pins are arranged along the periphery of a package, the pin count usually governs the size of the package. For instance, a 40 pin dual in-line pin (DIP) package occupies almost three in.^{2} of circuit board space, though the chip inside is very unlikely to exceed 0.3 in. on a side. Thus, the pressure to increase the pin count is accompanied by pressure to reduce the effect of pin count on the overall dimensions of the package. Extensive work in this area is bearing fruit. As shown in the next section, there now exist commercially available packages with pin counts in excess of 150 that are not much larger than 40-pin DIPs.

28.2. INTEGRATED CIRCUIT PACKAGES

Major work is now going on to refine IC package designs and bring others to market; the VHSIC (very high speed integrated circuit) program in particular has led to the development of a number of special purpose packages that may find application in some form in other markets. Other designs will certainly be along, however, the ones listed here are those likely to be in use in the normal range of military and commercial electronics in the near future.

Figure 28.4 shows the five major types of packages in current use. The TO can is an early package, used originally for single transistors. Today it is also used to package ICs, though its low number of leads generally limits its use to linear (analog) circuits.

The DIP package has become the symbol for digital electronics and ICs. It takes its name from the two rows of pins which are on 0.100 in. (100 mil) centers lengthwise, and of various widths. The DIP is used to package ICs from a few gates to complete one-chip computers and 65,536 bit memories. It is rugged, easily handled by manual or automatic insertion techniques, and supported by a mature system of tools, procedures, and experience. Its major disadvantage is size with large pin counts.

Figure 28.5 compares several packages with respect to pin count and occupied circuit board area. As shown, the widest DIPs occupy about three times as much area as chip carriers with the same number of pins.

The flat pack is the contemporary of the DIP. It is heavily used in U.S. military applications, having been proven to be a very reliable package due at least in part to the low stresses imposed on the flat array of leads. Flat packs are produced with lead counts to 100, and experimental versions with 150 to 200 leads have been tested or planned, making them suitable for LSI (large scale integration) and VLSI (very large scale integration), but there is an inherent disadvantage in area requirements compared with newer packages.

The pin-grid array is a multi-layered chip çarrier with a grid or matrix of pins on 100-mil centers. The pins may be arranged either in two or three peripheral rows, as shown in the figure, or in a fully filled matrix, without a hollow center. Commercial units have been fabricated with as many as 324 pins, and one supplier has reported an experimental 1024-pin model. The ceramic body of the pin-grid array may contain more than one chip, but normally only one is mounted.

TO-Packages

Dual in-line pin
(dip) packages

SO-Packages
(mini-dips)

Flat packs

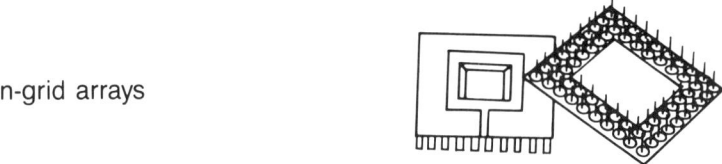

Pin-grid arrays

Chip carriers

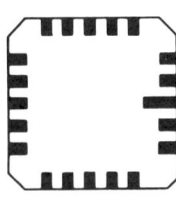 

FIGURE 28.4. Representative integrated circuit packages.

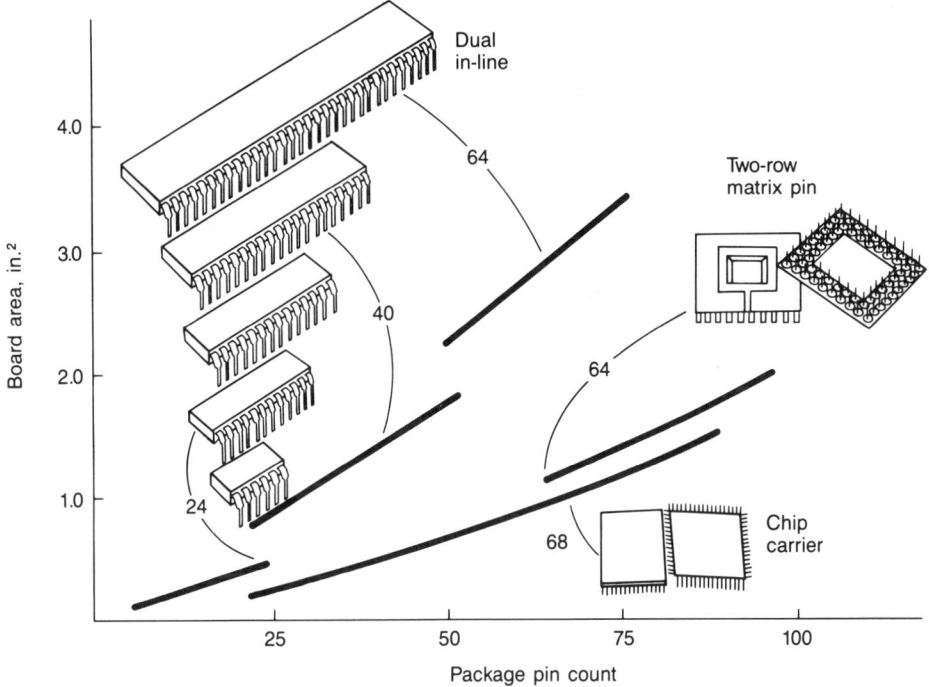

FIGURE 28.5. Package pin count vs. area trend for various IC packages. (*Source:* Plastic chip carriers, their advantages and applications by John W. Orcutt, Texas Instruments Semiconductor Group.)

the pin-grid array may contain more than one chip, but normally only one is mounted.

Access to the inner pins for soldering and inspection is the main disadvantage of any pin-grid device. In addition, the device is extremely difficult to remove and replace, especially in the field. On the other hand, no other package offers such high pin counts.

Chip carriers are likely to be the most common military package for ICs for the next few years. This is a ceramic package in which the chip is mounted; external connections are contacts or leads placed on 40- or 50-mil centers around the periphery of the package. The arrangement details are endless: cavity up, cavity down, multi-layer chip carrier to accommodate a variety of signal paths from the chip to the leads, square or rectangular shape, and several different spacings of leads. Although chip carriers are somewhat more expensive than DIPs or flat packs, they require little board area and are very reliable, and so have found many military and aerospace applications.

The major problem with leadless chip carriers is a requirement for circuit boards that match the thermal expansion characteristics of the ceramic chip carrier bodies. Otherwise, the solder connections tend to fracture due to lack of flexibility. Conventional epoxy glass boards are not suitable, and the required composite materials are quite expensive.

The leaded chip carrier is a fairly recent innovation. In this device, the leads are folded down toward the circuit board and back up in the tight "J" shape. When mounted, the leads accommodate differences in thermal expansion between the chip carrier and the circuit board.

Note that chip carriers are mounted on the surface of the circuit boards, not in through-holes. The device is first glued to the board to hold it in place, and then soldered either by wave soldering or by one of the reflow processes.

For the time being, it seems that leaded chip carriers in pin counts to 80 or 90, and pin-grid arrays above that will take over from the DIP and flat pack.

However, there is continuing development on several fronts. A DIP with 50-mil pin spacing is now available commercially. These devices take up much less board area than their 100-mil predecessors. For example, a 40-pin 50-mil DIP takes up half of the board area required by a 40-pin 100-mil package. The improvement is better at higher pin counts; at 64 pins, the 50 mil package takes one-third the area of its 100-mil counterpart. Special sockets are available to lead the 50-mil pins out to staggered rows of leads on 100-mil centers, to accommodate existing board technology and standard 100-mil spacing for bed-of-nails test fixtures. Even when these sockets are used, the 50-mil packages require much less area.

Similarly, work is going on to develop closer spacing for chip carriers. The current standard is 50- or 40-mil spacing; Table 28.2 shows the effects of the proposed 25- and 20-mil spacings. The problem with such close spacing is the fabrication of suitable circuit boards, which is also being pursued. It is clear that further increases in circuit packaging densities are well on the way.

TABLE 28.2. Standard and Fine-Pitch Chip-Carrier Families

Ceramic Size (in.)	Terminal Count		
	50-mil Centers	25-mil Centers	20-mil Centers
0.400	24	44	60
0.450	28	52	68
0.560	—	68	92
0.650	44	84	108
0.720	—	92	124
0.750	52	100	124
0.950	68	132	164
1.150	84	164	204
1.350	100	196	244
1.650	124	244	388

28.3. INTEGRATED CIRCUIT PACKAGING PROCESS

The following process describes chip mounting in ceramic carriers, as this is becoming the most common package in the military, high-reliability commercial market.

1. *Convert the Customer's Specification to the Packager's Specification.* Customers normally specify IC packages either by citing mil-specs or by providing their own specifications, which are usually in mil-spec format. The packager must then develop specific manufacturing procedures to meet these requirements using his equipment and practices.

2. *Order parts, Both Chips and Packages.* The ordering quantity is based on the expected yield, which includes loss and damage due to the packaging process. For instance, typical yields of usable IC chips are in the range of 60 to 70%. If a customer specifies tighter electrical tolerances than a chip manufacturer normally meets, the packager can meet the customer's requirements by ordering a large number of chips and selecting satisfactory chips from the larger lot. This can greatly reduce the yield and increase the final cost per device.

3. *Dicing.* Chips are normally supplied in wafer form, that is, about a thousand chips on a single piece of silicon. A wafer is roughly four inches in diameter, while an individual chip is about 0.1 in. square. Dicing is accomplished either by scoring and breaking the silicon substrate or by sawing with a diamond saw.

4. *Incoming Inspection.* Random samples of the ceramic carriers are inspected for dimensional tolerances and gross physical defects such as chips and cracks.

All chips are inspected visually, primarily for gross physical defects incurred during dicing such as chips, cracks, and breakage. In addition, for high reliability applications four chips per wafer are examined with a scanning electron microscope to verify the chip fabrication process.

At present there are two conflicting trends with regard to the visual inspection of chips. First, automatic (computerized) visual inspection is on the horizon. On the other hand, as chips become more and more complex, visual inspection becomes less useful; a trend is developing toward 100% dynamic electrical testing in place of visual inspection.

5. *Make Up Production Kits of Carriers and Dice.*

6. *Chip Attach.* In this step, the chip is bonded to its carrier. The normal bond is epoxy, polyimide, or eutectic gold.

7. *Wire Bond.* Tiny (0.001 in. thick) gold wires are thermosonically bonded to the chip and to pads on the chip carrier which are connected to the external pins. Wire bonding is normally a manual process, although N/C machinery is being developed.

8. *Ultrasonic Cleaning.* The chips in carriers are cleaned thoroughly prior to being sealed.

9. *Pre-Sealing Visual Inspection.* A visual inspection is made of every chip before sealing to check for damage incurred during the chip attach, wire bond, and cleaning steps.

10. *Vacuum Bake.* Chips are baked in a vacuum to remove any contaminants prior to being sealed.

11. *Lid Sealing.* Lids are delivered with solder rings in place. The lids are placed over the chip carrier cavities, and the chips are conveyor-feed through a furnace to melt the solder and form the seal. The process is accomplished in an inert atmosphere (nitrogen).

12. *Marking.* The cases are marked with device identification, manufacturer, certification, lot identification, and any other required information.

13. *Final Inspection and Test.* In accordance with customer requirements. (See below.)

Test and inspection contributes a major portion of the cost of IC packaging. Most military devices must conform to the requirements of MIL-STD-883B. This standard includes screening criteria, lot acceptance criteria and testing procedures. The primary methods that apply to ICs are Methods 5004 and 5005.

Essentially all devices for military use must undergo Method 5004 screening. Table 28.3 lists the tests and inspections; 100% qualification is required for each of 12 procedures, including visual inspections, seal leak tests, electrical tests, acceleration tests, and thermal cycling.

Method 5005 describes several levels of lot acceptance testing over and above Method 5004 screening. These are shown in Tables 28.4 through 28.7.

Group A testing is primarily a set of electrical tests. These include static, dynamic, switching, and functional electrical tests conducted at the highest and lowest rated operating temperatures and 25°C.

Group B testing is a set of physical and chemical tests, including physical dimensions, resistance to solvents, solderability, wire bond strength, and internal water vapor content.

Group C testing includes a 1000-hour-life test, with measurement of end point electrical parameters, temperature cycling, and acceleration testing.

Group D testing is a more extensive set of mechanical and life tests.

Method 5005 tests are expensive to conduct; also, since most of them are destructive, additional devices must be purchased and packaged solely for the purpose of guaranteeing adequate lot size. A discussion of the costs of packaging and testing follows.

TABLE 28.3. MIL-STD-883B Method 5004 Screening

Screen	Method	Requirement
1. Internal visual	2010, condition B	100%
2. Stabilization bake	1008, condition C, minimum; 24 hours minimum	100%
3. Temperature cycling	1010, condition C	100%
4. Constant acceleration	2001, condition E (min) Y_1 orientation only	100%
5. Visual inspection		100%

6. Seal
  a. Fine
  b. Gross
	1014	100%

7. Interim (pre-burn-in) electrical parameters	Per applicable device specification	
8. Burn-in test	1015 160 hours at 125°C minimum	100%
9. Interim (post-burn-in) electrical parameters	Per applicable device specification	100%

10. Final electrical test
  a. Static tests
    1. 25°C (subgroup I), Table 1, 5005)
	Per applicable device specification	100%

2. Maximum & minimum rated operating temp. (sub-groups 2, 3, Table I, 5005)		100%
b. Dynamic tests and switching tests 25°C (subgroups 4, 9 Table I, 5005)		100%
c. Functional test 25°C (subgroup 7, Table I, 5005)		100%
11. Qualification or quality conformance inspection test sample selection		Sample[a]
12. External visual		100%

[a]Does not include method 5005 qualification or quality conformance inspection. Therefore, samples are selected only when method 5005 conformance inspection has been specified.

TABLE 28.4. Method 5005 Lot Acceptance Testing Group A[a]

Includes:

Complete verification of all production and 883B 100% screening procedures.

Internal and external visual guaranteed 100%.

Electrical guaranteed per following table.

Subgroup[b]	Class B LTPD
Subgroup 1	
Static tests at 25°C	5
Subgroup 2	
Static tests at maximum rated operating temperature	7
Subgroup 3	
Static tests at minimum rated operating temperature	7
Subgroup 4	
Dynamic tests at 25°C	5
Subgroup 5	
Dynamic tests at max. rated operating temperature	7
Subgroup 6	
Dynamic tests at minimum rated operating temperature	7
Subgroup 7	
Functional tests at 25°C	5
Subgroup 8	
Functional tests at maximum and minimum rated operating temperatures	10
Subgroup 9	
Switching tests at 25°C	7
Subgroup 10	
Switching tests at maximum rated operating temperature	10
Subgroup 11	
Switching tests at minimum rated operating temperature	10

[a]The specific parameters to be included for tests in each subgroup shall be as specified in the applicable procurement document. Where no parameters have been identified in a particular subgroup or test within a subgroup, no group A testing is required for the subgroup or test.

[b]A single sample may be used for all subgroup testing. Where required size exceeds the lot size, 100% inspection shall be allowed.

TABLE 28.5. MIL-STD-883B Method 5005 Lot Acceptance Testing Group B

Test[a]	Method	Condition	Class B LTPD
Subgroup 1 Physical dimensions[b]	2016		2 devices (no failures)
Subgroup 2 Resistance to solvents	2015		3 devices (no failures)
Subgroup 3 Solderability[e]	2003	Soldering temp of 260 ± 10°C	15
Subgroup 4 Internal visual and mechanical	2014	Failure criteria from design & construction requirements of applicable procurement document.	1 device (no failures)
Subgroup 5 Bond strength[d] 1. Thermocompression 2. Ultrasonic or wedge 3. Flip-chip 4. Beam lead	2011	 1. Condition C or D 2. Condition C or D 3. Condition F 4. Condition H	15
Subgroup 6 Internal water-vapor content[c]	1018	1,000 ppm maximum water content at 100°C	3 devices (0 failures)[f] or 5 devices (1 failure)

[a]Electrical reject devices from the same inspection lot may be used for all subgroups.
[b]Not required for qualification or quality conformance inspections where group D inspection is being performed on samples from the same inspection lot.
[c]This test is required only if the package contains a desiccant.
[d]Test samples for bond strength may, at the manufacturer's option, unless otherwise specified, be randomly selected immediately following internal visual (precap) inspection. Unless otherwise specified, the LTPD sample size for condition C or D is the number of bond pulls selected from a minimum number of 10 devices, and for conditions F or H is the number of dice (not bonds) (see method 2011).
[e]All devices submitted for solderability test must have been through the temperature/time exposure specified for burn-in. The LTPD for solderability test applies to the number of leads inspected except in no case shall less than 3 devices be used to provide the number of leads required.
[f]Test three devices; if one fails, test two additional devices with no failures.

TABLE 28.6. MIL-STD-883B Method 5005 Lot Acceptance Testing Group C

Test	Method	Condition	LTPD
Subgroup 1			
Steady state life test	1005	Test condition to be specified (1,000 hrs at 125°C) as specified in the applicable device specification	5
End-point electrical parameters			
Subgroup 2			
Temperature cycling	1010	Condition C	15
Constant acceleration	2001	Condition E minimum Y_1 orientation only	
Seal	1014	As applicable	
a. Fine			
b. Gross			
Visual examination		Per visual criteria of method 1010 or 1011 as specified in the applicable device specification	
End-point electrical parameters			

TABLE 28.7. MIL-STD-883B Method 5005 Lot Acceptance Testing Group D

Test	Method	Condition	LTPD
Subgroup 1			
a. Physical dimensions	2016		15
b. Internal water-vapor content	1018	5,000 ppm maximum water content at 100°C	3 devices (0 failures) or 5 devices (1 failure)[d]
Subgroup 2[a]			
Lead integrity	2004	Test condition B2 (lead fatigue)	15
Seal	1014	As applicable	
a. Fine			
b. Gross			
Subgroup 3[b]			
Thermal shock[b]	1011	Test condition B minimum, 15 cycles minimum	15
Temperature cycling	1010	Test condition C, 100 cycles minimum	
Moisture resistance	1004		

TABLE 28.7. *(Continued)*

Test	Method	Condition	LTPD
Seal a. Fine b. Gross	1014	As applicable	
Visual examination		Per visual criteria of Method 1004	
End-point electrical parameters[c]		As specified in the applicable device specification	
Subgroup 4[b]			
Mechanical shock	2002	Test condition B minimum	15
Vibration variable freq	2007	Test condition A minimum	
Constant acceleration	2001	Test condition E minimum Y_1 (orientation only)	
Seal a. Fine b. Gross	1014	As applicable	
		Per visual criteria of Method 1010 or 1011	
End-point electrical parameters		As specified in the applicable device specification	
Subgroup 5[a]			
Salt atmosphere	1009	Test condition A minimum	15
Seal a. Fine b. Gross	1014	As applicable	
Visual examination		Per visual criteria of Method 1009	

[a]Electrical reject devices from that same inspection lot may be used for samples.
[b]Devices used in subgroup 3, "Thermal and Moisture Resistance" may be used in subgroup 4, "Mechanical."
[c]At the manufacturer's option, end-point electrical parameters may be performed after moisture resistance and prior to seal test.
[d]Test three devices; if one fails, test two additional devices with no failure.

28.4. TYPICAL COSTS

Integrated circuit packaging is a rapidly changing business; in addition, pricing is very market dependent. As a result, it is most difficult to give accurate cost estimating data. The following information is valid as of late 1982, for the U.S. military electronics market.

Packaging costs are in the general neighborhood of $5.00 to $7.00 per device, not including testing, for typical 20-pin ceramic chip carriers in lots of 500 to 1000. However, higher prices are not at all uncommon for more complex devices. Obviously, a device with 80 or 120 pins will require considerably more

INTEGRATED CIRCUIT PACKAGING

```
Job _____
No. devices req'd _____   Testing req'd _____
Wafer description _____   Dicing yield _____ %
Package description _____   Pkg & test yield _____ %
```

Wafers/Dice

		_____	Dice per wafer
	×	_____	Dicing yield
	=	_____	Net dice/wafer
		_____	Wafers req'd = No. devices req'd ÷ net dice/wafer (round up to whole number)
	×	$ _____	Per wafer cost
= $		_____ ¹	Cost of wafers

Packaging

	$	_____	Unit pkg cost/device
	×	_____	No. devices req'd
+ $		_____ ²	Raw cost of packaging
= $		_____ ³	Raw cost of packaged devices (1 + 2)
	÷	_____	Pkg & testing yield
= $		_____ ⁴	Cost of packaged devices

Lot Screening (5004)

	$	_____	Per device 5004 screening cost
	×	_____	No. devices required
	÷	_____	Pkg & testing yield
= $		_____ ⁵	5004 Screening cost (Minimum $750)

Lot Acceptance Testing (5005)

	_____	($1000) Group A — Electrical/temp
+	_____	($1500) Group B — Physical
+	_____	($1500) Group C — Life, end-point electrical
+	_____	($2500) Group D — Mechanical/life
= $	_____ ⁶	5005 Lot acceptance cost

$	_____	Total lot cost (4 + 5 + 6)

FIGURE 28.6. Integrated circuit packaging cost estimating worksheet.

labor than a relatively simple 20-pin one. At the high end of the scale, radiation immune devices may have prices in the $300 range, due to the very low yields involved. A great many devices must be packaged and tested to obtain a single radiation-qualified unit.

Testing is normally priced separately. Method 5004 screening typically adds $5.00 to $10.00 per device, with a minimum of about $750 per lot. Costs for

TYPICAL COSTS

Method 5005 lot acceptance testing are independent of lot size for quantities to 1000. Approximate costs per lot are:

Group A (electrical/temperature)	$1000
Group B (physical)	$1500
Group C (life test, end point electrical)	$1500
Group D (mechanical/life)	$2500

Thus, the complete range of Method 5004 screening and Method 5005 lot acceptance testing adds about $7250 to the price of a lot of integrated circuits. Note that pre-test costs are also higher when extensive testing is specified. Increased testing lowers the yield of a process; yields may range from 30% to 80%. Since more devices are rejected during test, more units must be purchased and packaged to obtain the required delivery quantity. The cost of the additional material and packaging is amortized as higher unit prices on the delivered devices. The higher unit prices are in addition to the direct cost of conducting the test.

The cost estimating worksheet given in Figure 28.6 is a guide for making an estimate of raw wafer cost, lot size, and unit packaging costs. In addition, yields must be known or estimated; these are normally part of the process specifications. The four intermediate sections of the worksheet are used to estimate the cost of raw wafers, packaging, Method 5004 lot screening, and Method 5005 lot acceptance testing. The final section is a total for a lot. Note that the bulk of these costs are by lot rather than by piece. Premium piece prices are often the result of small lot size.

SECTION TWENTY NINE

MATERIAL DISCOUNT CURVES

29.1. Uses of Material Discount Curves 362
29.2. Application of Material Discount Curves 363

Material cost reductions can be projected with log-log curves in the same manner as labor hours. The unit costs of 100 transistors is less than for 10, and the cost in a 1000 lot is less than a 100 lot. When the total material cost per system is plotted on a log-log graph at different buy quantities, the trend line will be very close to a straight line. Thus, the normal 10 or 20% quantity discount can be projected relative to the quantity at which it applies. A 95% material cost curve means that if the quantity of units is doubled, the cost of the material will be reduced by 5%. A 90% curve projects a 10% reduction when the quantity is doubled.

29.1. USES OF MATERIAL DISCOUNT CURVES

The curves in this section can be used as either estimating shortcuts, or as a check on vendor prices. As a shortcut they can be used to factor a known material cost either up or down to compensate for changes in buy quantities. This technique is not to be construed as a substitute for detailed material pricing, but it is a very handy tool for the estimator to have at his disposal.

When checking vendor prices, the curves serve as a norm from which to appraise either individual item quotations or total system material cost. Ven-

APPLICATION OF MATERIAL DISCOUNT CURVES

dors experience the same cost reduction with increases in quantity as the electronics manufacturer. The reduction may be at a lower rate but it is still there. The curves can be used to compare a particular product's discount rate for reasonableness.

29.2. APPLICATION OF MATERIAL DISCOUNT CURVES

The bulk of electronic component prices will follow a 95% curve. Fabricated parts, according to their degree of newness or innovation will follow a 90% or as steep as a 75% curve. The Learning Curve Selection Chart of Table 25.2, is helpful in selecting the appropriate curve slope.

Figure 29.1 gives curve slopes from 95 to 75%. They are applied by the following procedure.

The unit cost, based on a known quantity, is given for a particular material or product. What would be the unit price for the same product for a new quantity?

First, select the appropriate fabrication complexity curve on Figure 29.1. This selection should be guided by an understanding of the product's complexity either from its description or prior knowledge. Based on the definition of the discount curves from Table 25.2 of Section 25, select the one that most nearly matches the characteristics of the product. Next, identify the relative cost factor by matching up the original quantity of units given off the selected discount curve. Using the same curve now find the new relative cost factor based on the different quantity of units.

Calculate the rate of change of the relative cost factor due to the new quantity, from the relative cost factor of the original quantity by dividing cost factor (new) by cost factor (original). Multiply this factor times the unit price of the original quantity to arrive at the unit price for the new quantity.

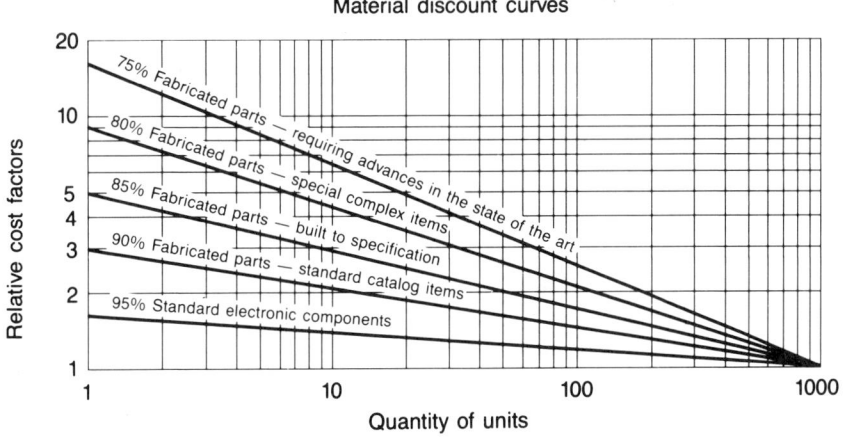

FIGURE 29.1. Material discount curves.

MATERIAL DISCOUNT CURVES

```
Item _____
Date _____                    By _____

┌─────────────────────────────────────────────────────┐
│ Basic Data                                          │
│                                                     │
│   Quantity of units quoted                   _____ │
│                                                     │
│   Unit price at this quantity                _____ │
│                                                     │
│   Degree of fabrication complexity                  │
│     (discount curve Figure 29-1                     │
│     that best matches product description)   _____ │
│                                                     │
│   New quantity.                              _____ │
│                                                     │
│ Procedure                                           │
│                                                     │
│   From selected discount curve of                   │
│     Figure 29-1, determine:                         │
│                                                     │
│     Relative cost factor (original quantity) _____ │
│                                                     │
│     Relative cost factor (new quantity)      _____ │
│                                                     │
│ Calculate                                           │
│                                                     │
│        _____ (Cost factor new quantity)          │
│     ÷  _____ (Cost factor original quantity)     │
│     =  _____                                     │
│     ×  _____ (Unit price original quantity)      │
│     =  _____ (Unit price new quantity)           │
│                                                     │
└─────────────────────────────────────────────────────┘
```

FIGURE 29.2. Material discount estimating worksheet.

This procedure is depicted by the worksheet in Figure 29.2.

Figure 29.3 illustrates the application of this procedure using the following example.

A linear array of eight opto-electronic diodes cost $3.00 each for a quantity of 25. What would be the unit cost for a quantity of 200 devices? It is known that this is a standard catalog item; however, it is not a stocked item and must be fabricated to customer order. Therefore, the discount curve best fitting the product description is 90% (a 95% discount would be more appropriate if it was a standard *stocked* item).

APPLICATION OF MATERIAL DISCOUNT CURVES 365

```
Item    8 - Optoelectronic Diode Array
Date    6/22/82                          By    H.
```

Basic Data

Quantity of units quoted	25
Unit price at this quantity	$3.00
Degree of fabrication complexity (discount curve Figure 29-1 that best matches product description)	90%
New quantity	200

Procedure

From selected discount curve of Figure 29-1, determine:

Relative cost factor (original quantity)	1.8
Relative cost factor (new quantity)	1.3

Calculate

```
         1.3    (Cost factor new quantity)
    ÷    1.8    (Cost factor original quantity)
    =    .722
    ×    $3.00  (Unit price original quantity)
    =    $2.17  (Unit price new quantity)
```

FIGURE 29.3. Material discount estimating worksheet—completed.

Following the worksheet procedure, the 90% selected discount curve on Figure 29.1 gives us a relative cost factor of 1.8 for the original quantity of 25 and a 1.3 relative cost factor for the new quantity of 200. The ratio of the new and original cost factors times the original unit price of $3.00 provides a discounted price of $2.17 based on a quantity of 200 units.

SECTION THIRTY

SPECIAL TOOLING AND TEST EQUIPMENT

30.1. Tooling, Test Equipment, and Production Planning Program Cost Ratios 367
30.2. Comparative Tooling and Manufacturing Methods 368
30.3. Tooling Cost Factors 368
30.4. Special Test Equipment Cost Factors 372

Special-purpose tooling and test equipment are those equipments that have no use other than to support the specific requirements of the job or program for which they were designed. This category can be a significant cost element because the full cost of the design, development, and purchase must be absorbed by the particular job. General-purpose tooling or test equipment, on the other hand, may be applied to any program, and is acquired from capital resources with costs distributed over all jobs as a part of the fixed burden.

Standard catalog items of test equipment may be classified as general or special purpose according to the manner in which they are used. A rule of thumb is that if a standard piece of test equipment is modified and permanently connected as part of a specially designed test system that is limited to a particular product, it becomes special purpose in nature. It is no longer available to other jobs in the shop; therefore, its costs are attributable only to the job for which it is wired. Similar reasoning applies to mechanical fixtures and tools when their design characteristics have been dictated solely by the product they support, severely limiting their use for anything else. In spite of these considerations, it is often not clear as to whether a piece of equipment can be classified as special or general purpose, and judgment must be employed. On government contracts this question is usually resolved during negotiations.

Detailed estimates of tooling and test equipment are generally made by specialists in their respective fields. The cost data in this section are designed primarily for estimating overall program requirements.

30.1. TOOLING, TEST EQUIPMENT, AND PRODUCTION PLANNING PROGRAM COST RATIOS

The cost of major manufacturing programs is usually segregated into start-up costs and recurring costs. The elements of this paragraph represent the start-up costs for such a program. Table 30.1 shows typical ratios of start-up to recurring program costs. All costs are in terms of fully burdened labor and material. They are used as follows:

100 Production Quantity	
Recurring cost/unit	$ 1,500
Recurring cost/lot	150,000
Composite start-up cost/lot	
(Medium implementation at 9%)	13,500
Total program	$ 163,500
1000 Production Quantity	
Recurring cost/unit	$ 1,000
Recurring cost/lot	1,000,000
Composite start-up cost/lot	
(Medium implementation at 4.8%)	48,000
Total program	$1,048,000

The degree of implementation is varied by the following equipment types and delivery schedules:

High—Ultra precision electro-mechanical instrumentation type systems, or highly tooled and planned programs for maximum monthly delivery rates.

Medium—Moderate circuit and design requirements and moderate delivery schedules.

Low—Simple circuitry and standardized designs, low monthly delivery rates, or high proportion of subcontract units.

TABLE 30.1. Manufacturing Start-Up Cost Ratios

		Percent of Recurring Manufacturing Costs			
		Lot Quantity			
Cost Element	Degree of Implementation	10	100	1,000	10,000
Production planning	High	20%	6%	1.7%	0.50%
	Medium	10	3	0.8	0.25
	Low	5	1.5	0.4	0.12
Special tooling	High	10	6	3.5	2
	Medium	5	3	2	1
	Low	3	1.5	1	0.5
Special test equipment	High	10	6	3.5	2
	Medium	6	3	2	1
	Low	3	1.5	1.0	0.5
Composite total	High	40%	18%	8.7%	4.50%
	Medium	21	9	4.8	2.25
	Low	11	4.5	2.4	1.12

30.2. COMPARATIVE TOOLING AND MANUFACTURING METHODS

Costs of special tooling and manufacturing methods for cast, stamped, and extruded parts is presented in Table 30.2. The data provides ranges of surface finishes and dimensional tolerances. The costs are average 1982 values. These equipments are typical of one-of-a-kind design, useful only to the product that they support.

30.3. TOOLING COST FACTORS

The production quantities of special electronic test equipments seldom require more than the minimum type of construction. Standard parts are used for punch and die sets wherever possible; rubber strippers; single action, rather than compound action; etc. Fabrication and assembly are manual operations, conducted in a model shop environment. The following operations are typical.

Drill Template	Shop Hours
Etched circuit board type, 50–500 holes.	
Photo hole pattern onto steel plate.	0.5
Center punch holes with toolmakers' microscope and drill holes.	
Constant time/template	1.0
Add per hole	0.03

TABLE 30.2. Comparative Tooling and Manufacturing Methods

Method	Tooling Cost Range (In Thousands of Dollars)	Optimum Quantity	Finishing Cost	Minimum Wall (In.)	Tolerance (In.)	Surface Finish	Notes
Sand casting	0.1–1.0	Up to 1000	High—clean and Machine	0.100	±0.030	Poor	Broad choices of Configuration
Permanent mold casting	0.4–4.0	200 Up	Medium—wash and Machine	0.100	±0.015	Fair	Repeatable dimensions
Die casting	1.0–10	1000 and up	Low—usually no machining	0.010	±0.002	Good	Repeatable dimensions
Investment casting	1.0–10	100–1000	Low—usually no machining	0.030	±0.005	Excellent	Best method for some configurations
Frozen mercury casting	2.0–20	100–1000	Low—little or no machining	0.010	±0.003	Excellent	Hollow parts
Stamped and formed parts	0.5–10	1000 and up	Low—trim or roll edges	0.003	±0.001	Good to Excellent	Good size and Material Range
Powdered metal parts	0.5–10	10000 and up	Low—usually no machining	0.030	±0.003	Fair	Best for small parts
Spinnings manual	0.05–0.2	1–750	Low—trim or roll edges	0.030	±0.008		
Mandrel	1.0–10	500–1000		0.020	±0.003	Good to excellent	May polish in place
Screw machine parts	0.1–0.5	1000 and up	Low—tumble deburr	0.030	±0.001	Excellent	May polish in place
Molded Plastics	5.0–20	1000 and up	Low—remove flash by tumbling or shearing	0.025	±0.005	Excellent	Wide variety of properties and colors
Drawing	2.0–8	750 and up	Trim or roll ears	0.025	±0.002	Fair	May need to polish to remove draw marks

Router Template
Etched circuit board type (aluminum)

	Shop Hours		
	Approximate Size (in.)		
Type	4 × 4	6 × 8	6 × 12
Simple	1	2	3
Medium	3	3.5	4
Complex	6	7	8

Blank-and-Pierce Die	Shop Hours
Per in. of blank circumference	1.5
Per standard punch	0.5
Per in. circumference of an irregular hole	2.0

Form Die	Shop Hours
Per in. of form line	
Very simple	0.8
Simple	1.0
Medium	1.2
Complex (make two single dies)	

Typical proportions of tooling cost elements:

Cost Element	Percent of Hours	Percent of Cost
Direct labor		
Tool Engineer	15	
Drafting and Design	25	
Fabrication	55	
Inspection	5	
Total direct labor	100%	
Burden		
Burdened labor		95
Material		5
Total tooling cost		100%

TABLE 30.3. Test Equipment Design and Drafting Time

Type Design	Description		Manual Operation		Computer Aided Design (CAD)	
	Standard Drawing Size	Ft.^{2} Drawing	Hours/ Ft.^{2}	Hours/ Drawing	Hours/ Ft.^{2}	Hours/ Drawing
Original concept	C	2.5	15	38	7.5	19
	D	5.0		75	(PR = 2.0)^a	38
	H	9.0		135		68
	J	11.0		165		82
Layout	B	1.0	10	10	5.0	5
	C	2.5		25	(PR = 2.0)^a	12
	D	5.0		50		25
	H	9.0		90		45
	J	11.0		110		55
Detail or Copy	A	0.7	3	2.1	0.8	1.7
	B	1.0		3.0	(PR = 3.7)^a	2.4
	C	2.5		7.5		6.0
	D	5.0		15.0		12.0
	H	9.0		27.0		21.6
	J	11.0		33.0		26.4

^{a}Productivity ratio—improvement due to use of CAD. (See also Section 32.3, "CAD/CAM Benefits and Costs.")

TABLE 30.4. Typical Proportions of Test Equipment Cost Elements

Cost Elements	Percentage of Hours	Percentage of Cost
Direct Labor		
Test Engineer	25	
Drafting and Design	20	
Chassis Fabrication	10	
Assembly and Wire	30	
Inspect and Check Out	15	
Total Direct Labor	100%	
Burden		
Burdened Labor		60
Material		40
Total Test Equipment Cost		100%

30.4. SPECIAL TEST EQUIPMENT COST FACTORS

Test equipment design and drafting labor is presented in Table 30.3 for various types of drawings and standard sizes. The complexity of the product and the amount of design effort required is represented by the size of drawing selected. Because computer aided design is becoming more and more prevalent as an engineering tool, a separate column provides the labor CAD estimates for equivalent types of designs and drawing sizes. CAD savings are significantly manifested once the design has been composed at the CRT terminal and data has been entered into the system. CAD savings are then realized for each successive iterations of drawings and perspectives developed from the initial design information as well as subsequent changes and revisions.

The cost elements associated with typical manufacturing test equipment developed are listed in Table 30.4. The proportion of labor for test engineering and drafting with respect to the other cost elements would likely be less if CAD is used. However, all the percentages are influenced significantly by the characteristics of the design, and materials and packaging concepts. For example, test equipment involving a high degree of logic design would be fairly engineering intensive, but relatively less costly to build if integrated circuits are used. On the other hand, a series of analog circuits for conditioning and switching of signal measurements may require only a modest amount of engineering, but result in a significant assembly and test effort if discrete components are used along with a lot of hard wiring. The same considerations also affect the burdened labor and material ratios.

As in every application of cost data in this book, judgement should always be used to modify the cost guideline to fit the particular situation.

SECTION THIRTY ONE

MANUFACTURING ENGINEERING

31.1.	Basis for Estimating Manufacturing Engineering	374
31.2.	Small Production Quantities—Minimum Manufacturing Engineering	374
31.3.	Moderate Production Quantities—100% Manufacturing Engineering	375
31.4.	Cost Comparison Example	379

Manufacturing engineering has many other names, including production engineering, production planning, and so on. As used here, manufacturing engineering is the production planning and operations analysis, both before and during production, that is applied to a specific project. This is in contrast to the general type of production engineering, which develops overall manufacturing techniques, facilities, and processes. The time a production engineering group spends to redesign a prototype unit in accordance with manufacturing or customer requirements is excluded here. Time spent designing tooling and test equipment for the specific project is also excluded, as it is usually estimated and charged as part of the cost of the special tools and test equipment. In any given company or production facility, however, the estimator must prepare his or her work in accordance with the way these functions are segregated in that facility.

A certain amount of manufacturing engineering or production planning is inevitable. If it is not done by anyone else, it will be done by the production workforce, with an accompanying increase in the direct labor requirements. The total cost of a job includes both the direct labor charges and the manufacturing engineering time. For small jobs, extensive manufacturing engineering often does not pay, whereas for large production runs the money spent on planning will be recouped in the form of higher efficiency and a steeper learning curve.

Although a range of intensities is possible, manufacturing engineering tends to be either a minimum "make per print" type of planning or a full, "100% planning," including fabrication, assembly, and test planning prior to production. It also includes sustaining engineering during the production cycle for the purpose of correcting design errors and developing more efficient manufacturing methods. Sustaining manufacturing engineering is a major contributor to the learning curve later in the production run.

In the following sections, sample estimates are developed for both levels of manufacturing engineering, and then the two are compared over a range of production quantities, to show where each would be appropriate. In practice, similar calculations would be useful in choosing the best level of manufacturing engineering.

31.1. BASIS FOR ESTIMATING MANUFACTURING ENGINEERING

It is very difficult to accurately predict what amount of manufacturing engineering will be required for any given program. By nature, extensive start-up and sustaining manufacturing engineering are problem-solving activities, and the time required is just as unpredictable as the problems themselves. Nevertheless, an estimate must be made.

The standard hour of production labor is used here as the basis for estimating manufacturing engineering labor. Judgement and comparisons with other projects are often used, and can serve as useful checks on the general range of results, but by themselves the judgement methods tend to give inconsistent results when used by different estimators. Basing the estimate on the standard labor content has several advantages. Since it is based on the fully developed process, when the operator has reached peak efficiency and all tools and jigs are available and proven, it is constant and independent of the quantity to be produced. It varies with the complexity of the unit, as does the amount of management engineering time required. And, it is already available as a part of the basic cost estimate.

31.2. SMALL PRODUCTION QUANTITIES—MINIMUM MANUFACTURING ENGINEERING

An average ratio for estimating manufacturing engineering for production quantities in the 1- to 20-unit range is about 15% of the fabrication and assembly labor estimate. This is adequate for only minimal planning of the "make per blueprint" type, plus planning the shop routing. The detailed production method is left to the production shop, so there will be higher direct labor time required than if extensive planning were done.

The 15% ratio is reasonably constant. If only one unit is to be made, it is designed to be built in a minimum of subassemblies and work orders, so minimal planning is required. For quantities in the 10 to 20 range, the units are normally designed to be produced in more subassemblies and modules, so more planning is required. The extra engineering and planning make the production process easier and reduce the overall time requirements.

31.3. MODERATE PRODUCTION QUANTITIES—100% PLANNING 100% MANUFACTURING ENGINEERING

More extensive manufacturing is typical for larger production quantities; the extra cost here is more than justified by reduced direct labor. Start-up manufacturing engineering affects fabrication, assembly, and test.

Fabrication planning includes:

Prepare operation sheets for each part, including operation sequence, material, machine, feed, and speed.

Prescribe standard and special tooling. Write-up tool order for design and construction of special tooling.

Apply standard time data to operation sheet. Maintain liaison with production and design engineering.

Assembly planning includes:

Prepare operation sheets for each assembly.
Build first sample unit.
Itemize assembly sequence and location of each part.
Write up tool order for design and construction of special jigs and fixtures.
Develop exact wire and component lengths.
Build harness lacing jib board.
Apply standard time data to operation sheet.
Balance the time cycles of final assembly line work stations.
Maintain liaison with production and design engineering.
Set up the material and layout of each work station per the operation sheet.
Instruct operator in construction of the first unit.

Test planning includes:

Determine the overall test method required to meet performance and acceptance specs.

Break total test effort into positions by function and desired test time cycle.

Prepare a test equipment list and schematic for each position.

Prepare a test equipment design order for the design and construction of special purpose test fixtures.

Prepare a step-by-step procedure for each position.

Maintain liaison with production and design engineering.

Set up test position and check out.

Instruct test operator on first unit.

Manufacturing engineering is also required after a job starts production. Sustaining manufacturing engineering has two phases. After a job is initially planned and released to the shop a debugging process of both manufacturing methods and engineering design data must be accomplished. This requires considerable effort in the beginning of a production run.

Later the object of sustaining manufacturing engineering is to apply more efficient manufacturing methods throughout the life of the production run. The reduction in labor cost per unit as forecast by the learning curve technique does not just "happen" by itself. It is the result of a *planned* and *continuous* cost reduction program on the part of management and manufacturing engineering. If a job is put into production without the follow-up effort, the anticipated reductions due to the learning process will never be realized.

Typical estimating ratios for manufacturing engineering for moderate production quantities are given in Table 31.1. Figure 31.1 is a cost estimating worksheet. Figure 31.2 shows a sample calculation for a unit with a typical breakdown of 40% fabrication and 60% assembly time. Note that about 21.4 hours of manufacturing engineering are required to properly set-up for each hour of standard time in such a unit.

Subjective Recheck:

The "values" established by the above methods should be given a final judgement check in comparison to:

1. Completeness of the unit design
2. State-of-the-art of the design
3. Available time period for production planning
4. General forecast by week or by month of the number of engineers required over the production planning period

Description of unit:

Production quantity: _____

Start-up Mfg. Eng. Hrs:

	Manufacturing Engineering Ratio	Standard Hours Per Unit		Manufacturing Engineering Hours
Fabrication	13	x _____	FAB =	_____
Assembly	15	x _____	ASSY =	_____
Test	12	x _____	ASSY =	_____
Total start-up mfg. eng. hrs.				_____

Sustaining Mfg. Eng. Hrs.:

```
            _____   Fab & assy std hrs.

    ×       _____   Learning curve cum. av. cost factor

    =       _____   Est. Hrs./Unit
                      7%  20-100 Units
                      3%  500   Units
    ×       _____   2%  1000  Units

    =       _____   Sustaining mfg. eng. hrs./unit
```

FIGURE 31.1. Manufacturing engineering cost estimating worksheet for moderate production quantities.

Description of unit:	Military Communications Reciever, Designed to Existing State of the Art.				

Production quantity: __100__

Start-up Mfg. Eng. Hrs:

	Manufacturing Engineering Ratio	Standard Hours Per Unit		Manufacturing Engineering Hours
Fabrication	13	x __4.0__	FAB =	__52__
Assembly	15	x __6.0__	ASSY =	__90__
Test	12	x __6.0__	ASSY =	__72__
Total start-up mfg. eng. hrs.				__214__

Sustaining Mfg. Eng. Hrs.:

       __10.0__   Fab & assy std hrs.

×   __2.2__   Learning curve cum. av. cost factor

=   __22.0__   Est. Hrs./Unit

                 7% 20-100 Units
                 3% 500 Units

×   __.07__   2% 1000 Units

=   __1.54__   Sustaining mfg. eng. hrs./unit

FIGURE 31.2. Manufacturing engineering cost estimating worksheet for moderate production quantities—sample calculation.

TABLE 31.1. Ratios for Estimating Manufacturing Engineering Labor for Moderate Production Quantities

Phase of Manufacturing Engineering	Manufacturing Engineering Hours/Fabrication Standard Hours/Unit		
Start-up			
Fabrication			
Fabrication methods engineering	13		
Assembly			
Assembly methods engineering	13		
Set-up and instruct	2		
Test			
Test procedure engineering	9		
Set-up and instruct	3		
Sustaining	Number of Units in Lot		
Percentage of realized	20–100	500	1000
Fabrication and Assembly hours/unit	7%	3%	2%

31.4. COST COMPARISON EXAMPLE

A natural question is "At what quantity of units does it pay to expend the effort for 100% planning?" Table 31.2 develops comparative cost data for a unit requiring one standard hour of fabrication and assembly labor per unit. It takes into account the effect of the two types of planning on the learning curve rate of cost reduction as well as the actual planning effort. The graph of Figure 31.3 shows the theoretical breakeven points for both labor hours and labor dollars. Some gross assumptions have been made, but the general indication is clear. A similar calculation for a specific job would be of value in planning the program.

Figure 31.4 graphically portrays the totals of manufacturing engineering labor to productive labor for different lot quantities. A 15% ratio is average for the 1-to-20 quantity range. Thereafter, the 100% planning is more economical even though the ratio of engineering to productive labor is higher. The graph is based on the separate start-up and sustaining engineering ratios as applied above to one standard hour of production labor.

Note that the curve shows a step down in total hours at 20 units, but since engineering hours are more expensive than production hours, the total cost reduces more smoothly.

TABLE 31.2. Manufacturing Engineering Cost Comparison Example

	Production Quantities							
	1		10		100		1000	
	Hour	Cost	Hour	Cost	Hour	Cost	Hour	Cost
100% Planning								
Production Labor at 85% Learning Curve								
Standard hour at 100th unit	1.0		1.0		1.0		1.0	
Production hours/unit— Cumulative average	6.7		3.8		2.2		1.3	
Production hours/lot	6.7	$67	38	$380	220	$2200	1300	$13,000
Planning hours/lot	21.0	$420	21	$420	21	$420	21	$420
Subtotal	27.7	$487	59	$800	241	$2620	1321	$13,420
Adding 10% tooling[a]								
Total per lot	27.7	$536	59	$880	241	$2882	1321	$14,762
Minimum Planning								
Production labor at 95% Learning Curve								
Production hours/unit— Cumulative average	6.7		5.5		4.6		3.8	
Production hour/lot	6.7	$67	55	$550	460	$4600	3800	$38,000
Planning at 15% × production hours	1.0	$20	8	$160	Negligible		Negligible	
Total per lot	7.7	$87	63	$710	460	$4600	3800	$38,000

Labor dollars are calculated as follows to give relative values:

	Production	Planning
Hourly rate	$4.00	$8.00
Burden at 150%	6.00	12.00
Total burdened rate per hour	$10.00	$20.00

[a] Production planning plus special tooling and test equipment are required to increase production efficiency at an 85% rate. Ten percent of total burdened labor is a planning purpose average.

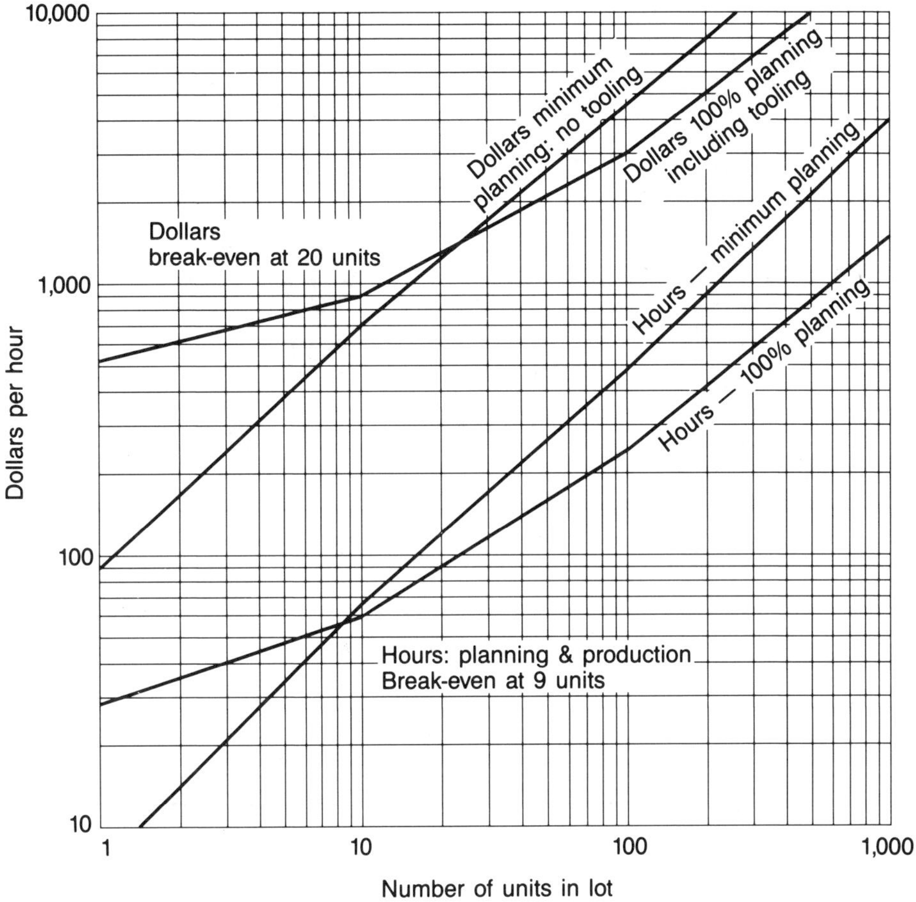

FIGURE 31.3. Manufacturing engineering cost comparison example by total direct labor hours and burdened labor cost per lot.

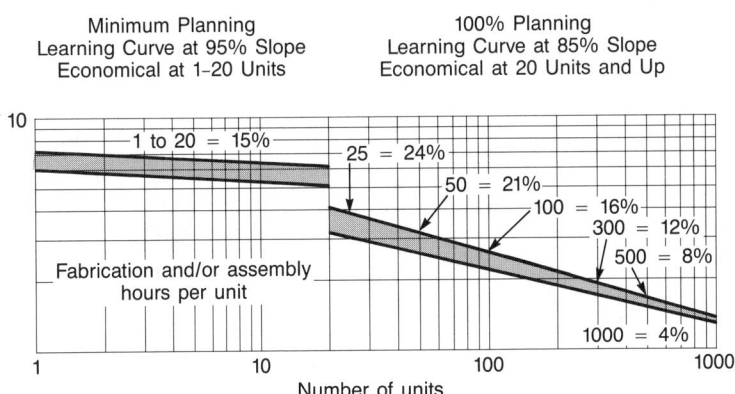

FIGURE 31.4. Manufacturing engineering composite start-up and sustaining effort as a percentage of production labor.

SECTION THIRTY TWO

COMPUTER AIDED DESIGN AND MANUFACTURING

32.1.	Computer Aided Design	385
32.2.	Computer Aided Manufacturing	389
32.3.	Benefits and Costs	390

CAD (Computer Aided Design). A process that uses a computer to assist in the creation or modification of a design.

CAM (Computer Aided Manufacturing). A process that uses computer technology to manage and control the operations of a manufacturing facility.

The basics of CAD/CAM will be presented in this section. The above definitions are quite broad; the technology is usually perceived in a much narrower sense. CAD pertains to automated drafting and drawing preparation, while CAM concerns numerical control, or computer aided machining. However, CAD/CAM has come to include much greater capabilities, the breadth of which requires detailed knowledge of the subject.

As shown in Table 32.1, CAD/CAM has a wide variety of applications, ranging from heavy manufacturing to integrated circuit production. Figure 32.1 shows the distribution of CAD/CAM among the major fields of applications. The projected growth of CAD/CAM is enormous, as Figure 32.2 attests.

The key attributes of CAD/CAM are twofold: first, that descriptions of parts and processes can be reduced to numerical data; second, that modern computers can process the resulting information in many different ways.

TABLE 32.1. Categories of CAD/CAM Applications

PC Board Design, Documentation
  Board Geometry
  Component Placement
  Signal Routing
  Editing
  Checking
  Documentation

Integrated Circuit Design
  Circuit
  Process
  Mask

Wiring Diagram Generation
  Charts, Diagrams, Wire Tags, Component Labels
  Labor Rate Setting

Two- and Three-Dimensional Design

Computer Aided Plant Design

Generation of Numerical Control Data

Manufacturing Process Design

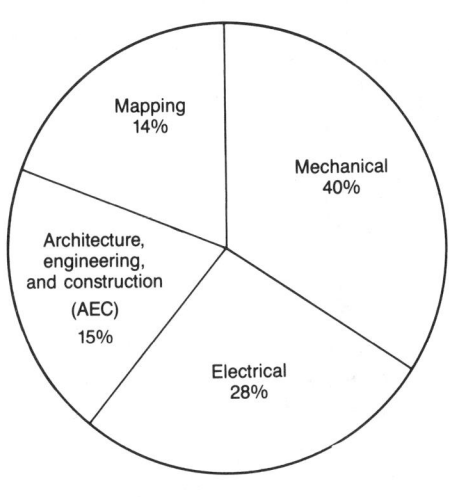

FIGURE 32.1. Worldwide CAD/CAM applications.

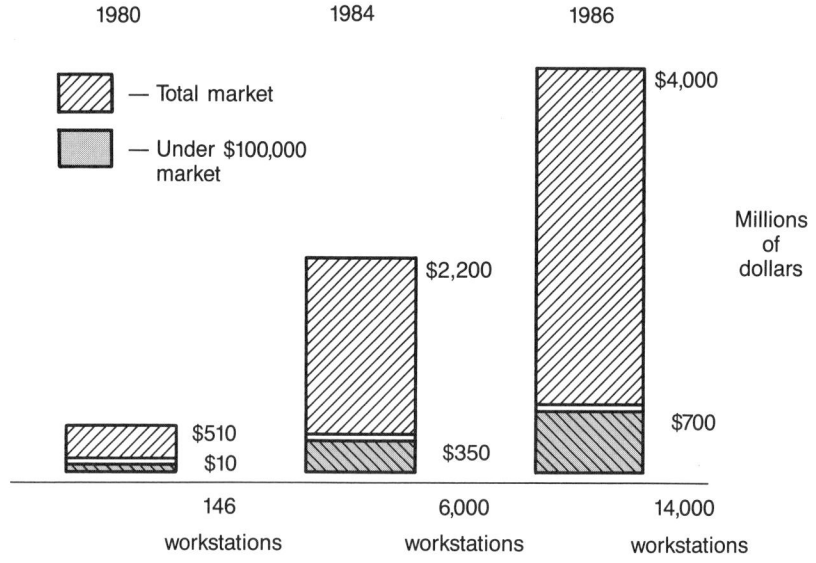

FIGURE 32.2. Worldwide CAD/CAM market growth. (*Source:* Summagraphics Corp.)

The "computer aided" in CAD/CAM refers to the use of a computer to store and process information about a product or process. The CAD/CAM revolution results from the fact that modern computers can process and distribute information rapidly and reliably. As will be shown later, a major benefit of CAD/CAM is the reliability of information transfer, quite apart from the increases in speed of information manipulation that are possible.

It is not possible to manually process part descriptions in numerical form, beyond a very restricted sort of numerically controlled machining. On the other hand, once part data is in a computer, it can easily be manipulated.

For instance, the simplest form of numerical processing is to code the dimensions of a part onto punched tape, which can be done manually, without the help of a computer. The tape can then be used to control automatic production machines accurately and reliably. But a given tape can be used to produce only one shape of part. However, if the part data is placed in a computer file, then it can be modified to produce similar parts. Also, existing parts can be used as building blocks for larger assemblies.

Other things can also be done with the part data. For instance, mold dimensions can be calculated automatically from the dimensions of the part to be molded, taking material shrinkage into account as necessary.

Sophisticated analyses and computations on part data can obviously help the design process. The existence of a detailed database and the necessary computer power can also help the management of the manufacturing process. For instance, the data file which is adequate for drilling a printed circuit board obviously contains all the information necessary to calculate the production process time. Such times constitute a source of accurate inputs to planning and scheduling. The integration of information about the different aspects of a large operation offers perhaps the greatest benefits in the form of smooth, predictable work flows.

The following sections discuss computer aided design and computer aided manufacturing individually, aspects of integrated manufacturing, and briefly, the economics of CAD/CAM.

32.1. COMPUTER AIDED DESIGN

Computer aided design began as computer aided drafting. The emphasis was on the speed, accuracy, and reliability of machine drawing. Not only could drawings be produced quickly, in any desired view and scale, but the chance of copying error was eliminated, since each drawing was produced from the same master file of numerical data.

But computer aided design in the fuller sense offers much more than faster, more-accurate drafting. The numerical data that defines the part is available for other uses, such as weight and moment calculations, stress analysis, tooling design, and so forth. And, once the design of the part is completed, the part data can be used to generate numerical control tapes or other N/C files to drive production machines.

It is important to realize that CAD is more than a picture on a screen or a machine-made drawing. It is primarily the precise geometric data that has been captured in the computer memory, combined with ability of the computer to manipulate that information quickly, in a variety of ways.

The simplest "manipulation," of course, in reproduction. In a manual drafting room, a great deal of time is spent simply redrawing parts that have already been drawn because of changes not being completely worked through the drawing set.

With a CAD system, however, there is only one description of the part: the numerical database. All screen presentations and drawings are derived from the same data. It is impossible for views not to agree. And, revision of the data base automatically accomplishes the revision of all related views and drawings made at any time thereafter; obtaining an up-to-date revision is as simple as obtaining a current drawing from the system. It is not necessary to manually trace the effects of a change through every drawing and view, from part detail through system assembly drawings.

Other manipulations are easier with a CAD system as well. Manually generating a new view from a paper drawing takes a significant amount of time. The CAD system, on the other hand, can do so in seconds, to show any point of view, any magnification, or any cross section.

386 COMPUTER AIDED DESIGN AND MANUFACTURING

Figure 32.3 shows typical applications of "traditional" CAD/CAM. The isometric views, with and without the hidden lines, help the designer visualize the art. Once the designer is satisfied that the part is properly described, the system takes care of the details of preparing the required drawings.

There are a variety of modeling techniques for describing smooth three-dimensional objects. Figure 32.4 shows a technique known as sculpturing, where the surface is defined at first very roughly, and then in finer and finer detail.

CAD/CAM now is capable of going well beyond line drawings. Solid modeling and color graphics now make possible very realistic pictures of the

Dimensional Part Graphics

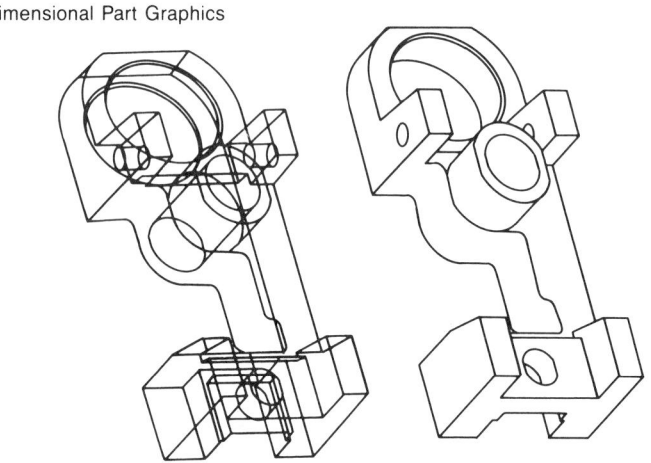

Translation into Drawings

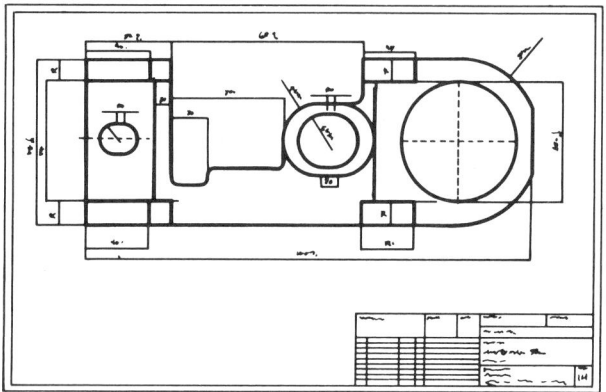

FIGURE 32.3. Traditional CAD/CAM. (*Source:* Geometric modeling for design and technological planning, by Dr. F. L. Krause—MIT CAD/CAM Conference, 1982.)

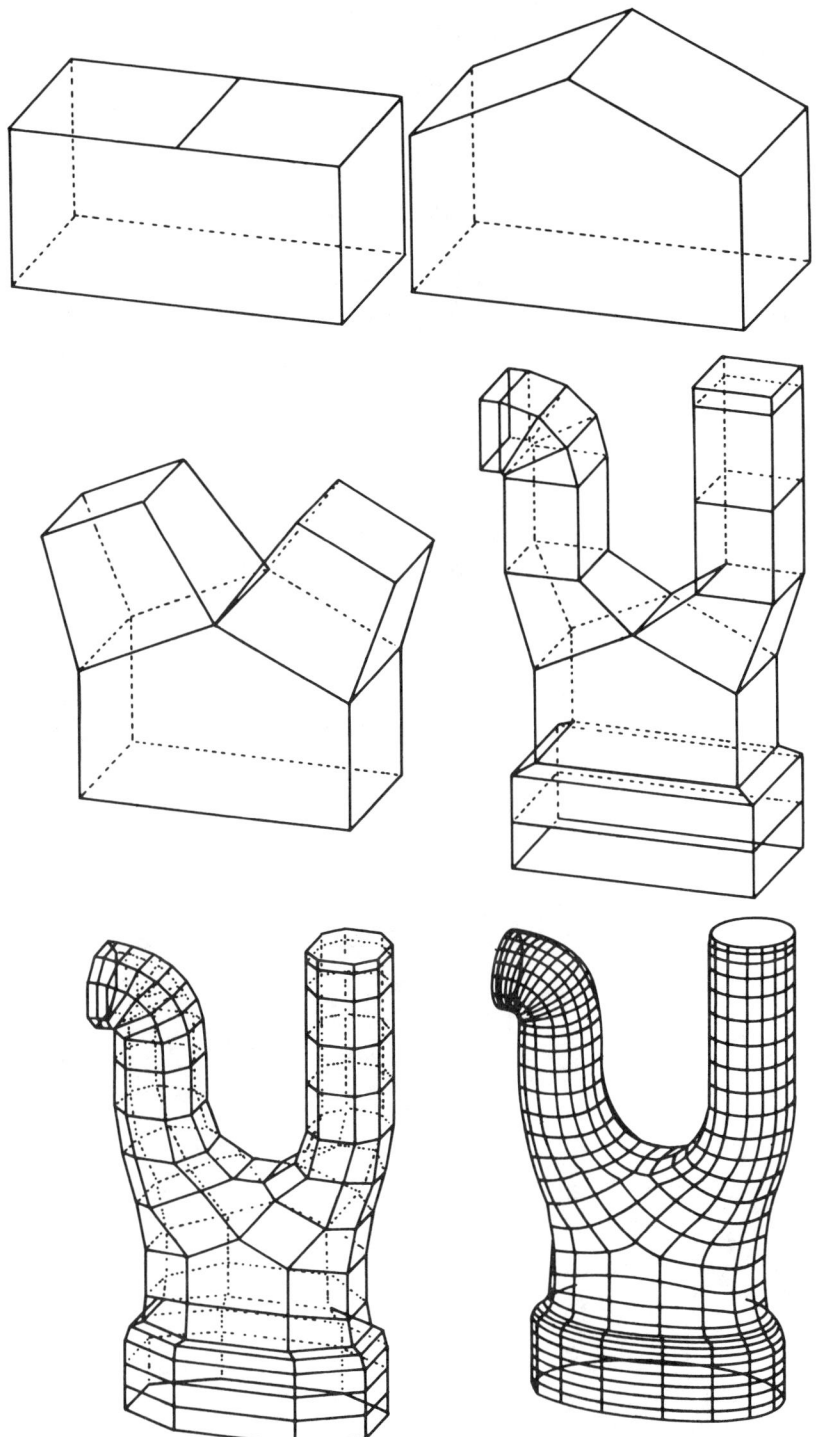

FIGURE 32.4. Sculptured surfaces—texturing. (*Source:* The design of sculptured surfaces using recursive subdivision techniques, Peter Veenman—MIT CAD/CAM Conference, 1982.)

388 COMPUTER AIDED DESIGN AND MANUFACTURING

finished parts. Some systems can properly show shadows and shading, as well as accurately portraying the surface texture of the finished part, as shown in Figure 32.5.

As well as manipulating the data for display, a CAD system can perform calculations to support various design functions. For instance, design-rule checking permits the designer to formulate a design rule for fit/interference tolerances, and then have the computer system check everywhere that the parts satisfy the specified conditions. In PC board layouts, it is possible to specify a minimum spacing between circuit paths and have the computer check the entire board, reporting on any violations of this tolerance.

Programs are available to help the designer lay out PC boards and wire-wrap panels. Given the device printouts and required signal connections, a CAD system will attempt to automatically route PC board traces. The IC designer can define portions of chip geography, such as gates or larger logical elements, and use the system to copy the elements into the required locations on the chip, rather than drawing the elements over and over again. Once the individual chip is laid out, the CAD system is used to copy the chip into the many separate positions on each of the masks required for fabrication.

The design database can also be used to provide input for engineering analysis programs, such as those used to analyze stress or thermal expansion. Again, there is considerable benefit in the use of a common database, both to avoid additional time and effort, and to avoid the introduction of errors.

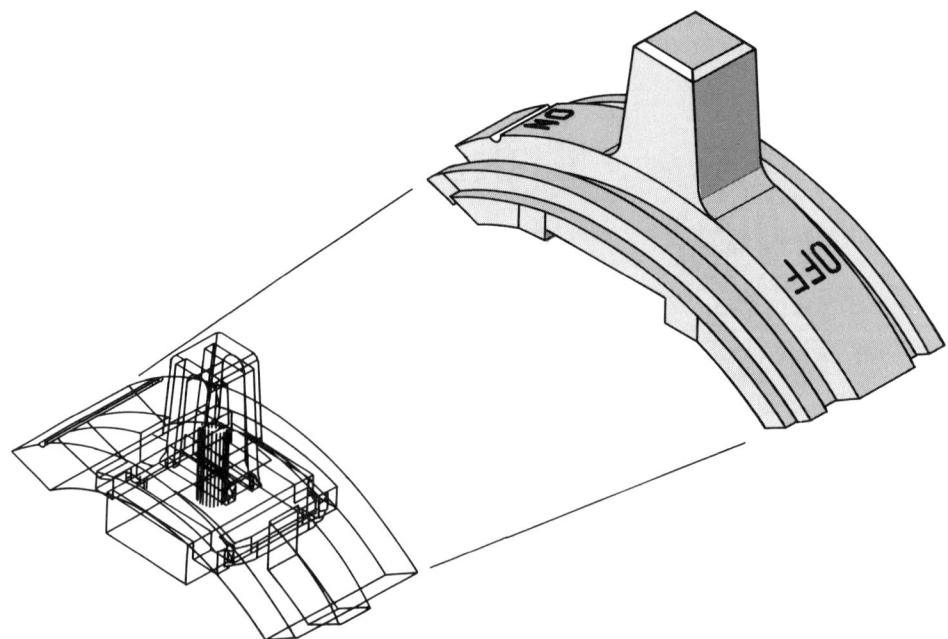

FIGURE 32.5. Solid modeling, with shadows and shading. (*Source:* The mechanical design process using solids modeling, Martin D. Schussel—MIT CAD/CAM Conference, 1982.)

32.2. COMPUTER AIDED MANUFACTURING

Although the line between CAD and CAM is often loosely defined, in general, CAD is used to define a part and CAM to prepare for its production.

The output of the CAM system may simply be a machine-produced drawing for a manual production process. More typically, it is a tape to drive N/C machinery. Or, the information may be used to produce tooling which will be used to fabricate the end product.

Given the part data created by a CAD system, the computer can do several things to assist the production process. These are all based on the existence of the single, central design database of part information.

Most directly, the CAM system can help prepare the N/C tapes or other data files needed for the production machines. Although there are many different N/C machines and formats, the designer need not be concerned with the details. With proper programming the system will automatically convert the numerical data in the CAD database to the format required by the target machine.

Conversion may be more or less complex, depending on the production process involved. For instance, for cutting machines, the required tool path is different from the part profile. The CAM system can take into account the shape of the cutter in defining a tool path as well as accommodating machine dependent requirements such as cutter changing. Some systems take tool wear into account in defining the cutting-tool path, others will automatically change tools after a prescribed operating time.

A relatively recent development is the ability to completely simulate the machining or production process. With the use of color graphics and solid modeling, the picture on the screen looks like the part, complete with shading and shadows. This has been coupled with the ability to simulate the tool and cutting action, so that it is now possible to set up simulated raw stock in a simulated fixture and watch a simulated tool take simulated cuts, arriving eventually at the desired part shape. This capability has lead to an enormous reduction in N/C tape errors. According to one estimate, the probability of having no errors in a manually produced tape is about 10%. Where computer assisted parts programming is used, the probability of no errors goes to 55%. When simulation of the cutting process is used, 75% of the resulting tapes have no errors on the first use. These benefits of computer assistance clearly go well beyond those of more-rapid production of tapes and drawings.

Another use of CAM on the horizon is the development of adaptive robotics in the production process. In many processes, including metal cutting and PC board manufacture, the part can be adequately produced by specifying simple tool actions. The production machine can be a blind, numb robot. However, in processes such as welding, forging, inspection, and sheet metal bending, phenomena such as spring-back, and uncontrollable material and process nonuniformities, make it impossible to produce adequate parts by simply specifying tool actions. The robot must be given some additional dimension of sight or feel to be of value.

Considerable computer power is required to implement such adaptive control. Lower computer costs have stimulated interest in developing the required software.

The results of this adaptive approach should show up in the near future in the electronics industry in such areas as the automatic insertion of components, inspection, and test. Prior work on automatic component insertion, for instance, has been devoted to careful control of the uniformity of the printed circuit boards and the components. However, there remain significant problems in the area of circuit board dimensional control, as parts become smaller and circuit densities increase. Adaptive systems may be a more economical way of solving the problem.

Beyond the gains from CAD/CAM improvements discussed above, computer assistance offers major benefits in the form of improved coordination among the various production areas. Information in the CAD/CAM system can be used to drive other elements in the manufacturing process as well, such as material resource planning, process planning, inventory control, and parts classification and coding.

32.3. BENEFITS AND COSTS

The economics of CAD/CAM are very difficult to quantify in the abstract. What is important, of course, is the comparison of costs and benefits rather than the absolute value of either. The actual economics of an installation can only be based on specifics. However, some historical data for benefits can be cited, and some "ball-park" costs provided.

The greatest benefit of a CAD/CAM system is increased productivity through reduced labor requirements. The labor reductions occur not only at the design stage, but also in manufacturing. Major systems have been justified on the basis of error and rework reduction rather than reduced design time. However, data on production cost reductions are very application-specific.

Design/drafting productivity improvements are hard to quantify. The following list (Table 32.2) provides examples of drawing types of various complexity levels (simple to complex) and their associated productivity levels as determined from limited industrial sources. The general range of values agrees with other published data.

The productivity improvement ratio of 10 to 1 for diagramatic revisions reflects a basic feature that CAD/CAM has in common with word processing; that it is possible to create the basic information file once, and from then on make only the necessary changes rather than completely redoing the design information for every change. This is no small item. For instance, in PC board design even a small change requires considerable checking and clerical work, much of it duplicating work previously done. These are noncreative, time-consuming, error-prone tasks.

The benefits from reduced design drafting time can be estimated in the

TABLE 32.2. CAD/CAM Productivity Improvements by Drawing Type

Drawing Type	Expected Productivity Improvement Ratio
I. *Simple Logic Drawings*	4.5–5.0
Single-line drawings	3.5–4.0
Wiring diagrams	3.0–3.5
Piping and instrumentation Diagrams	3.0–3.5
Revision of diagrams	10
II. *Stock Drawings*	4.3
Assembly/detail	3.7
Sheet metal drawings	3.7
Extrusion drawings	3.2
Numerical control (tape preparation)	2.7
Detail aircraft drawings	2.4
Layout drawings (tool design, etc.)	1.7–2.2
III. *Structural Steel*	1.5–2.0 (Est.)
Piping layout	1.25–1.75 (Est.)

following way. First, estimate the time required to perform drafting work by conventional means. In the absence of data for a particular situation, a value of 7.3 work-hours per ft.^{2} of drawing area may be used. This was developed by a major manufacturer of military electronics by collecting time data for one year. The 7.3 work-hours per ft.^{2} total is made up of elements shown in Table 32.3.

Next, for each type of tasks, apply the expected productivity improvement ratios to find the expected man hours using CAD/CAM. The values in Table 32.2 may be used in the absence of better information. An illustration of this procedure is presented in Table 32.4. The equivalent number of CAD terminals is included in the table to assist in the cost comparison. This is based on an 80% utilization factor in two-shift operation, for a total of 3200 hours per year availability.

TABLE 32.3. Conventional Drawing Work-Hours per Square Foot

Task	Hours/Ft.^{2}
Preparation for drawing	1.0
Drawing	2.2
Checking	2.7
Administration	1.4
Total	7.3

TABLE 32.4. Design Drafting Work-Hours Comparison with and without CAD/CAM

Drawing Task	Conventional Work-Hours	Expected Productivity Ratio	CAD/CAM Terminal Work-Hours	Equivalent Terminal —Years (2-shift, 80%, utilization)
1. Stock drawing	632	4.3	146	0.1
2. Assemblies and detail drawing	18,940	3.7	5,167	1.6
3. Sheet metal drawings	7,945	3.7	2,175	0.7
4. Extrusion Drawings	1,873	3.2	577	0.2
5. NC tape preparation	1,780	2.7	668	0.2
6. Other (layouts, tool design, etc.)	6,000	2.0	3,000	0.9
Total	37,170	3.2	11,733	3.7

With respect to the cost of CAD/CAM, there is a wide range of equipment available, with prices ranging from $50,000 to well over $500,000 and capabilities varying at least as much. In addition, advancing hardware and software technology is driving the cost per function down, and increasing the available capabilities.

Figure 32.6 is a block diagram of a typical installation. The number and capabilities of the terminals, the size of the central processing unit, and the variety of software packages in the system contribute to the broad range of cost.

The following cost data was developed by the Computervision Corporation for a "typical, four-station" CAD/CAM system. The numbers are not adequate for cost justification of a system; they are ball-park values for illustrative purposes only.

For a system to be amortized over five years, operating four consoles two shifts per day, with an 80% utilization factor:

(a) Hardware: includes CPU, displays, digitizers, light pens, printers, plotters, mass storage, and so forth. $450K
(b) Software to drive the CAD/CAM systems and special applications. $ 50K
(c) Average interest cost or equivalent $150K

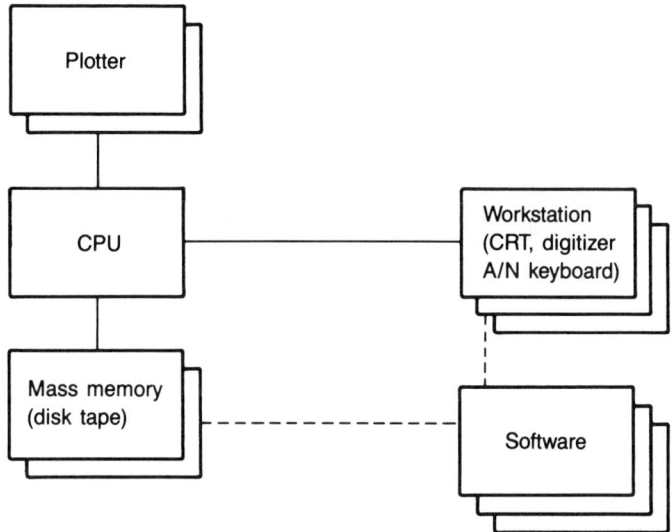

FIGURE 32.6. Block diagram of stand-alone CAD system (turnkey).

(d) Special operations and system operation (two-shift operation). $500K
(e) System maintenance. $250K
(f) Site preparation and special equipment. $ 40K
(g) Special training and administration. $300K
 Five-year total $1740K

This works out to $348K per year, or $27.19 per terminal hour, in a two-shift basis for four terminals.

Comparing benefits and costs, 37,170 work-hours of coventional drawing preparation would reduce to 11,733 work-hours plus 11,733 terminal hours with the CAD system. The differences are 25,500 fewer work-hours compared with 11,733 more terminal hours. With the terminal hour costing $27.18, the two methods have equivalent costs if the saved work-hours are at the rate of $12.50. In other words, if the fully burdened labor rate for draftsmen is above $12.50, the CAD system will save money in the drawing room.

Note that this does not take into account any savings due to reduced errors, lower production rework, and more rapid response time for initial design and for engineering change orders and revisions. As mentioned above, these factors can far outweigh the direct savings of drawing time.

From a more distant perspective, it has been estimated that a minimum-size company that can justify a CAD/CAM system is one having $20–25 million sales or 15–20 people in the drafting department.

SECTION THIRTY THREE

HOURLY LABOR AND BURDEN RATES, EARNINGS

33.1.	Hourly Labor Rates	394
33.2.	Burden Rates	395
33.3.	Earnings	396

Hourly labor rates, burden rates, and earnings or mark-up rates are the three final factors that build labor hours and material dollars into a product selling price. Some general examples and methods can be given here, but specific company policy, product, and business conditions will dictate the individual cases.

33.1. HOURLY LABOR RATES

A forecast of average hourly rates should be developed for bidding on new business. According to the normal time lapse between bidding and performance of the work, a forecast of 3 months to 3 years may be required. If the period of performance is in one year or more, it is advisable to apply an economic factor to compensate for the current inflationary trend. The next round of wage increases can be used for a union shop, or an average based on the Consumer Price Index (CPI) can be added as a general inflationary factor.

The Bureau of Labor Statistics monthly publication *Employment Earnings* serves as a guide in determining current average hourly earnings for the electronic components and accessories industry such as presented in Table 33.1.

33.2. BURDEN RATES

A burden rate is applied to direct labor costs for the purpose of recovering those costs that are not easily identifiable with the product. The burden rate is merely a means of allocating expenses that otherwise are not practical or economical to segregate as a direct cost.

It is not the purpose of this book to go into the many different methods of accumulating and allocating indirect costs. This is the function of the accounting profession. However, it is important that the person applying these burden rates have an understanding of their purpose and derivation. It must be remembered that what is not itemized as a direct cost for a product must be included as a burden cost. If it is not included in either category, a legitimate cost element is being left out. If it is included in both categories it is being charged twice. There are a number of fringe areas such as low value and bulk material items, inspection and manufacturing engineering labor, and so on, which are classified as direct by one company and indirect by another company. It is the estimator's responsibility to ferret out these fringe elements and make certain that they *are* charged once, but not twice.

The simplest burden structure is that used by small job shops. It is very common for the small job shop to price its services in terms of $15, $20, or $25 per hour. This single rate includes the cost of the basic labor, all indirect costs, and a profit allowance. Corporation-type accounting methods have developed a somewhat more complex system of cost allocation and burden rates. A typical example is given here:

Overhead Rate by Cost Center. A cost center may be an entire plant or a department within the plant. The appropriate facility expenses, management salary expenses, and so on, are allocated as a percentage of the direct labor

TABLE 33.1. Average Hourly Earnings: Components and Accessories

Industry Category	Average Hourly Earnings($)
Electronic Components and Accessories	
Electronic tubes	8.68
Semiconductors and related services	7.95
Electronic components	6.48
Miscellaneous electrical equipment and supplies	8.89
Communications equipment	9.21

Source: Bureau of Labor Statistics Employment & Earnings (April 1982).

dollars. The cost center technique allows the burdened labor rate to reflect the difference in facility cost utilized by different departments. It is not uncommon that heavy equipment departments such as punch press, machine shop, and paint finishing will run at a 175% to 225% burden rate. The high burden ratio is further amplified by the fact that the use of high-production machinery not only increases the facility cost per hour but also decreases the labor cost. Thus, the burden rate appears higher than if only the facility cost was increased. Even with the high burden ratio, the high-machinery, low-labor combination is still the least-cost method of production. The lower facility costs of bench assembly operations are reflected by the fact that they typically run at a 100% to 150% burden ratio. The least costly operations in terms of facility and supervisory effort will require a 60–80% burden rate.

Material Burden Rate. A further refinement in the allocation of indirect expenses is to segregate the cost of the material function and allocate it on the basis of the dollar value of material procured. The material function is usually defined as the purchasing, shipping, and receiving, and storeroom activities. The material control or production control activity that is maintained by the production department is not normally included. The cost of this function is usually 5 to 20% of the value of the material procured and handled. Procurement departments serving prototype-production operations will be in the upper range while the procurement department serving a quantity-production activity will be in the lower range.

General and Administrative Burden Rate. This cost element is characterized by the multi-cost center operations of corporations. The burdened labor and burdened material of the above two elements make up the "factory cost." The cost of the corporate administrative activity is allocated over the various factory operations in terms of a G and A percentage. The typical range is 5 to 15%.

A definition of the various cost elements and their relationship is presented in Figure 33.1.

33.3. EARNINGS

In building up a product sales price, we are now at the point of "total manufacturing cost." Even this point of total cost must be qualfied by the possibility that the cost of royalties, warranties, and so on must be added. Assuming that this has been done, the next question is one of earnings or gross mark-up. Table 33.2 provides a rough breakdown of the channels of distribution through which a product might be sold, and the associated earnings or mark-up percentages. A cost plus fixed fee contract with a government agency is rated as the least-effort, least-risk method of making a sale. It warrants the lowest earnings.

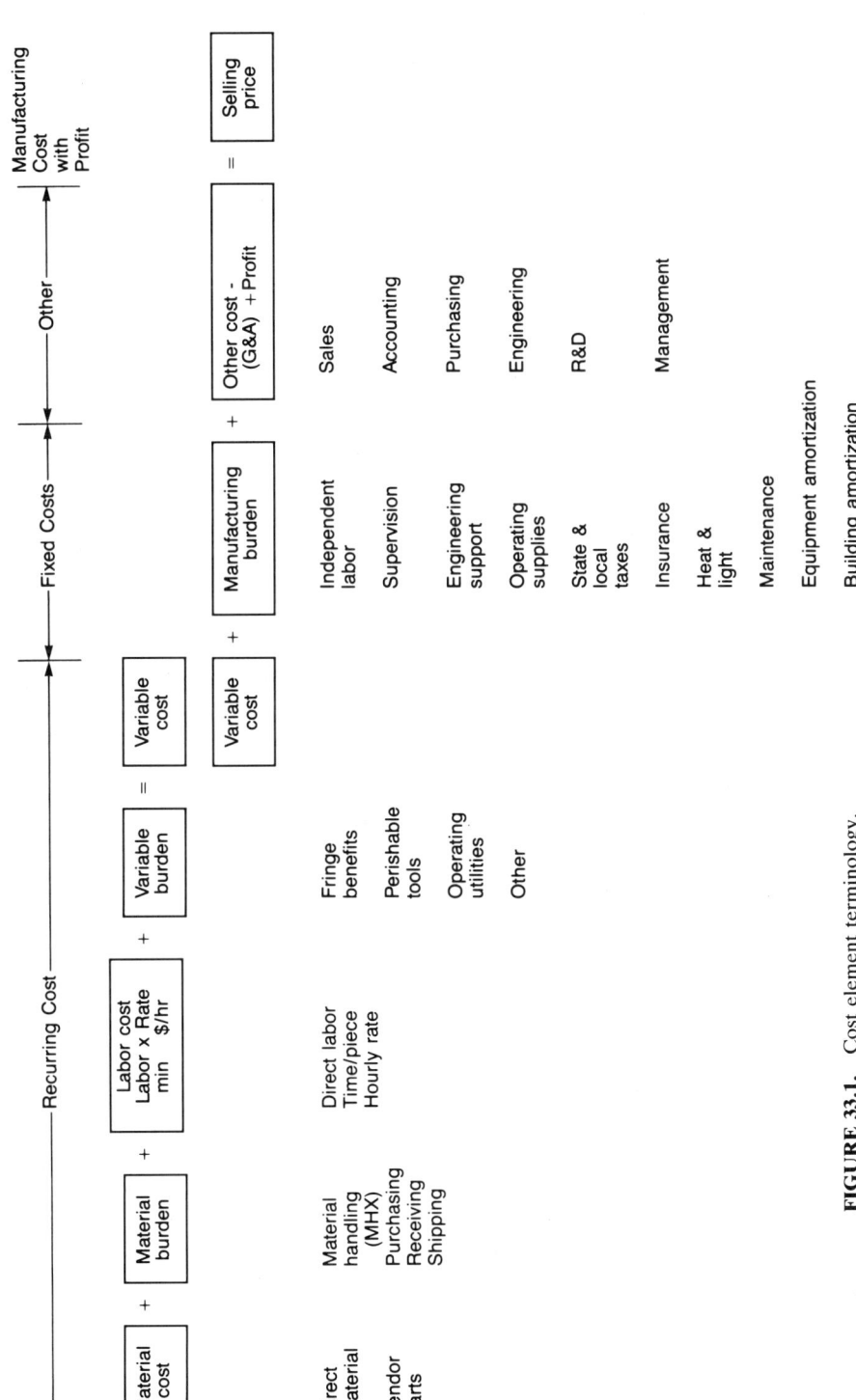

FIGURE 33.1. Cost element terminology.

TABLE 33.2. Product Earnings—Mark-Up Chart by Channels of Distribution

Government Sales		Commercial Sales		
CPFF	Fixed Price			
Manufacturing	Manufacturing	Manufacturing	Manufacturing	Manufacturing
↓	↓	↓ 20%	↓ 20%	↓ 20–25%
User	User	Manufacturing sales	Manufacturing sales	Manufacturing sales
		↓ 10%	↓ 10%	↓ 10–15%
		User	Distributor	Distributor
			↓ 20%	↓ 20–30%
			User	Retailer
				↓ 30–50%
				User
		Typical Total Markup		
5–8%	9–12%	30%	50%	100%

As more and more marketing activities are required to sell the product, the mark-up must be increased to pay for these functions, and for the risk involved. (The earnings and mark-up percentages on the chart are given as relative values only. They could not possibly reflect all products or all business situations.)

A final and perhaps the most important technique for pricing commercial products is to determine what price the customer is willing to pay for the desired product, and then, working backwards, determine if it can be designed, manufactured, and marketed for that price.

INDEX

Acceptable quality level (AQL), *see* Inspection
Allowances, 34, 35, 333–338
  bid, 34, 338, 339, 343, 344
  labor, 34, 334–338
  material, 35, 343, 344
  Standard Allowance Selector, 337
  standard hours:
    labor, 34, 334–338
    multipliers, 34, 338, 339
  trouble shoot and retest, 280
American Machinist's Manufacturing Cost Estimating Guide, 68
Annealing, 21, 111, 112
Anodizing, 128, 138, 139
Arc welding, *see* Welding, Arc or Gas
Assembly:
  costs, concept estimating, 63, 64
  labor, ratios, 376, 379
Assembly methods, manufacturing engineering:
  planning, 37, 374–376, 379–381
  ratios, 376–381
Assembly operations, 27–29, 198–200
Assembly times:
  connector/plug to harness, 30, 31, 227–229
  direct labor proportions, 49
  ground lead to coaxial cable, 225
  identification tape to cables, 227
  mechanical, *see* Mechanical assembly
  Plug or connector to harness assembly, 227, 228
  RF connector to coaxial cable, 226, 227
  wiring and component insertion, 235–247
  see also Manufacturing engineering
Automatic test equipment (ATE), 281
  component, 282–289, 294, 295

  printed circuit boards, 24, 25, 170, 171, 173, 175, 176, 293, 301–303
  types of, 282, 283–289

Baking, 134
Blanking, 19, 20, 102, 103, 106, 107
Boring, tolerances, 94, 96
Broaching, 17, 81, 82
  tolerances, 96
Burdened labor, cost elements, 371
Burden expenses, classifications, 50, 51
Burden rates:
  definition, 394–396
  general and administrative burden rate, 396
  factory cost, 396
  material burden rate, 396
  material control, 396
  material function, 396
  overhead rate by cost center, 395, 396
  special tooling and test equipment, 366
Bureau of Labor Statistics:
  Employment And Earnings, 395
  Producer Prices And Price Index (Materials), 340–342
Burring, sheet metal operations, 21, 22, 115, 116, 118
Bus wire, 222, 240, 245, 246

Cable:
  coaxial, 222, 225–227, 242, 243
  shielded, 223, 224, 241, 242
CAD/CAM, *see* Computer aided design/computer aided manufacturing
Capacitors, 222, 223, 240, 241
Cement, use of, 207

399

INDEX

Channels of distribution, mark up chart by, for product earnings:
   commercial sales, 398
   government sales, 398
Charts:
   learning curve selection, 324–329
   manufacturing phasing, schedule, 41, 42
   mark-up, product earnings, by channels of distribution, 398
   phasing, manufacturing, 41, 42
Coil winding operations, 210–217
   analysis for, 210–217
   detail values, application of, 211, 212
   general time values, 30, 211–217
   run time analysis, 213–217
   set-up analysis, 212, 213
   summary table, 30
   terms, 210, 211
Comparative tooling and manufacturing methods, 368, 369
Component counts, 57–63
Component insertion, 235–247
Compression molding, 156–159
Computer aided design/computer aided manufacturing (CAD/CAM), 382–393
   applications, 382–390
   benefits, 390–393
   block diagram of, 393
   costs, 392, 393
   definition, 382
   manufacturing, 389, 390
   market forecast, 384
   numerical control, 383, 389, 390
   special test equipment, 371, 372
Concept estimating, *see* Estimating, concept
Conductors, wire preparation, 224, 225
Connectors, wire preparation, 30, 31, 226–229
Consumer Price Index, 394
Contracts, government:
   cost plus fixed fee, 397, 398
   fixed price, 397, 398
Cost allocation, accounting methods:
   corporations, 395, 396
   small job shops, 395, 396
Cost control, internal, 320
Cost estimating, *see* Costs
Cost ratios, manufacturing start-up, *see* Ratios, cost, manufacturing start-up
Cost reduction, projections, materials, 362
Costs:
   American Machinist Cost Estimating Guide, 68
   burden, 395, 396
   cost estimating worksheets:
      automatic testing, 303
      bid allowance, 338
      chemical surface treatment, 136, 137
      coil winding, 213–217
      electroplating, 136, 137
      engraving, 197
      integrated circuit packaging, 360
      labor allowance, 338
      manufacturing engineering, 377, 378
      material discounts, 364
      packing and packaging, 312, 313
      painting, 188
      plastics, 161
      printed circuit boards:
         double-sided, 171
         multi-layer, 176
      silk screen printing, 196
      standard hour multipliers, 338
      terminal boards, 178
   engraving, 194–197
   estimating, 49
      concept, 56
   estimating forms, 8–13
   estimating methods, 4
   estimating process, 6, 7
   fiber optics, 271
   fixed, 50, 51
   inspection, 272
   integrated circuits, 58–61, 359–361
   labor, *see* Direct labor; Indirect labor
   learning curve, 317, 329
   machine cost index, 99
   machining, 68, 69
   manufacturing, *see* Manufacturing costs; Manufacturing engineering; Manufacturing start-up costs
   material cost summary, *see* Material costs and allowances
   packing, 306–316
   painting, 188
   plastics, 160–163
   printed circuit boards, 170, 171, 175, 176
   raw material, miscellaneous, 35, 340–342
   silk screen, 193, 194
   special tooling and test equipment, 366–372
   terminal board fabrication, 177, 178
   testing, 171, 176, 303, 360
   variable, 50, 51
Crimping:
   definition, 236
   obstruction allowances, 244–246
   terminal, detail values, 236–238
   wiring values, 30, 31, 240–246, 248, 249
   see also Wiring

INDEX

Curves, learning, *see* Learning curves
Cutting, raw material, machine
    ship operations, 15, 68–70

Decals, 27, 194
  assembly, 194
  silk screen method for making, 189–194
Delivery schedules, 45–48
  learning curve, 46–48
  program breaks, 47
Design, *see* CAD/CAM
Designing, CAD/CAM, 382–388, 390–392
Dimple (or joggle), 108
Direct Costs, *see* Costs, labor
Direct labor:
  burden expense ratio, 50, 51, 395, 396
  costs (inspection), 272
  hourly rates, 394–397
  proportions, 49, 50
  special tooling and test equipment, 371
Discounts, *see* Material discount curves
Distributions, product earnings, 38, 396–398
Drafting (drawings), labor hours, 370, 371
Drilling:
  dynamic balance in, 207–209
  machining shop operations, 17, 79–81
  NC tapes, 167, 171, 173
  printed circuit boards, 165–169, 172–174
Dynamic balance operations, 29, 207–209

Earnings, 396–398
  gross mark-up, 396–398
  mark-ups by channels of distribution, 38, 396–398
Electronic equipment manufacturing:
  concept estimating, 56–64
  direct labor proportions, 49, 50
  direct labor requirements, 49–51
  indirect labor proportions, 49–51
Electroplating and chemical surface treatment of metals, 23, 124–148
  analysis for summary time values, 124–137
    cleaning and baking parts, 132–135
    handling time, 132, 133
    inspection, 134, 135
    labor times, 135–137
    man dipping time, 130–132
    masking, 132
    number of batches, 129, 130
    parts handling time, 132, 133
    parts processing time, formula for, 127
    sign-in time, 128
    time variables, 126, 127
    printed circuit boards, 166–169, 171–174

  process descriptions, baths and times per batch, 23, 128, 138–148
  summary table, 23, 128
  analysis for, 124–148
Engineering change notices, 43, 45
Engine lathe, 16, 73–76
Engraving, 27, 123, 194–196
  analysis for summary table, 194–197
  costs, 194–197
  lacquering, 195, 196
  pantograph, use of, 195, 196
  sheet metal, 123
  summary table, 27
Estimating:
  concept, 56–64
    applications, 63, 64
    block diagram estimate, 58
    component counts, 57–59
    costs, 63, 64
    definition, 56, 57
    Parameters For Hardware, 57–63
    Ratios, 61–63
  methods, 4
  personnel and planning ratios, 49–55
  process, 6, 7
  special tooling and test equipment, 366
Etched circuit boards, *see* Printed circuit boards
Expenses:
  burden, classification, 50, 51
  indirect, 395–397
  see also Costs
Eyelet operations, 201

Fabrication:
  costs, concept estimating, 63 64
  engineering ratio, 379
  harness 30–31, 231–234
  manufacturing engineering:
    methods, 374–376
    ratios 376–381
Facility planning (learning curve), 321. *See also* Personnel and facility planning ratios
Fault detection and isolation, 296–300
Fiber optics:
  applications, 269, 270
  cable, 260–262
  characteristics, 254, 255
  connectors, 263, 264
  costs, 271
  detectors, 264
  emitters, 264, 265
  fibers, 260–262
  market projection, 256–259

402 INDEX

Fiber optics (*Continued*)
optoelectronics, 267–269
receivers, 265, 266
test equipment, 266
transducers, 266, 267
transmitters, 264, 265
Forming, *see* Press operations
Forms, *see* Estimating
Functional testing, 284, 286, 288–291, 296–303

Gas welding, *see* Welding, gas
Gauging frequency, 275
Gauging times, machine shop, 274
Gear hobbing *see* Hobbing, gear
Glyptol, use of, 28, 207
Grinding:
centerless, 18, 82–84
in-feed method, 83, 84
through-feed method, 83
external cylindrical, 18, 84–87
formula for grinding time, 85
internal cylindrical, 18, 87, 88
speeds and feeds, 94
summary tables 18–19
analysis for, 87–89
tolerances and surface finishes, 94–100

Hardware:
concept estimating, 56–64
prices, 59, 60
Harness:
assembling connectors to, 31, 227–229
fabrication, 231–234
nail board, 31, 234
spot tie, 31, 230, 231
Hobbing, gear, 89–94
speeds and feeds, 93
time estimating formula, 90–94
Hobbing formula, 92–94
basic, 92
example, 92, 93
Holes:
printed circuit boards, 165–169, 172–174
sheet metal:
drilling, 104–106
techniques for producing, 103–106
time values, 20, 105
turret punch, 19, 104
Hourly rates, 394, 395

In-circuit testing, 284, 286, 288–291, 294, 296–303
Indirect costs, *see* Costs, labor
Indirect labor:
burden expense ratio, 50, 51

costs (Inspection), 272
rates, 395–397
Industrial work, *see* Silk screening operations and engraving
Injection molding, 156–158
Inspections, 32, 272–276
acceptable quality level (AQL), 276
cost, labor:
burden rate, 272
estimating ratios, 272, 273
percentage ratios, 272, 273
electroplating, 134, 135
estimating ratios, 32, 272, 273
types of inspection, 32, 273
gauging frequency, 275
gauging times, machine shop, 274
labor, 272
printed circuit boards, 165–169, 171–174
record keeping, 273, 274
sampling, random, 276
summary table, 32
analysis for, 272–276
time standards, 272
time values, elemental, 274, 275
visual inspection, 274
Integrated circuits:
CAD/CAM, 383
concept estimating, 57–64
costs, 59, 60, 359–361
joggle (dimple), 108
market projection, 345–347
MIL-STD-883-B, 354–359
packages, types, 349–352
packaging process, 353–359
technical projection, 346, 348
testing, 354–359
see also Hardware

Labor:
allowances, 34, 333–338
direct, 49–51
hours, summary sample, 34. *See also* Inspection
indirect, 49–51
learning curve, 34, 333, 339. *See also* Learning curves
proportions, 49
standard hour multipliers, 34, 338, 339
Labor allowances and multipliers, 34, 333–339
standard hour allowance, 333–338
design growth, 335
engineered change, 335
engineered prototype, 335

INDEX **403**

measured labor variance, 334
miscellaneous, 336
rework and repair, 335
standard hour multiplier, 34, 338, 339
summary table, 34
analysis for, 333–339
total bid allowance, 34, 338, 339
Labor costs:
inspection, 272
learning, 321
test equipment design, 371
Labor hours summary, sample estimating, 34
Labor learning curves, 34
analysis for, 317–332
Labor rates, 394, 395
electronics industry, 395
unions, 395
Lacquering, *see* Engraving
Lathe, machine shop operations:
engine, 16, 73–76
turret, 15, 16, 70–73
Learning curves, 34, 317–332
aircraft industry, 317
applications, 320, 321
completion *vs.* expenditure, 331, 332
cumulative average cost, 322–324
definition, 317
delivery schedule, 45–48
formulas for, 321, 323
industrial learning, 318–320
debugging engineering data, 319
improvements, 319
management learning, 319
operator, 319
labor, 376–380
log log curves, 322, 327, 328
manufacturing engineering, 373–381
packing and packaging, 306–311
percentages, 34, 322–324, 331, 332
principles, 318
selection chart, 326
shipbuilding industry, 320
summary table, 34
analysis for, 317–332
time reductions, 318–320
time requirements, 329–331
uses, 329–331
variables:
individual *vs.* total program progress, 326, 329
job shop vs. mass production, 326, 329
product complexity, 325, 326
product newness, 325, 326, 328
Learning curve selection chart, 326
Line stock ratio, 342, 343

Log log curves, 322, 327, 328

Machine shop operations, 15–19, 67–100
broaching, 81, 82
cut raw material, 68–70
drilling, 79–81
gear hobbing, 89–94
grinding:
centerless, 82–84
external cylindrical, 84–87
internal cylindrical, 87, 88
surface, 88, 89
inspection, *see* Inspection
lathe:
engine, 73–76
turret, 70–73
machining operations, *see* Machining operations
milling, 77–79
speeds and feeds, 94
summary table, 15–19
analysis for, 67–100
time values, 68, 69
tolerances and surface finishes, 94–100
Machining costs, 99
Machining data handbook, 68, 94
Machining operations:
electronic equipment, direct labor proportions, 49–51
inspection, gauging times, 274, 275
summary table, 15–19
analysis for, 67–100
Manpower requirements, forecasting, 321
Manufacturing costs:
start up, 367
total, 397, 398
Manufacturing engineering, 37, 42, 44, 373–381
assembly, 374–376
cost comparison, 379–381
costs:
moderate production quantities, 375, 376
small production quantities, 374, 375
see also Burden rates; Hourly rates; Plastics
definition, 373
labor hours, estimate, 379, 380
learning curve, 376–380
planning:
assembly, 376
fabrication, 376
100%, Moderate Production, 375, 376
minimal, Small Production, 374, 375
schedule Determination, 41, 42, 44
test, 376

INDEX

Manufacturing engineering (*Continued*)
  plastics (fabrication methods), 155–159
  production quantities:
    moderate, 375, 376
    small, 374, 375
  ratios:
    burden, 50, 51
    direct labor proportions, 49–51
    moderate production quantities, 375, 376
    small production quantities, 374, 375
    start up and sustaining, 37, 376–379
  research and development funding, 51–55
  summary table, 37
    analysis of, 373–381
  test and inspection, 41, 42, 44, 367, 368
  tooling, 367, 368
Manufacturing engineering CAD/CAM, 382, 383, 389–393
Manufacturing start-up costs, 367, 368, 376–379
Masking, 132, 250–253
Material:
  allowances, 35, 343, 344
  cost elements, special tooling, 371
  manufacturing start-up costs, 367, 368
  packing price list, 314
Material costs and allowances, 35, 340–344
  line stock, 342, 343
  plastics, 160–163
  raw material prices, 35, 306, 314, 316, 340–342
  raw material ratios, 343
  summary table, 35
    analysis for, 340–344
  total bid allowance, 35, 343, 344
Material discount curves, 35, 362–365
  applications, 35, 363, 364
  cost reductions, projections, 362
  electronic component prices, 363
  log log curves, 363
  packaging, 306, 308
  packing, 306, 308
  summary table, 35
    analysis for, 362–365
  uses of, 362, 363
Mechanical assembly:
  cement and glyptol, 28, 207–
  dynamic balance, 29, 207–209
  miscellaneous operations, 28, 29, 204–206
  parts handling, 27, 198–200
  rivet, stake, eyelet, 28, 201
  screw and nut operations, 27, 202–204
  tape and tags, 29, 206
  tool handling values, 27–29, 206
  walking, 29, 207
  see also Riveting and mechanical assembly
Metal treating:
  analysis of time values, man time, 124–137
  process descriptions, 138–148
  summary table, 23, 128
    analysis for, 124–148
  see also Electroplating and metal treating
Milling, 16, 17, 77–79
  cutters, types of, 79
  machine shop operations, 67–100
  speeds and feeds, 94
  tolerances, 94–100
MIL-STD-883B, integrated circuits, 354–359
Multipliers:
  standard allowance selector, 337
  standard hour. 34, 338, 339

Notching, 19, 103
Numerical Control (NC):
  CAD/CAM, 383, 385, 389–392
  printed circuit boards, 24, 167, 171, 173, 176

Operations:
  assembly, miscellaneous, 28, 29, 204–206
  automated, 50
  coil winding, *see* Coil winding operations
  dynamic balance, 29, 207–209
  machine shop, *see* Machine shop operations; Machining operations
  manual, 50
  painting, *see* Painting operations
  press, *see* Press operations
  rivet, eyelet, stake, 28, 201
  screw and nut, 27, 202–204
  sheet metal, *see* Sheet metal operations
  soldering, *see* Soldering
  wiring, *see* Wire preparation; Wiring
Optoelectronics, 267–269

Packing and packaging, 32, 33, 304–316
  definition, 304
  labor requirements, 306, 307, 311, 312
  learning curves, 306–311
  material costs, 33, 306, 308, 312, 313, 314, 316
  material discount curve, 306–311
  methods:
    methods of preservation (MIL-P-116), 304
    moisture protection (Methods 1A8, 1C1, 11B, 11C, 1A14), 305, 306
  physical protection:

INDEX

405

maximum (Methods 1A14, 11B), 305, 306
minimum (Method 111), 305, 306
moderate (Methods 1A8, 1C1, 11C), 305, 306
price list material, 314
reusable containers, 316
selection of 61, 62
summary table, 32, 33
  analysis of, 304–316
time and material requirements, 32, 33, 306, 307
Painting operations, 26, 179–188
  costs, 188
  detail values, miscellaneous, 26, 184–187
  enamel, 26, 183
  fungicide, 26, 184
  lacquer, 26, 182, 183
  masking, 26, 180
  plastic protective film, 26, 183, 184
  primer, 26, 181, 182
  set up time per paint type, 26, 181
  sheet metal, 122
  summary table, 26
    analysis for, 179–188
  surface preparation, 23, 26, 180
  surfacer, 26, 182
  time values, 26, 180
  time variables, 179, 180
  vinyl, 26, 183
Pantograph, use of, for engraving, 195, 196
Parts handling, 27–29, 198–200
  detail, elemental table, 199, 200
  electroplating, 132, 133
  part-types, explanation of, 199, 200
Personnel and facility planning ratios, 49–55
  burden expense, 50, 51
  direct labor proportions 49, 50
  indirect labor, 50, 51
  sales and facilities, 51–55
  space planning, 51
  use of, in estimating, 49
Phasing, manufacturing, 41–45
  chart, 41, 42
  design, 43
  manufacturing, 44
  manufacturing test engineering, 44
  material procurement, 44
  production planning, 44
  test and inspection, 44
Photographic operations, 27, 189–193
  equipment, basic, 191
  film:
    development, 191–193

exposure, 191–193
labor times, 192, 193
machine times, 192, 193
silk screen stencil, fabrication, 27, 191–193
summary table, 27
  analysis for, 189–193
Planning ratios:
  direct labor proportions, 49, 50
  estimating, 374–379
  indirect labor and burden expenses, 50, 51
  sales and facilities, 51–55
  space, 51
Plastics, 149–163
  applications, 150–155
  costs, 160–163
  fabrication methods, 155–159
    compression molding 156–159
    injection molding 156–158
  types, 150–155
    new materials, 155
    thermoplastic, 151–153
    thermosetting 153–155
  use in electronics industry, 150
Plating, *see* Electroplating
Press operations, sheet metal, 106–113
  annealing, 21, 111, 112
  blank and pierce, 20, 106, 107
  brake form, 108, 109
  deep draw, 110, 111
  dimple or joggle, 108
  hydroform, 21, 112, 113
  roll form, 21, 109, 110
  time values, 19–23
Prices per pound:
  plastics, 162
  raw materials, 341
Pricing, 394–398
  Bureau of Labor Statistics, producer prices and price index, 342
Printed circuit boards, 24, 25, 164–176
  CAD/CAM, 383, 388
  double sided costs, 169–171, 175, 176
  fabrication:
    drilling, 165, 166–169
    electroplating, 165–169
    etching, 165–169
    inspection, 165–169
    numerical control, 24, 167, 171, 173, 176
    pre-production, 24, 166, 167
    silk screening, 166, 169
    subtractive process, 165
  multi-layer:
    costs, 175, 176

406 INDEX

Printed circuit boards (*Continued*)
    fabrication, 24, 25, 171–175
    summary table, 24, 25
    analysis for, 164–176
    testing, electrical, 167, 173, 175, 176
Printing, silk screen, 27, 193, 194, 196
    costs, 196
    decals, 194
Product earnings, 38
    commerical sales, 398
    government sales, 398
Production design, 44, 45–48
Production discontinuities, 47
Production engineering, 373. *See also* Manufacturing engineering
Production planning, *see* Manufacturing engineering
Production planning program cost ratios, 367
Product selling price, determination of, 396–398
Profiling, 21, 113–115
Program breaks, *see* Delivery schedules
Program planning:
    personnel and facility planning ratios, 49–55
    schedule determination, 41–48
Progress rates, 325, 326

Rates:
    burden, *see* Burden rates
    delivery, 45–48
    hourly, *see* Hourly rates
    production, 46, 48
Ratios:
    burden expense, 50, 51
    cost, manufacturing start-up, 367, 368
    direct labor, 49–51
    estimating:
        active *vs.* support components, 61, 62
        inspection, 32, 272, 273
        manufacturing engineering, 376–379
        packaging, 62, 63
        test, 32, 277, 278
    facilities, 51–55
    indirect labor, 49–51
    manufacturing engineering:
        moderate production quantities, 375, 376
        small production quantities, 374, 375
        start up and sustaining, 37, 376–379
    material, 35, 343, 344
    planning:
        facilities and sales, 51–55
        personnel, 49–51
    space, 51
    sales, 51–55
Raw material prices, 35, 162, 306, 311, 312, 340–342
Raw material ratios, 35, 342, 343
Recurring to start-up costs, manufacturing, 367, 368
Research and development funding, 51–55
Resistors, 31, 222, 223, 240, 241, 244–246
Riveting and mechanical assembly, 27–29, 198–209
    detail tables, elements, 199–207
    operations 23, 27–29, 120, 198–207
    parts handling, 27, 198–200
    summary table, 27–29
    analysis for, 198–209
    see also Mechanical assembly
Routing, 21, 113–115

Sales:
    commercial, 396
        mark up chart for, 398
    government, 396
        mark up chart for, 398
    ratios, 51–55
Schedule determination, 41
    delivery schedule 45–48
    manufacturing phasing charts, 41, 42
    master production schedule, 41, 42
Screw and nut operations, 27, 202–204
    elemental table for nuts 203, 204
    elemental table for screws, 202
Sensors (fiber optics), 266, 267
Sheet metal operations, 19, 101–123
    blanking, 19, 102, 103
    burring, 21, 22, 115, 116
    engraving, 123
    holes, 19, 20, 103–106
    notching, 19, 103
    painting, 122
    press operations, 20, 21, 106–115
    riveting, 23, 120. *See also* Riveting and mechanical assembly
    silk screening, 123
    summary table, 19–23
    analysis for 101–123
    surface preparation and protection, 121, 122
    trim, profile, rout, 21, 113–115
    welding:
        arc or gas, 22, 116–118
        spot, 22, 118, 119
Silk screen, printing and engraving, 27, 189–197

INDEX

costs, 196, 197
operations:
   engraving, 27, 123, 194–197
   photographic processes, 27, 189–193
   printed circuit boards, 166–169
   printing, 193, 194, 196
   sheet metal, 123
   stencil fabrication, 27, 189–193
   terminal boards, 177, 178
   summary table, 27
     analysis for, 189–197
Sleeving:
   preparation and assembly, 241
   wiring operations, 222, 224, 225
Soldering, 30, 31, 248–253
   brazing, nonapplicability, 248
   buss wire, 240
   cable, 241–243
   definition, 248
   dip, 250–253
   etched circuit boards, 250–253
     cleaning, 251–253
   insulated wire, 236–239
   masking, 250–253
   seam, 31, 253
   silver, nonapplicability, 248
   summary table, 30, 31
     analysis for, 248–253
   terminals, 243–246, 248, 249
   time values, 30, 31, 248, 249
   wave, 250–253
   wire, 249, 250
   see also Wire Preparation, Wiring
Space planning ratios, 51
Special tooling and test equipment:
   comparative tooling and manufacturing methods, 368, 369
   computer aided design, 371
   cost factors:
     tooling, 368–370
     special test equipment, 372
   cost ratios:
     manufacturing start-up, 367, 368
     recurring, 367, 368
   definition, 366
   design time, 370–372, 373
   drafting, 370, 371
   estimates
     implementation, degree of, 367
     manufacturing methods, 368
     manufacturing start-up cost ratios, 367, 368
     production planning costs, 367

summary table, 36, 37
   analysis for, 366–372
   test equipment costs, 371, 372
   design and drafting time, 371
   tooling costs, 368–370
Spot welding, *see* Welding, spot
Speeds and feeds:
   band saw, 70
   broach, 82
   drilling:
     drill, 80
     drill, tap, countersink, 80, 81
   engine lathe:
     taper, 76
     thread, 75
     turn and bore, 74
   gear hobbing, 90–94
   grind:
     centerless:
       in-feed method, 83
       through-feed method, 83
     external cylindrical, 86, 87
     internal cylindrical, 88
     surface, 89
   milling:
     corners, grooves, slots, 79
     profile or end mill, 78
     side, straddle, slothing, 79
     surface or face mill, 78
   power hack saw, 70
   turret lathe:
     tap, 72
     taper, 73
     thread, 72
     turn and bore, 72
Staking, 28, 201
Stamping, *see* Press operations; Sheet metal
Standard hour allowance, 34, 334–338
   assembly cycle, 336
   design growth, 335
   engineered change, 335
   engineered prototype, 335
   measured labor, variance from, 334
   miscellaneous, 336
   normal rework and repair, 335
Standard hour multipliers, 34, 338, 339
Standards, time, 34
   allowances for, 334–338
Standard time data, learning curves, 4, 318–320, 338, 339
Start-up costs, 367, 368
Stencil, *see* Silk screen, printing and engraving, operations

408 INDEX

Summary tables:
  coil winding, 30
    analysis for, 210-217
  electroplating and chemical surface:
    analysis for, 124-148
    treatment, 23, 128
  hourly rates, burden rates and earnings, 38
    analysis for, 394-398
  inspection and test, 32
    analysis for, 272-276, 277-280
  labor allowances and multipliers, 34
    analysis for, 333-339
  learning curves, 34
    analysis for, 317-332
  machine shop operations, 15-19
    analysis for, 67-100
  manufacturing engineering, 36
    analysis for, 373-381
  material costs and allowances, 35
    analysis for, 340-344
  material discount curves, 35
    analysis for, 362-365
  packing and packaging, 32, 33
    analysis for, 304-316
  painting operations, 26
    analysis for, 179-188
  plastics, 24
    analysis for, 149-163
  printed circuit board fabrication, 24, 25
    analysis for, 164-176
  riveting and mechanical assembly, 27-29
    analysis for, 198-209
  sheet metal operations, 19-23
    analysis for, 101-123
  silk screen printing and engraving, 27
    analysis for, 189-197
  soldering, 31
    analysis for, 248-253
  special tooling and test equipment, 36, 37
    analysis for, 366-371
  terminal board fabrication, 25
    analysis for, 177, 178
  wire preparation, 30, 31
    analysis for, 218-234
  wiring, 30, 31
    analysis for, 235-247
Surface finishes, 94-100
  effects on machining costs, 98-100
  tolerances, rules of thumb, 99
Surface preparation:
  painting, 26, 180
  sheet metal, 121, 122
Sustaining ratios, 37, 376-379

Tags, identification, 206
Tape, 206
  electrical, 206
  identification, 227
  masking, 206
  pressure sensitive, 206
Taper, machine shop operations, 15, 73, 75, 76
Taper pin:
  crimping, to wire, 243-246
  staking, to wire, 221, 223
  termination, 238, 239
Terminal board fabrication, 25, 177, 178
  costs, 178
  operations, 178
  summary table, 25
    analysis for, 177, 178
Terminals, soldering, 248, 249
Test equipment:
  adjustments, 277, 278
  automatic, 281
  computer aided design, 371, 372
  cost ratios, 367-372
  fiber optics, 266
  types of, 282, 283-289
    costs, 367-372
    labor, 371, 372
Testing (automatic), 32, 277-280, 281
  algorithm based, 289-293
  comparison testing, 289-291, 296-302
  component, 282-289, 294, 295
  estimating ratios, labor, 32, 277, 278
  definition, 277, 278
  fault detection and isolation, 296-302
  functional, 284, 286, 288-291, 296-303
  in-circuit, 284, 286, 288-291, 296-303
  in-system, 289, 290
  integrated circuits, 354-359
  planning, 375, 376
  printed circuit boards, 25, 167, 173, 175, 176
  schedule determination, 42, 44, 45
  summary table, 32
    analysis for, 277-280
  test programs, 293, 301-303
  time standards elemental test, 279, 280
    adjustment (equipment), 279, 280
    definition, 277
  trouble shoot and retest allowance, 280
Thermoplastics, 151-153
  costs, 162
  summary data, 24
  types:
    ABS, 151, 152

INDEX 409

acrylic, 151, 152
  costs, 162
  nylon, 151, 152
  polyacetal, 152, 153
  polyamide-imide, 152, 153
  polycarbonate, 152, 153
  polyester, 152, 153
  polyethylene, 152, 153
  polyphenylene sulfide, 152, 153
  summary data, 24
Thermosets, 153–155
  costs, 162
  summary data, 24
  types:
    allyl, 152, 154
    aminos and urea, 152, 154
    epoxy, 152, 154
    fluoroplastic, 152, 154
    phenolic, 152, 154
    silicone, 152, 155
Time requirements, production quantities, 329, 330
Time standards, elemental test, 277, 279, 280
Time values:
  CAD/CAM, 391, 392
  coil winding, 30, 211–217
  drafting, 371, 372
  electroplating, 23, 124–137
  inspection, 32, 274, 275
  machine shop operations, 15–19, 67–100
  packing and packaging, 32, 306, 307, 311
  painting, 26, 180–187
  plastics, 24, 160–163
  printed circuit boards, 24, 25
  record keeping, 273, 274
  riveting, mechanical assembly, and fastening, 27–29, 199–209
  sheet metal oeprations, 19–23, 101–123
  silk screen, 27, 192–196
  soldering, 31, 248, 249
  special tooling and test equipment, 36, 37, 367–372
  terminal boards, 25, 178
  test, 32, 277, 279, 280, 293–303
  wire preparation, 30, 31, 218–234
  wiring, 30, 31, 236–240, 243–247
Time variables, electroplating, 126, 127
Tolerances:
  gauging, 274, 275
  machine shop operations, 94–100
  machining costs, effect of tolerances, 98
    steel parts, 99
  recommended operation machine at metal parts, 95, 96
  surface finishes, 97
Tooling:
  comparative and manufacturing methods, 368
  cost factors, 368–370
  cost ratios, 367, 368
  definition, 366
  special and test equipment, 366–372
  summary table, 36, 37
  analysis for, 366–372
  tool handling:
    screw and nut operations, 202–204
    values, riveting and mechanical assembly, 206
    wiring, 246, 247
Transistor, 240, 241, 243–246
Trimming, sheet metal, 21, 113–115
Turret lathe, 15, 70–73

Walking times, mechanical assembly, 29, 207
Welding, sheet metal operations:
  arc or gas, 72, 116–118
  spot, 22, 118, 119
Wire dressing values, 246
Wire gauge and diameter, 249, 250
Wiring, 30, 31, 235–247
  bus wire, preparation and assembly, 31, 240, 246
  butt solder, 236–238
  CAD/CAM, 383
  capacitor, 31, 240, 244–246
  coaxial cable, 31, 242, 243
  component insertion, 235–247
  conductor, 30, 31, 241–243
  crimping, 31, 236–238, 240–246, 248, 249
  detail values, terminal, 243–246
  elemental breakdowns, 236, 243–246
  estimating values, applications, 236–240, 241–246
  insulated:
    butt solder, 236–238
    crimp and solder, 31, 240–246
    "lace wire into harness," 30, 236–239
    "lay in u-channel," 30, 236–239
    operations, 236–241
    pneumatic wrap termination, 30, 239
    "point to point," 236–239
    preparation and installation or assembly, 30, 31, 219–221
    taper pin terminations, 30, 221, 222, 238, 239
  preparation, *see* Wire preparation
  resistor, 31, 222, 240, 241, 243–246

Wiring (*Continued*)
  shielded cable, 31, 223–225, 241, 242
  sleeving, 31, 222, 241
  soldering, 31, 236–238, 240–246, 248, 249
  solderless, 30
    pneumatic wrap termination, 30, 239, 243–246
    taper pin termination, 30, 238, 239, 243–246
    summary table, 30, 31
    analysis for, 235–247
  transistor, 240, 241, 243–246
  values:
    average, 218
    terminal, 243–246
    tool handling, 246, 247
    wire dressing, 246
  wire list, 31, 247
  wire selection, 246
  work station, 31, 247
  see also Wire preparation
Wire preparation, 30, 31, 218–234
  bus wire, 222
  buzz wire, 229
  cable, 30, 31, 223–230
    coaxial, 223, 225
    shielded, 223, 224
  capacitors, 222, 223
  conductors, 224, 225
  connectors, 31, 226–229
  cutting, 218–221, 222–224
  ground lead, 225
  harness, 31, 227, 230–234
    fabrication, 231–234
    form and lace, 30, 231–233
    spot tie, 31, 230
  identification tape, 227
  insulated wire, preparation:
    hand, 219–221
    machine, 220
  operations, 218–234
  plug or connector assembly, 227–229
  resistors, 222, 223
  sleeving, 222
  stamping, 218–220, 227
  summary tables, 30, 31
    analysis, 218–234, 235–247, 248, 249
  taper pin staking, 221, 222
  times per operation, 218–234
  time values, elemental, 218–234
  tubing, 230
  twisting, 218–221, 230
  see also Wiring
Work sheets, *see* Cost estimating